Bayonets before Bullets

Indiana-Michigan Series in Russian and East European Studies
Alexander Rabinowitch and William G. Rosenberg, general editors

Advisory Board

Deming Brown
Jane Burbank
Robert W. Campbell
Henry Cooper
Herbert Eagle
Ben Eklof

Zvi Gitelman
Hiroaki Kuromiya
David Ransel
Ronald Grigor Suny
William Zimmerman

Bayonets Before Bullets

THE IMPERIAL
RUSSIAN ARMY,
1861–1914

Bruce W. Menning

Indiana University Press
BLOOMINGTON & INDIANAPOLIS

This book is a publication of
Indiana University Press
601 North Morton Street
Bloomington, IN 47404-3797 USA

http://www.indiana.edu/~iupress

Telephone orders 800-842-6796
Fax orders 812-855-7931
E-mail orders iuporder@indiana.edu

Maps and diagrams designed and produced by the University of Kansas Map Associates, David Halliday, Mark Lippincott, John B. McCleary, and George F. McCleary, Jr., with John Brewer, George Dalton, Mary Catherine Prante, and Bill Skeet.

First reprinted in paperback in 2000
© 1992 BY BRUCE W. MENNING
All rights reserved

No part of this book may be reproduced or utilized in any form or by any means, electronic or mechanical, including photocopying and recording, or by any information storage and retrieval system, without permission in writing from the publisher. The Association of American University Presses' Resolution on Permissions constitutes the only exception to this prohibition.

The paper used in this publication meets the minimum requirements of American National Standard for Information Sciences—Permanence of Paper for Printed Library Materials, ANSI Z39.48-1984.

MANUFACTURED IN THE UNITED STATES OF AMERICA

Library of Congress Cataloging-in-Publication Data

Menning, Bruce.
    Bayonets before bullets : the Imperial Russian Army, 1861–1914 / Bruce W. Menning.
    p. cm. — (Indiana-Michigan series in Russian and East European studies)
    Includes bibliographical references and index.
    ISBN 0-253-33745-3 (alk. paper)
    1. Russia. Armiia—History—19th century. 2. Russia. Armiia—History—20th century. 3. Military doctrine—Russia—History—19th century. 4. Military doctrine—Russia—History—20th century. 5. Russo-Turkish War, 1877–1878—Campaigns. 6. Russo-Japanese War, 1904–1905—Campaigns. I. Title. II. Series.
UA770.M467 1992
335.3′0947—dc20   92-8233

ISBN 0-253-21380-0 (pbk : alk. paper)

2 3 4 5 6 05 04 03 02 01 00

FOR TONY AND ALICE

CONTENTS

Acknowledgments	ix
Introduction	1
1. The Army of D. A. Miliutin and M. I. Dragomirov	6
2. Russo-Turkish War, 1877–1878	51
3. The Army of P. S. Vannovskii and A. N. Kuropatkin	87
4. The Legacy of G. A. Leer and M. I. Dragomirov	123
5. Russo-Japanese War, 1904–1905	152
6. Theory and Structure, 1905–1914: Young Turks and Old Realities	200
7. Dilemmas of Design and Application, 1905–1914	238
Conclusions	272
Notes	*279*
Bibliography	*309*
Index	*329*

Illustrations

D. A. Miliutin	37	A. N. Kuropatkin	93
M. I. Dragomirov	37	N. P. Mikhnevich	212
N. N. Obruchev	37	A. A. Neznamov	212
G. A. Leer	37	A. F. Rediger	223
I. V. Gurko	77	V. A. Sukhomlinov	223
M. D. Skobelev	77	M. V. Alekseev	223
P. S. Vannovskii	93	Iu. N. Danilov	223

Maps and Diagrams

Russian Army Corps, 1877–1878 (Transitional)	25
Regimental Combat Formation (Offensive), 1876	43
War Plans, 1877	54
Military Actions in the Balkan Theater, 1877–1878	59
Military Actions in the Caucasian Theater, 1877–1878	79
Russian Army Corps, 1904	110
Russian War Planning, 1887: Offensive Variant	118
Infantry Regiment in the Offensive, 1904	140
Far Eastern Theater, 1904–1905	156
Port Arthur	162
Liaoyang	177
Sha-ho	182
Mukden: Initial Deployments	188
Mukden Operation	190
Russian Army before the Reforms of 1910	224
Russian Army after the Reforms of 1910	225
Russian Army Corps, 1912	229
Initial Strategic Deployments and Offensive Plans, 1914	244

Acknowledgments

It is a pleasure for me to recognize assistance in the completion of this manuscript for publication. In this respect, three colleagues stand out from the small crowd who inevitably figure in larger research and writing projects. First, my thanks go to Major General William A. Stofft, USA, who, in 1983–84 as Director of the Combat Studies Institute at the U.S. Army Command and General Staff College, provided initial impetus for me to spend part of a student year engrossed in a study of the Imperial Russian Army. Next, Dr. Jacob W. Kipp, a long-time friend and mentor, gave unselfishly of his precious time and insight at critical stages along the way from thesis to book. After him, I owe much to Professor Andre L. de Saint-Rat, who willingly shared his expertise on the Imperial Army with me as his junior colleague at Miami University in Oxford, Ohio. The collection bearing his name at King Library, Miami University, furnished many of the research materials for this book.

In addition, a small host of scholars, friends, and professional acquaintances offered advice and assistance at various stages of research and writing. They include Dr. Robert F. Baumann of the Combat Studies Institute; Dr. Christopher Bellamy, correspondent for the *Independent*; Dr. Philip J. Brookes, Director of Graduate Degree Programs, U.S. Army Command and General Staff College; Professor John Erickson of Defence Studies, University of Edinburgh; Professor William C. Fuller, Jr., of the U.S. Naval War College; Dr. Robert F. Ivanov of the Institute of History, USSR Academy of Sciences; Professor David R. Jones of Dalhousie University; Professor John L. H. Keep, formerly of the University of Toronto; Professor George F. McCleary, Jr., of the University of Kansas; Professor Walter M. Pintner of Cornell University; Professor Dennis E. Showalter of Colorado College; Dr. James J. Schneider of the School for Advanced Military Studies, U.S. Army Command and General Staff College; A. G. Kavtaradze of the Institute of History, USSR Academy of Sciences; and Mr. Carl Van Dyke, private scholar. Others who offered professional support in the all-important task of running down rare research materials include Mmes. Helen Ball and Frances McClure of Special Collections at Miami University; Mme. Irina Kuk, Librarian at the Hoover Institution on War, Peace and Revolution; Mr. Charles Potts of Whitehall Library, United Kingdom Ministry of Defence; Dr. Timothy Sanz of the Soviet Army Studies Office; and Mme. Maria Alexander of Watson Library, University of Kansas.

Institutions which provided varying degrees and kinds of sustenance include the Combat Studies Institute, U.S. Army Command and General Staff College; the Department of History, Miami University; the Department of History, University of Kansas; the Ford Foundation; the Russian and East European Studies Center, University of Illinois; the Russian and East European Institute, Indiana University; and the Soviet Army Studies Office, U.S. Army Combined Arms Command.

Above all, I owe a great deal to my wife, Margaret, and my children, Anitra and Tony. Their sacrifices made this book possible, and it was they who were forced to accept my absence during innumerable weekends and evenings as research and writing exacted a debt which words alone can never repay. Over the longer term, my wife and partner of twenty-five years has given freely of her time and freedom, invariably meeting with grace and wit the challenges of our combined interests in Russian and Soviet studies. Her love, dedication, and self-sacrifice have been inexhaustible sources of strength for me.

It goes without saying that any errors are my own.

Bayonets before Bullets

Introduction

THIS STUDY traces the organization and military art of the Imperial Russian Army through two wars and two phases of the industrial revolution. The story begins with the soul searching which followed Crimean defeat, describes the principal changes of the reform era, and provides an operational and tactical narrative of the Russo-Turkish War of 1877–78. The book goes on to depict the Russians' response both to their own wartime experiences and to perceptions of other conflicts, to show how the Imperial Russian Army adapted to smokeless powder weaponry and to reveal why that army failed to meet the test of Far Eastern combat in 1904–5. In 1905–14, the story closes as it begins, with a round of profound introspection and military reform in anticipation of possible future war.

This survey focuses on the Imperial Russian Army as a military institution. It is a description of how that army organized, trained, and armed itself to fight. It is a depiction of war and combat experience. It is an account of tacticians, inventors, teachers, politicians, organizers, and dreamers. Above all, it is the story of how a military institution struggled mightily to accommodate itself to change in a Darwinian world which dealt harshly with failure and the inability to adapt.

Part of this study is organizational and doctrinal, and the other part is combat. Modern military experience suggests that the two are bound together across time through peace and war, a realization which the historian Peter Paret has reaffirmed by asserting recently that all the essentials of action in war find their initiation and development in peacetime.[1] The current study acknowledges this unity by combining the martial preoccupations of traditional military history with the broader organizational and intellectual concerns of the new military history. From the beginning, the assumption was that wartime performance (the province of battle and operations analysis) could be fully understood only with reference to prewar preparation (the province of concept and infrastructure).

For reasons which will become apparent in this book, peacetime preparation for future war between 1861 and 1914 assumed unprecedented scale and importance, often prompting novel departures in thought and action, including preliminary excursions into the newly emerging discipline of military science. True to the intellectual traditions born of this era, Major General S. N. Kozlov, a Soviet officer and military theorist, would observe a century later that "the most significant task of military science has always been determining the character of future war." He further noted that an understanding of the character of future war is predicated on an understanding of three separate—but related—phenomena: the level of technology of a given era, the experience of past wars, and the general political milieu.[2]

General Kozlov's categories proved a useful point of departure in establishing a convenient framework for imposing order on and making some sense of the Russian army's attempts to change itself in anticipation of future combat. What made the period in question particularly challenging and complex was the rapidly accelerating pace of technological change. During the span of several generations, the Russians moved from flintlock muzzle loaders to small-caliber repeating rifles, from smoothbore cannon to rifled, rapid-firing field pieces. Without going into detail, it is sufficient to assert that the pace and scale of technological change were notable enough to become dominant influences on military development throughout the era.

If technology emerged to become perhaps the single most important arbiter of change in military affairs, politics in the broadest sense of the word established the context in which military figures and institutions debated the nature of change and orchestrated appropriate responses. During the period between 1861 and 1914, three successive Romanov sovereigns remained crucial players in an imperial military drama played out against a background both of new and old antagonisms and of traditional and novel interests. Tsarist political structures established the overall framework in which military change took place. For this reason, the soldier and the tactician grudgingly shared the military limelight with the politician and the reformer—even in a period when technological change appeared to enhance the authority of the specialist and military professional.

Besides technology and political context, precedent exerted a profound influence in imperial Russian military development between 1861 and 1914. In an effort to gain perspective on the perplexing effect of new technologies and attendant methods and means, the Russians as never before systematically studied their own military experience and that of their contemporaries. To this study the Russians and their latter-day Soviet counterparts owe both the origins of their modern military professionalism and the birth of institutional military history. The degree to which the Russians were able to enlist the past in service of the future forms an important part of the picture of military change and adaptation for the entire late imperial period.

The combat element forms a second major concern of this book. How did changing technology and institutions meet the test of war? For a variety of reasons, Western historians and military observers—despite occasional bursts of enthusiasm and study—have often either ignored or dismissed the lessons of the Russo-Turkish War of 1877–78 and the Russo-Japanese War of 1904–5. Yet, for the world of European-style, large-scale military establishments, these two conflicts represented the most important test of concepts and weapons between 1871 and 1914. These wars revealed in all their starkness virtually all the military challenges of the industrial age, including the mobilization of mass regular cadre and reserve armies, the command of large units across vast distances, the growing lethality of firepower, changing strategic and tactical designs, and the importance of the home front.

The portrait of the Russian army which emerges from its organizational and combat history is that of a complex, multi-faceted structure, the faces, junctures, and components of which changed over time to affect each other and the shape and dimensions of the whole. Changes in technology wrought changes in organization and structure. Meanwhile, societal and political context conditioned overall institutional responses to challenges of technology and adaptation. All of the parts interacted with each other within an environment that was itself changing; consequently, the functioning of the parts cannot be completely understood in isolation from either one another or the whole.

How well the parts worked together more often than not depended upon linkages—organic and mechanical, intellectual and physical, spiritual and moral. Once the army assumed its basic post-reform shape, linkages not only between past and present but also among soldiers, ideas, and weapons determined combat effectiveness. And, for various reasons, it was in the realm of linkages where the Imperial Russian Army most often came up short. M. I. Dragomirov was a case in point. For much of the period under consideration, he was the theoretician and practitioner whose thought and actions characterized prevailing attitudes about critical linkages between man and weapon. It was he who reigned over nearly a half-century of Russian tactical development, and it was he who popularized A. V. Suvorov's eighteenth-century maxim, "The bullet's a fool, the bayonet's a fine lad." Although 1904–5 largely discredited Dragomirov's tactical views, for reasons of other failed linkages Russian soldiers in 1914 would find themselves marching off to a war in which a shortage of bullets caused them once again to ponder the dilemmas inherent in combat reliance on will and cold steel.

An organizational and combat history of the Imperial Russian Army from 1861 to 1914 seemed a worthwhile undertaking for several reasons. One was the realization that for reasons of language and distance—in several senses of the word—important aspects of the Russian military past remain obscure to Western observers. Among these are significant continuities between the old imperial and the new Soviet regimes in a number of areas, ranging from weaponry to military thought. Some of the more striking aspects of the latter, as reflected in General Kozlov's remarks, include a preoccupation with defining the substance and limits of military science, a regard for the importance of military doctrine, and a search for utility in military history. Another vital aspect of the late imperial military legacy was consistent concern for development of a viable theory for the conduct of modern military operations. Given the enduring significance of these and other aspects of late imperial Russian military theory, an examination of the circumstances under which they first came to prominence would be sufficient justification in itself for the present study.

Several historiographical and practical considerations provided additional justification. The first was a perception that historians have tended to compartmentalize the experiences of the nineteenth and twentieth centuries, with the result that traditional treatments fail to illuminate important continuities over the entire period between the end of the Crimean War and the onset of World War I.[3] Still another factor has been the appearance over the last several decades of new materials and interpretations on various subjects related to Russian military history between 1861 and 1914.[4] A final—and not inconsiderable—justification was the idea that one more attempt to answer some of the knottier questions about the military past might yield greater insight into both that past and the Soviet present.

Historiographical considerations also helped set chronological limits and determine depth of coverage. John L. H. Keep's recent *Soldiers of the Tsar* and John S. Curtiss's earlier pioneering studies provide in-depth coverage of the Russian army's pre-reform institutional evolution.[5] However, there is no single volume in any language which links their treatments either with the recent penetrating analyses of the Russian army in 1914–17 by Norman Stone and I. I. Rostunov or with subsequent and voluminous military historiography on the revolutionary and Soviet periods.[6] The current volume is intended to fill this gap. That it can aspire to do so is in part thanks to the work of others whose solid, groundbreaking studies on discrete aspects of Russian military history form much of the grist for the mill of the present survey.[7]

Materials for analysis and argument in support of various observations and conclusions have been drawn from a combination of available primary and secondary sources. The more important primary materials have consisted of diaries, reports, and various collections of documents and writings, including those gathered to support compilation of official histories. The histories themselves, with all their strengths and weaknesses, were an important and voluminous source of information on the course of the two important conflicts in which the Russians participated during the period under examination.[8] In addition to secondary materials of the prerevolutionary period, another important source of information and interpretation has been Soviet materials, especially those published during the last several decades as Soviet military historians have devoted increased attention to the tsarist military past.

The historian E. H. Carr once observed that history is an extended dialogue between past and present, with contemporary concerns shaping the historian's approach to the past and conditioning the conclusions drawn from it. The present study is no exception. Its origins date to the author's attendance in 1983–84 at the regular course of the U.S. Army Command and General Staff College, where his instructors encouraged him to combine an interest in Russian military history with a renewed curiosity about interaction among doctrine, technology, and organization in the preparation for and conduct of modern war. Subsequently, occasional instruction at the same institution's School for Advanced Military Studies afforded the

Introduction / 5

opportunity both to extend the limits of the original study and to test many of its assumptions and conclusions on instructors and officer-students charged with peacetime preparation for future war. Their commentary and penetrating criticism provided valuable guidance for additional research and writing.

Some final words are in order on matters of transliteration, dates, and documentation. Throughout this study, the modified Library of Congress system of transliteration has been observed to render Cyrillic names and words in the Latin alphabet. The only exceptions have been personal and place names (e.g., Alexander instead of Aleksandr, and Moscow instead of Moskva) which have come into popular English usage through other renditions and forms of transliteration. On the matter of dates, this study has adhered to the Julian calendar, which remained in use in Russia until 1918. In the nineteenth and twentieth centuries, the Julian lagged behind the Gregorian respectively by twelve and thirteen days.

1.

The Army of D. A. Miliutin and M. I. Dragomirov

> Such were the requirements and fruits of the generally positive flow within the Russian Army after the Crimean War, that is, the urge to prepare for future military action not only whole units of the army but also each man individually, especially the leaders on whom new conditions of battle imposed especially responsible tasks, often demanding of junior officers conscious, bold, and rational initiative.
>
> Anonymous Imperial Russian Official Military Historian, 1901[1]

BECAUSE MASS military establishments are aggregates of many variables, only rarely do modern armies reflect the personality and outlook of individuals. The Imperial Russian Army of the 1860s and the 1870s was perhaps an exception to this rule. Special circumstances plunged that army into a period of rapid transition, enabling two officers, D. A. Miliutin and M. I. Dragomirov, to leave their personal imprint on one of the era's largest military machines. As War Minister after 1861, Miliutin presided over a series of far-reaching reforms which touched nearly every aspect of Russian military existence. As instructor of tactics and training methods, Dragomirov was instrumental in fashioning the tactical system which endowed Miliutin's reforms with battlefield significance. The army that charged across Romania to the Danube and beyond in 1877 and 1878 was in large measure the army of Miliutin and Dragomirov.

The Wellsprings of Change

The period between 1856 and 1877 was one of painful transition for both the Russian army and Russian society. Crimean defeat left the proud military edifice of Tsar Nicholas I (1825–55) in shambles and prompted a far-reaching round of public and private agonizing over the causes of failure and the measures necessary to avoid repetition. Proponents of change—social and military—were numerous and opinions diverse. On the social side, criticism fixed on the manifold evils of serfdom—economic, moral, and military—and following the Congress of Paris in 1856, one of the first steps taken by Tsar Alexander II (1855–81) was to set in motion the legislative machinery that would culminate in 1861 with the liberation of the serfs.[2] This act, in turn, helped inspire a series of liberal reforms which spilled over into the next decade, concluding in 1874 with the proclamation of universal military service.[3]

The impulse for military reform proceeded not only from social change but also from military imperative. Of considerable consequence in the calculations of would-be reformers was the perception that Russia had strayed from its true military heritage. That heritage was embodied in the spirit and actions of the previous century's heroes, including Peter the Great and A. V. Suvorov, both of whom had preached a pragmatic military organization and tactics oriented on enemy, terrain, and Russian resources.[4] However, in the years between the defeat of Napoleon and the fall of Sevastopol, their legacy had been engulfed in a torrent of what many observers called "paradomania." This epithet they used to describe the obsession of the monarch and his military retinue with rigid uniformity and blind adherence to drill regulations at the expense of flexibility and innovation. Huge in size, ponderous in movement, unimaginative in leadership, and bound to an outmoded socioeconomic system, the Imperial Army stumbled to humiliation in 1854–55 by failing to drive the enemy from its own homeland.[5]

Even before debacle and subsequent debate, many military observers had seen the need for change. Since the days of Paul I (1796–1801), influential segments within the Russian officer corps had perceived the drift to institutional rigidity and conservatism as detrimental to the military interests of Russia. In the wake of the Napoleonic campaigns their resentment and dissatisfaction had festered until December 1825, when a brief interregnum invited an ill-conceived coup. For the next two decades memories of the "Decembrist Uprising" figured prominently in the military and political calculations of Nicholas I, a tsar obsessed not only with the fear of foreign revolution but also with the dangers presented by malcontents within his own nobility and officer corps.[6]

However, the prevailing reactionary mood could not entirely stifle the voices of discontent, loyal and disloyal. Would-be rebels studied the lessons of December 1825 to emerge with "the algebra of revolution," while would-be reformers sought to express their sentiments in terms acceptable to the regime. As early as 1839, D. A. Miliutin, a young officer on Caucasian assignment and a recent graduate of the Nicholas Military Academy, openly extolled the virtues of Alexander Vasil'evich Suvorov, Catherine the Great's talented and victorious commander, in the pages of *Fatherland Notes*, a leading periodical of the era. After recounting Suvorov's many achievements, Miliutin concluded that the Russian generalissimo had stood so far in advance of his own century that in many respects his methods presaged the tactics and strategy of Napoleon.[7] Implicit in Miliutin's observations was the idea that Russia's military survival necessitated a return to the spirit of Suvorov, a sentiment shared by many officers, including several leading instructors at the Nicholas Academy. It was probably no accident that while Miliutin was serving in the Caucasus, his thoughts had openly turned to Suvorov, for conditions on the southern military frontier had traditionally encouraged daring departures from the accepted "paradomaniacal" practices of units located closer to St. Petersburg.[8]

Two other officers, A. I. Astaf'ev and F. I. Goremykin, argued for changes in approach, regulation, and training which would put the army in closer touch with both its true roots and new tactical realities. Writing in the 1850s, Astaf'ev, an officer of the General Staff, asserted the primacy of Suvorov over Napoleon and railed against the ossification of military thought and practice which had resulted from a slavish acceptance of military truth as interpreted by the latter-day disciples of Napoleon. He believed that many aspects of Napoleonic military art no longer corresponded with modern developments, including the introduction of rifled shoulder weapons which "would require changes not only in army organization, deployment, movement and action of troops in war but also in tactics, military administration, strategy, and the whole of military science."[9]

In 1849, Goremykin, a professor at the Nicholas Academy, published his *Handbook for the Study of Tactics in Their Fundamentals and Their Practical Application*. Like Astaf'ev, he condemned the drift to military pedantry and wrote that "it is time to leave behind the oppressing and strange forms in which military learning is sometimes wrapped."[10] However, while decrying the dead hand of Napoleon, Goremykin retained at least one conviction that was quite Napoleonic: the distinction between the material and moral factors in war. Goremykin would write that despite the importance of armaments and equipment, preponderant moral force was an absolute precondition for the successful employment of armies in combat. Interestingly, during Goremykin's tenure at the Academy, an outstanding graduate of that institution was M. I. Dragomirov, a young infantry officer destined to become one of the most influential minds in the post-1856 Russian army.

In contrast with his more tactically oriented contemporaries, Miliutin believed that mere changes in tactics and training alone would not force a return to the tradition of Suvorov. Miliutin argued that tactical flexibility, rapidity of calculation and action, and a thirst for decision were all qualities that required a new breed of officer, one schooled in accepting responsibility and seizing the initiative. Indeed, one of Miliutin's most telling diary entries dated to 1840, when he castigated the officer education system, writing that it was calculated to produce parrots who, like Polly, repeated orders but could do little on their own.[11]

Other critics of the system understood the necessity of carrying fundamental transformations into the ranks. In place of the serf recruit who was beaten into submission and forced into the mold of the goose-stepping automaton, rebels and reformers alike dreamed of training free men to act and obey not through fear but through conviction and sentiment. Russian military observers near and far in time and distance had witnessed the effects of the French Revolution on European armies and had watched enviously as it unlocked the military potential of the masses.[12]

Beyond these issues the prophets of change saw the need for fundamental alterations in recruitment, organization, and equipment. In reflecting on Crimean defeat, Miliutin cited the inflexibility of serfdom which prevented Russia from tapping its vast manpower resources and discouraged the maintenance of a large trained reserve. In addition, Miliutin perceived the need for a fundamental reorganization of Russian military administration. The War Ministry was overcentralized and too cumbersome to respond to the challenges of a changing military world. Traditional peacetime military deployments, which had been dominated by army and corps-style organizations since the era of Napoleon, had demonstrated themselves incapable of making a smooth transition to a wartime footing. In addressing these and other issues Miliutin and like-minded reformers proposed nothing less than a revolutionary transformation of the Russian army and its supporting institutions.[13]

After the Congress of Paris, several years dragged by before the patient proved itself capable of swallowing the required dose of medicine. In 1856, the new Emperor, Alexander II, appointed the cautious N. O. Sukhozanet as War Minister. Parsimony and palliative marked the tenure of this ostensibly decisive combat veteran, whose chief accomplishment was abolition of Russia's outmoded system of military colonies. Only after three years did Alexander and his closest advisers apparently conclude that Sukhozanet lacked the motivation and temperament to effect more far-reaching change. Their dissatisfaction mounted during the Piedmontese-Austrian crisis in 1859, when Sukhozanet informed the Emperor that concentration of four Russian corps on the Austrian border in response to urgent French requests would require five and one-half months! Although the Tsar and his reform-minded supporters were far from united in their convictions and programs, Miliutin was at last summoned to St. Petersburg, where on 30 August 1860 he received the portfolio of Deputy War Minister. His appointment came on the recommendation of the Tsar's boyhood friend, Field Marshal Prince A. I. Bariatinskii, now Commander-in-Chief of the Caucasus, whom Miliutin had been serving as chief of staff. The unlikely marriage between the Minister and his Deputy persisted until 9 November 1861, when the Emperor at last eased the ailing Sukhozanet from office with a transfer to viceroy duties in Warsaw, thus making room at the top for Miliutin.[14]

Miliutin had two decades in which to work his organizing genius on the army. During that time he displayed an impressive combination of foresight, persistence, astuteness, and good fortune in attempting to implement his vision of a more perfect military institution. He came to the task more a scholar than a fighting soldier. Born in 1816 into an impoverished noble family, Dmitrii Alekseevich Miliutin had finished his preparatory education with a silver medal at the Pension of Moscow University, then departed for St. Petersburg in 1833 to enter active service as an artillery junker in the Life Guards. He received his commission within six months, and three years later, after skipping the introductory course, he graduated—again with a silver medal—from the newly created Nicholas Military Academy. Following appointment to the Guards General Staff in 1837, Miliutin was posted to the Separate Caucasian Corps in 1839–40, where he was seriously wounded in the right shoulder during a skirmish with Muslim insurgents. After returning to St. Petersburg as a Guards Captain, then spending ten months on recuperative leave in Europe, Miliutin served two additional (and unhappy) years in the Caucasus, after which he obtained reassignment in 1845 as a lieutenant colonel to the faculty of the Nicholas Military Academy.[15]

Miliutin remained at the Academy more than a decade, a period fortuitously divided between reflection and limited engagement. His tenure as Professor of Military Geography encouraged him to ponder in depth the relationship between resources and military potential, an issue which had first attracted his interest in the Caucasus. He pioneered in the application of statistical analysis to military geography and established an enduring tradition for statistically based studies of possible adversaries and theaters of operations. Meanwhile, from the recently deceased military historian A. I. Mikhailovskii-Danilevskii, he inherited an incomplete study of Suvorov's campaigns in northern Italy and Switzerland. Completion of this project between 1848 and 1853 both deepened Miliutin's appreciation for Suvorov's generalship and led to publication of a prize-winning multivolume work, *The History of Russia's War with France during the Reign of Paul I in 1799*.[16] Finally, during the Crimean War, he appeared in the Imperial Suite as occasional adviser and served as special consultant to the War Minister, V. A. Dolgorukov. Now a major general, Miliutin also sat as a member of several important advisory bodies, including the Committee on the Improvement of the Military Aspect. By the spring of 1856, Miliutin had received his baptism to court and ministerial politics and had grown accustomed to submitting advisory memoranda directly to his military superiors and to the imperial throne. His writings and arguments marked him as a zealous advocate of military reform.[17]

One lesson Miliutin learned was that access did not necessarily translate into acquiescence. True, some of his views on military education found grudging acceptance and incorporation into statutes governing the limited reorganization of the officer education system, including the curriculum at the Nicholas Academy. Also, his third proposal to found a professional military periodical, modeled in substance on the more senior *Naval Collection* (*Morskoi sbornik*) and in format on the popular liberal journal *Contemporary*, would eventually bear fruit in 1858 with the appearance of the *Military Collection* (*Voennyi sbornik*). However, even in the reformist climate of the immediate post-Crimean period, Miliutin's clarion calls to thoroughgoing military change went largely unheeded, leaving him in the unenviable position of a prophet ahead of his own time.[18]

The result was that Miliutin found himself having to serve an additional apprenticeship before his views enjoyed more than indirect influence. When the conservative Sukhozanet replaced Dolgorukov, Miliutin toyed briefly with applying for extended leave, then reluctantly accepted an unexpected opportunity to serve as Bariatinskii's chief of staff in the Caucasus. A third Caucasian tour between late 1856 and mid-1860 enabled Miliutin to translate into action his earlier theories on pacification and decentralization of the war effort against the Tsar's Caucasian adversaries. By the time that Miliutin was recalled to the imperial capital in the summer of 1860, the Bariatinskii-Miliutin team had captured the celebrated rebel Shamil and set in place the political-military infrastructure that would facilitate successful conclusion of the Caucasian conflict four years later.[19] Meanwhile, Miliutin would endure an additional year in reformer's purgatory as Sukhozanet's deputy.

Foundations of Reform: Expansion, Stability, and Decentralization

On 15 January 1862, scarcely ten weeks after accession to full ministerial power, Miliutin presented the Emperor with a comprehensive set of reform proposals. Miliutin's program, based in large part on his earlier experiences, observations, and conclusions, established the objectives and tenor of his activity for the next two decades. Above all he was concerned with reducing state expenditures while simultaneously closing the gap between Russian military potential and the reality of Russia's international position. Since at least 1856 he had argued for a system that would permit peacetime reduction of the army to its smallest possible size while retaining a capacity for rapid wartime expansion. In his comprehensive report to the Emperor, Miliutin wrote that "under present conditions of the European powers' situation, when each of them has a significant standing army and the assured means of expanding its military forces in case of war, the relative political significance of Russia can be supported in no other way than by corresponding armed forces with the same foundation for proper expansion in wartime." He noted that in the event of war Prussia could expand its peacetime army by a factor of 3.4, Austria by 2.2, and France by 2.[20]

The Russian army was also capable of expansion, but only in theory. Miliutin asserted that on paper the Russian army numbered 765,532 men with a reserve that would nearly double that figure in the event of war. However, he called this capacity for expansion "imaginary" because it was based not on hard realities but on outmoded and inaccurate data and tables of organization. As a remedy to this and other discrepancies he proposed a series of reforms in organization, recruitment, and education, the sum of which would enable Russia in its own fashion to adopt and support the European system of cadre standing army and massive reserve backup force.[21]

In addition to tsarist support, Miliutin's ambitious program required a stable threat environment, but St. Petersburg would enjoy only partial success in winning the serenity and time necessary for thoroughgoing military reform. In this respect, Foreign Minister A. M. Gorchakov's avowed policy of *recueillement*, or calculated self-absorption, accorded well with the imperatives of domestic reform and Russia's diminished international status in the immediate post-Crimean period.[22] Over the longer term, however, the shifting European balance, together with emerging Panslavism in Russia and rising nationalism in the Balkans and East Europe, promised to keep the international pot boiling. Initially, the Russians distanced themselves from Austria and counted on the France of Napoleon III, which generally supported nationalist movements in the Balkans, to counter the growing influence of Prussia and Great Britain in Eastern European and Near Eastern affairs. Although Russia drew some Balkan advantage from support for the Serbian nationalist cause, the emergence in 1866 of a unified Romanian state under non-Russian auspices worked for a time against St. Petersburg's interests.[23]

Frustrated in the Balkans, Russian expansionism found outlets in the Far East and Central Asia, where *recueillement* could be conveniently overlooked because great power intervention appeared remote. In 1860, the Treaty of Peking secured Russian commercial privileges in China, confirmed Russian presence on the Amur, and granted the Tsar an immense stretch of territory between the Amur and Ussuri rivers and the Pacific Ocean. This latter concession made possible the development of a Russian port at Vladivostok. Meanwhile, farther to the west, Russian troops intervened in 1871 on the Russo-Chinese border to begin a decade-long occupation of Kuldzha, thus winning temporary control of the strategic Ili Valley with access to passes in the Tien Shan.[24]

In Central Asia, beginning in the late 1850s, the Russians wrung more apparent than real advantage from a series of minor military campaigns, which brought local nomadic and semi-barbarous tribes first under Russian protection, then under Russian hegemony. In 1864, Russian columns advanced from Orenburg and Semipalatinsk to capture Chimkent and envelop the Kazakh steppe, bringing tsarist troops into direct confrontation with Khiva, Bukhara, and Kokand. In June 1865, the adventurous General M. G. Cherniaev captured Tashkent. Three years later, General K. P. von Kaufman captured Samarkand. In 1869, Bukhara and Kokand became protectorates. In 1873, General von Kaufman reduced Khiva to the same status, followed in 1876 by the annexation of Kokand. Not to be outdone by their counterparts to the northeast, in 1869 Russian troops from the Caucasus established a foothold in Trans-Caspia, from which they conducted a series of campaigns against Turkmen tribes between 1873 and 1877, edging ever closer to Afghanistan.[25]

Because gains in Central Asia conferred little immediate practical advantage, the chief impact of their acquisition was to levy an additional drain on scarce military resources and to sow alarm among the great powers, especially Great Britain, which had interests in the area. Unlike the prolonged campaigns of the Caucasus and the southern steppe, Central Asia left few lasting impressions on the tactics and administration of the army. To the delight of enemy sharpshooters in later wars, the army would gradually adopt for its summer uniform the white duck tunics of troops serving in the warmer climates of Asian conquest. Conquest would also capture the imagination of more aggressive Russian nationalists who had never reconciled themselves with Gorchakov's policy of calculated self-absorption. Meanwhile, Central Asia would leave its stamp on many officers who had served there. For some, like Cherniaev, the "Lion of Tashkent," the exigencies of frontier warfare conferred license to exceed instructions, to build popular reputations in the newly unfettered press, and occasionally to challenge the wisdom of St. Petersburg.[26] For more junior participants, the same frontier experiences might instill bad military habits, which the so-called Tashkentsy would elsewhere unconsciously attempt to translate into combat against less tractable future foes.[27]

Closer to European Russia, other kinds of pacification levied incessant and large military requirements with their own complex set of international implications. Although the Caucasian wars ended in 1864, Russian administration in Tiflis presided over an uneasy peace, requiring the stationing of troops in key areas, especially those exposed to Turkish influence and attack. Poland broke into open rebellion in 1863, thanks to long standing grievances, rising national sentiment, and a ham-handed Russian attempt to impose military recruitment on a recalcitrant populace.[28] Although the Russian army moved quickly to quell the insurrection, French support for the Polish nationalists caused relations to cool between St. Petersburg and Paris. Berlin gradually provided compensatory warmth, but the price was steep: Russian neutrality during the wars of German unification. In 1873, even after the Franco-Prussian War had demonstrated the military threat inherent in a Prussian-dominated Germany, St. Petersburg joined Berlin and Vienna in adherence to the Three Emperors' League, thereby winning additional time for Miliutin's reforms.[29]

The cumulative impact of Russia's evolving external and internal circumstances was initially to impose somewhat greater military requirements, then to support modest reductions, and finally to justify a renewed emphasis on reform and expansion that would at last lay the foundation for a truly massive cadre and reserve army. Thus the combined effects of military operations in Central Asia, the Caucasus, and Poland precluded the sharp reductions in active forces originally envisioned by Miliutin. When reductions finally occurred, they took place against a background of increasing concern caused by the growing military potential of Prussia as revealed in its lightning victory of 1866 over Austria. The events of 1870–71 transformed concern into a genuine apprehension which would find reflection in revised threat assessment and a program which both reaffirmed and deepened the Tsar's commitment to military reform and expansion within the limits of fiscal and manpower constraints.

Throughout the reform period, domestic circumstances shaped possibilities and imposed frustrations. Universal military service was a major case in point. Liberation of the serfs in 1861 made mass conscription theoretically possible, but universal military service in the early years of a Miliutin-dominated War Ministry remained an idea ahead of its time. Assertion of traditional gentry interests, together with the sheer complexity of application, formed insurmountable obstacles. Consequently, Miliutin had to content himself with implementing measures that boosted conventional recruitment, furloughed trained personnel, reduced the number of noncombatants carried on the active rolls, and revised the structure of reserve units to make them genuine vehicles of training and sources of manpower for a regular army which was still only capable of very limited wartime expansion. Meanwhile, Miliutin worked strenuously to build support for universal military service, but more than a decade would elapse before it became a reality.[30]

Cadre army and reserve force were important parts of an overall reform vision which required development of a revamped support and administrative structure to eliminate the waste, corruption, and inefficiency of the old system. Before 1862 the army had relied on a system of deployed army and corps organizations to satisfy basic organizational and military requirements in peace and war. As John Erickson has written, these entities, including especially the 1st (or Field) Army and the Army of the Caucasus, "carried their support systems snail-like on their own backs," reporting not to the War Ministry but directly to the tsar.[31]

On the basis of his experience in the Caucasus, where Field Marshal Prince Bariatinskii had devised military districts as key units in effective military administration, Miliutin proposed the creation of a series of military districts to shoulder a major part of the army's support and administrative burden. He began in 1862 by abolishing the outmoded peacetime corps organization and dividing the 1st Army into three districts (Warsaw, Vilnius, and Kiev), to which he added a fourth, the Odessa, which was fashioned from V Corps and the Composite Cavalry Corps. In 1864, the Tsar decreed the establishment of districts throughout Russia, adding six to the four already in existence. Four more followed in 1865 for a total of fourteen.[32] Two decades later, the fruits of Central Asian conquest produced a fifteenth in Turkestan. With modification, for more than a century this network of military districts would serve the military establishments of both Imperial Russia and the Soviet Union.

The military districts embodied Miliutin's conception of rational decentralization to improve efficiency and ease the army's transition from peacetime to wartime footing. Each district resembled a miniature version of an improved War Ministry. Within the district the division replaced the corps as the army's basic tactical entity. At the head of the district stood the commander, who directed general affairs through a district council and military affairs through a staff. The district council included key representatives of support organizations, all of whom coordinated their activities with respective subdivisions of the War Ministry under the supervision of the district commander and his chief of staff. In wartime the district apparatus in theory could readily be converted into an active army staff. To complete the system Miliutin in 1874 created an administrative infrastructure of military chiefs (*voinskie nachal'niki*) at the province (*guberniia*) and district (*uezd*) levels. Their function was to assist in the maintenance of reserve forces, to facilitate the flow of recruits into the army, and to ensure conformity of practice at the regional and local levels with Imperial military regulations.[33]

Restructuring the War Ministry and Creating the Main Staff

The next step in reforming military administration was to restructure the War Ministry itself. Prior to 1862 it consisted of the Office of the War Minister, nine departments, and various other offices, staffs, and agencies, nearly all of which had grown haphazardly with little recent action devoted to coordination, rationalization, and unity of command. Some department heads reported directly to the tsar, while others remained needlessly buried in subdivisions which permitted little flexibility for healthy institutional development. To cut into this bloated and inefficient apparatus Miliutin implemented a series of organizational reforms which effected a gradual transformation between 1862 and 1869. After some trial and error, there gradually emerged a simplified and consolidated ministerial organization of five major divisions (Imperial Headquarters, Military Council, High Military Court, War Ministry Chancellery, and Main Staff), seven main administrations (Intendance, Artillery, Engineer, Medical, Military-Educational Institutions, Military Juridical, and Cossack Hosts), and several subsidiary inspectorates and main committees.[34] All divisions and administrations were subordinate to the war minister, who remained the undisputed chief military adviser to the tsar with vast authority in all military matters. The effects of the reform were dramatic: it simplified ministerial organization, clarified roles and functions, and reduced the Ministry's personnel by 1,000, and the administrative work load by 45 percent.[35]

Positive results aside, the overall arrangement was not completely satisfactory to all observers, including Field Marshal Prince Bariatinskii, who saw in his protégé's unfolding design the dashing of his own hopes for a reorganization of ministerial and staff functions on an alternative model—the Prussian. Adoption of the Prussian system would have handed the army's operational planning and direction over to an autonomous and highly centralized general staff, the chief of which reported directly to the sovereign. Meanwhile, the war minister would have been relegated to the position of an administrative figurehead, with his institution providing infrastructure and housekeeping support for a larger military establishment dominated by the general staff. Bariatinskii's aim, of course, was to elbow Miliutin aside and make himself master of Russia's military destiny as chief of a revamped general staff.[36] This was not an unnatural objective for an official who, as Viceroy of the Caucasus, had direct access to the tsar and combined in his own person vast civil and military powers. However, throughout the late 1860s, Miliutin's own adroit maneuvering and retention of imperial favor proved sufficient to ward off this thinly veiled challenge to his own authority and position.

In retrospect, the Miliutin-Bariatinskii controversy held far greater implications for the army than a simple contest of wills for domination of the military establishment. At the time, neither protagonist probably understood—or perhaps understood only imperfectly—that general staff development had reached a critical stage. Although in Russian usage the term "general staff" dated to at least 1763, its definition remained vague even in Miliutin's time. Thus official codes of the late 1850s noted that the term was appropriate to three situations: to auxiliary staff organs at division level and higher, to a sphere of military-scientific activities associated with the preparation for and conduct of war, and to a level of jurisdictional competence within the War Ministry for matters requiring either special oversight or a higher military education. Officers assigned to discharge these functions or responsibilities either in the War Ministry or with troops were carried on the rolls as Officers of the General Staff.[37]

Also in retrospect, aspects of all three of the above categories lay within the province of the evolving capital staff, a term which the historian Dallas Irvine has chosen to describe a central military organ assisting the supreme command authority in exercising command and administrative control and in intellectually determining and implementing higher directives. Indeed, the origins of the capital staff, of which Helmuth von Moltke's Prussian General Staff was the premier contemporary example, cut across organic and functional lines to emphasize two activities: systematic peacetime collection of information for the conduct of possible future operations and intellectual preparation for the conduct of future operations, either through the development of skills for contending with contingencies or through the elaboration of specific war plans, or both.[38] If these activities were combined with operational authority within a single institution and if the head of that institution assumed a direct military advisory role to the head of state, then something like the Prussian General Staff might emerge.

The Russian context, however, offered neither cognitive nor political justification for such a departure. Although Miliutin and many of his contemporaries realized that changing military organization and technology had rendered warfare more complex, they apparently saw little need to introduce anything other than compensating adjustments into their basic concept for ministerial and central staff redesign. Perhaps no one better than Miliutin understood that modern war now required both the systematic collection of military-geographical information and the capacity to train specialized staff officers who were suited to serve as assistants to field commanders and skilled in conducting war by maps. However, circumstances were such for Russia that the exigencies of modern war had not yet engendered the impulse to create an all-powerful central staff charged with preparing specific war plans. The time required at the outset of war for concentration of troops was still too great to make exact planning calculations, and possible enemy options during the same interval were too numerous to permit reliable forecasts. Consequently, grander operational schemes might exist, but they would remain hypothetical in nature, while more exact plans would be feasible only for finite theaters of military action.[39] With several notable exceptions, this situation would hold true for Russia until the 1880s.

Meanwhile, the varying pace of technological change imposed a varying degree of urgency on the necessity to evolve a capital staff organization. For Prussia, the early advent of railroads, which introduced speed of troop concentration and the exactitude of timetables, would dramatically alter fundamental mobilization and planning calculus and spur final evolution of a capital staff—even as Miliutin and Bariatinskii argued the merits of a Prussian-style system. For Russia, where railroad construction proceeded at a slower pace, there was no such immediate impulse for creating a centralized planning staff. Until the wars of German unification hit home in all their implications, the Russians might remain satisfied with a mixture of rationalization and incremental change in staff procedures and organization.

With no iron-clad intellectual justification for Bariatinskii's general staff, practical politics assumed the ascendancy, dictating Miliutin's assertion of the War Ministry's bureaucratic prerogatives over the personal preferences of a viceroy. Better from Miliutin's perspective to retain general staff–type functions within the War Ministry on the French model, thereby retaining for himself unified control not only over infrastructure but also over a rudimentary version of preparation and planning. It was against this background that the Imperial Russian Main Staff, an imperfect version of the ideal capital staff, made its debut.

The origins of the Main Staff's immediate predecessor dated to 1863, when Miliutin transformed the Department of the General Staff into the Main Directorate of the General Staff (GUGSh) under the Quartermaster General, who was himself subordinate to His Imperial Majesty's Main Staff, a traditional version of a command staff not to be confused with the later Main Staff. The new GUGSh inherited control of the Nicholas Academy, which in 1855 had been renamed the Nicholas Academy of the General Staff, and of the Military-Topographical Depot. Overall coordination came from the Consultative Committee, which conducted special investigations and exercised broad supervisory powers over the preparation of General Staff officers. Various sections within the GUGSh retained responsibilities for statistics, studying the coordination of transport facilities, educational activities, military-historical work, conducting surveys of potential theaters of operations, proposing deployments, and gathering intelligence.[40]

Nearly three years later, the iterative nature of the reform process dictated further transformation. Creation of the military districts had revealed the wisdom of combining closely related quartermaster (operational) and unit assignment (deployment) functions within district staffs. Common sense and Miliutin's own experience also indicated that the resulting broadening of horizons and functional cross-fertilization would help bridge the inevitable gap between line officers and officers of the General Staff, the latter of whom occupied corresponding staff positions both at the center and within the districts. By the end of 1865, the logic of combining analogous functional concerns and diversification of job experience at the highest level justified merging the GUGSh with the Inspectorate Department of the War Ministry to form the Main Staff. Thus Miliutin attained the overall objective of combining within a single organization on the French pattern both the operational concerns of the Quartermaster General and the order of battle concerns of the Inspectorate. The resulting change also reduced confusion by standardizing functional areas of competence within both the Main Staff and analogous staffs at the district, army, and wartime corps levels.[41]

The negative side of reform was that Miliutin had created within the War Ministry a sprawling secondary empire, the multiple and compartmentalized concerns of which prevented single-minded concentration on fundamental linkages between strategy and mobilization. The new Main Staff possessed broad military-administrative powers, including jurisdiction over statistical information, personnel, and recruitment and matters related to the structure, deployment, and service of troops. According to the 1869 table of organization, the Main Staff consisted of six subdivisions: Military-Topographical Section, Committee for the Movement of Troops by Rail and Water, Military-Academic Committee, Nicholas Academy of the General Staff, the Corps of Officers of the General Staff (144 generals, 242 field-grade officers, and 109 company-grade officers), and the Corps of Military Topographers and Couriers. Only in 1875 was a Committee for Preparation of Data for Troop Mobilization belatedly added to the complement of the Main Staff.[42] Meanwhile, as one of six sections within the Main Staff, the Officer Corps of the General Staff lacked the mandate and resources to play any role other than that of a reservoir of superior technical talent within the War Ministry.[43]

This overall arrangement resulted in divided focus and fragmentation of effort. Because the Chief of the Main Staff was also Chief of the General Staff, his administrative span was simply too great to permit devoting single-minded attention to complex information-gathering and planning processes. At the same time, these activities remained divided among too many personnel and agencies to promote true staff integration and economy of effort. It was no accident during the 1870s that embryonic versions of Russia's first war plans came not from any collective "brain of the army" but from an individual brain, that of Nicholas Nikolaevich Obruchev. Obruchev was a talented but stigmatized Miliutin protege who had replaced his mentor as instructor at the Staff Academy, had served as assistant editor of the

Military Collection, then had improbably become involved during the early 1860s with Land and Liberty, a revolutionary organization. After falling under a cloud in 1863, when he refused to serve in the suppression of the Polish insurrection, Obruchev had only barely managed to save his career, thanks possibly to Miliutin's intervention. By 1867 Obruchev had recovered his fortunes sufficiently to sit as Chief of the Military-Academic Committee.[44]

Field Administration and Dilemmas of Mobilization

Organizational and personal growing pains reflected not only politics but also the difficulties of anticipating and managing complex transitions from peace to possible future war. Other thorny problems of infrastructure included field administration and recurring shortcomings in mobilization. Abolition of peacetime deployments based on order of battle required provision for the wartime creation of command and staff instances higher than division level. In 1868, after more than five years' study and trial, Miliutin's answer was a new *Regulation on the Field Administration of Troops*. It detailed the command and staff composition of a field army and its constituent corps and outlined relevant duties and responsibilities. An army's field administration consisted of four principal components (field staff and separate directorates for intendance, artillery, and engineering) and six secondary elements (army chaplain; directorates of military police, military medicine, and field post; and inspectorates of military communications and military hospitals). The principal components were directly subordinate to the commander-in-chief, who "represented the person of the Tsar and was invested with Imperial authority," his commands having the power of laws signed by the Emperor. The six secondary functional areas were immediately subordinate to the field army chief of staff, who acted as the commander's immediate assistant on all issues related to field administration. Meanwhile, an assistant chief of staff assumed responsibility for quartermaster and deployment functions. The field staff also included a military-topographical section and an officer charged with intelligence collection. An analogous but reduced version of the same scheme for field administration applied to corps organization within a field army and to separate corps.[45]

Although the Regulation of 1868 represented a distinct advance over its 1846 predecessor, the solution to one set of problems tended to create others. Bariatinskii, who had by now become Miliutin's chief critic, was only too happy to point out that the very essence of the new scheme for field administration permitted the War Ministry and its bureaucrats wide latitude for interference in matters heretofore lying within the jurisdiction of field commanders. With a thinly veiled reference to Miliutin, Bariatinskii held that "he who has not himself commanded a regiment cannot know the greatness of our army or understand the meaning of its command."[46] Worse still if the field commander were a member of the royal family or even the tsar himself, in which case there might arise the issue of whether authority lay either with the bureaucracy or with the person of the commander.

At the same time, an unintended consequence of Miliutin's arrangement was to overload the commander and his chief of staff with too much paperwork and too many administrative activities, thus diverting them from their primary activity, the conduct of operations.[47] Another major shortcoming which neither Bariatinskii nor other critics discerned was that of weak linkages between fighting front and supporting rear. Although the reorganization provided for simplified administrative and logistical structures and procedures, it failed to spell out in detail relationships between the leaner field staffs and their corresponding (and heavier) elements within the district staffs from which field formations would ultimately draw support and sustenance. The Russo-Turkish War of 1877–78 would lay bare this shortcoming, but at least for a time Miliutin could count on the Tsar's support to contend with lesser criticisms of the new scheme for field administration.[48]

In contrast, shortcomings inherent in the mobilization process presented a far more immediate and serious challenge. Thanks primarily to improvements in administration, by the time of the Polish insurrection in 1863, the entire Russian army could be brought to a full wartime footing within a span of two and one-half months, a figure comparing most favorably with Sukhozanet's lamentable projections of 1859. By 1867, additional improvements meant that the period required for mobilization varied from a minimum of twenty-five days within the Kiev Military District to a maximum of 111 days within the Caucasus Military District. By 1872, the same figures had been reduced to as little as nine days and as many as thirty-nine days.[49]

What marred this otherwise solid achievement was realization that Russia's potential adversaries had made even more impressive leaps, not only in their ability to mobilize but also—thanks to the railroad—in their ability to concentrate mobilized troops for field operations. N. N. Obruchev, Miliutin's replacement at the Nicholas Academy, had written extensively throughout the 1860s to champion Russia's construction of strategic rail lines, but to little avail. By 1870, the possibilities opened by rail mobility and the rude entry of a Prussian-dominated Germany onto the central European scene suddenly confronted Russia with a new threat that was too disturbing to ignore. Even as the Franco-Prussian War raged, Obruchev pondered the lessons inherent in the mobilization capacities of the major European powers and concluded that "our army can scarcely be considered adequate for defense and still less adequate for offense."[50]

The Russian army faced a crisis of speed and mass, with no easy solutions in sight. In the beginning of 1873, Major General Obruchev presented Miliutin with a special report, "Thoughts on the Defense of Russia," which outlined in some detail the consequences and implications inherent in the changed European military scene. Obruchev's analysis was significant for two reasons: it made clear in digested form what a number of separate studies had been saying since 1870 and it represented a kind of war plan, not the detailed excursion so characteristic of the Prussian General Staff but a general discourse on options available to Russia in the event of war with Germany or Austria-Hungary, or both.[51]

Obruchev's conclusions were alarming. He fully expected that future war in the west might find Russia confronting not a single adversary but very possibly a coalition including both Germany and Austria-Hungary. Worse, in the event of full mobilization for war, both likely adversarial armies now separately possessed numerical superiority over the Russian army, the more so if all trained reserves entered calculations. Still worse, Russia now confronted a mobilization gap stemming from its own vast expanse and failure to keep pace with either Germany or Austria-Hungary in strategic railroad construction. According to Obruchev, the Russian army would require from fifty-four to fifty-eight days for mobilization and concentration against Germany, while the requisite span for Germany against Russia was twenty to twenty-three days. Similarly, Russia would require sixty-three to seventy days against Austria-Hungary, while the Austrians would require only thirty to thirty-three days. Short of full wartime mobilization, the Russians might count on fleeting numerical superiority only if they opted for timely concentration of the largest part of their active peacetime army in a single theater of war.[52]

After citing other shortcomings, including insufficient engineering preparations and an inadequate distribution of horses, Obruchev concluded that in the event of war in the west the Russian army would be able to retain a foothold on the Vistula, where it would be forced to conduct defensive operations before going over to the offensive against either Germany or Austria-Hungary. Obruchev's design retained viability only if measures were taken to strengthen defenses between the Vistula and the Narew and Bug rivers. Further, Russia would have to construct additional strategic railroads to serve these defenses, while simultaneously enlarging its army and adjusting troop deployments to protect the vulnerable flanks of the central fortified region in Poland. Adoption of these measures would require spending at least forty million rubles in fortress construction, building nearly 7,000 kilometers of railroads, and creating an army of 820–840 infantry battalions in place of the current 516 in European Russia.[53]

Inherent in Obruchev's thinking was a changed pattern of deployments, which, depending upon perspective, might bring either comfort or consternation. Heretofore, Russia had rested its western defenses on an aging and incomplete network of fortresses backed by covering forces in varying degrees of combat readiness arrayed in depth across the main approaches to the interior. Now, however, the time-distance-mass calculus caused Obruchev to prescribe a greater forward concentration of active forces in the Kingdom of Poland, thereby calling for changed peacetime deployments, additional engineering requirements, and a substantially altered mobilization and reinforcement regime. Only by pushing a heavier screen forward could the Russians hope to win time both to disrupt a potential adversary's advance and to cover their own cumbersome mobilization process. At the same time, these very actions might increase apprehensions in Berlin and Vienna, where military planners would logically view heavier Russian forward deployments as threatening. More ominously for St. Petersburg, subsequent Russian military planners might view forward deployments as a springboard for lightning offensive operations against either Germany or Austria-Hungary (or both) in the event that acceptance of calculated risk seemed to outweigh prudence and caution. Then too, the forward positioning of substantial forces in and adjacent to Poland created a situation with a logic of its own to support and protect a suddenly more vulnerable portion of Russia's military establishment. Obruchev thus laid the foundations for a mixed legacy: on one hand, his design would give rise to significant distortions in deployment and mobilization schemes, while simultaneously creating apprehensions and potentially ephemeral possibilities; on the other hand, the overall design would retain during his own lifetime a solidly defensive emphasis.[54]

From Crisis to Conscription: The Miliutin System on Trial

Obruchev's prescription for 300 additional battalions in European Russia was hopelessly at odds with the realities of Russia's force structure and trained pool of military manpower. Despite Miliutin's best efforts since 1861, Russia's rudimentary cadre and reserve system could deliver by 1872 a fully mobilized force of only 1,358,000, a figure which fell short of the War Minister's own requirements by more than a half-million. As early as the autumn of 1870 Miliutin had once again pressed for the imposition of universal conscription to contend with the numerical side of the Western military challenge. The Tsar responded by creating two deliberative bodies, the Organizational Commission and the Commission on Terms of Service, to study the closely related problems of conscription and military organization. Although both commissions had completed their work by the summer of 1872, the prospect of universal conscription incited a storm of public controversy, with the liberal press either ambivalent or supportive of Miliutin and the conservative press registering varying degrees of disapproval.[55] To retain initiative and momentum, Miliutin importuned the Tsar in the summer of 1872 to launch a more thoroughgoing examination of Russia's military requirements.

The Tsar acceded by convoking the secret Special Conference early in 1873, but the substance and tenor of deliberations briefly and unexpectedly placed Miliutin's entire reform program in jeopardy. Representation at the Conference, running between 28 February and 31 March 1873, included not only the Ministers of War, Navy, Foreign Affairs, and Finance, but also several prominent figures openly hostile to Miliutin's reform agenda, most notably Field Marshal Bariatinskii and the influential Count P. A. Shuvalov, head of the Emperor's security apparatus. The range of potential opposition extended to other participants, including the Crown Prince, Alexander Aleksandrovich, whose sympathies lay with Bariatinskii, and the Grand Dukes Michael Nikolaevich and Nicholas Nikolaevich. The Conference began smoothly enough, with Miliutin determining the agenda and the Tsar himself presiding. The War Minister opened discussion with a survey of Russian military development within European context since the Crimean War. He recounted the progress of his reforms, outlined the current threat in terms resting heavily on Obruchev's "Thoughts on the Defense of Russia," and concluded that additional resources were required to assure the continued security of the Empire.

Although no one seriously faulted Miliutin's threat assessment, the discussion quickly became acrimonious when it focused on the organization of the army, the quality of its composition, and finances. Critics of the system compared it unfavorably with that of Prussia. Bariatinskii rose to assail Miliutin with charges that the War Minister's reforms had both bureaucratized the army and led to waste and unnecessary expenditures. Count M. Kh. Reitern, the Minister of Finance, followed with a report asserting that additional military expenditures would imperil the fiscal survival of the state.[56] Amid a welter of voices expressing similar sentiments, Miliutin alternated between stony silence and aggressive counterattacks grounded in fact and statistics. Meanwhile, the Tsar sought balm through compromise and additional study, eventually designating a special subcommittee to investigate ways of reducing military expenditures. To Miliutin's rage and despair, the Tsar named Bariatinskii as chairman of the subcommittee, with the clear implication that its findings might provide justification for undoing much of what Miliutin had accomplished in the previous dozen years.[57] Suddenly a man who in Miliutin's words "did not know how to count money, even in his own pocket," stood in judgment of the 160 million ruble budget of the War Ministry.[58]

The majority of the War Minister's apprehensions ultimately proved groundless. Despite the weight of the opposition and debate sharply critical of Miliutin, the Tsar eventually came down on the side of his reformer, affirming the basic soundness of Miliutin's program and reaffirming the throne's support for universal conscription. With Bariatinskii slowed by gout, his subcommittee delayed circulating its findings until mid-1873. When the highly critical report finally appeared, Miliutin relied on information provided by the Main Staff to brush aside his critics' superficial analysis, even gloating in his diary about their ineptitude.[59] Thus, with the opposition intellectually discredited and Bariatinskii increasingly relegated to the sidelines by illness, Miliutin emerged from the trials of early 1873 with enhanced stature and a stronger mandate to press forward in the struggle for universal military service.

This did not mean that Miliutin escaped completely unscathed. The Conference adopted modifications to basic military organization, including both the resurrection of corps- and brigade-level command instances in the western military districts and alterations in regimental organization to provide for a larger complement of four battalions—albeit at a reduced peacetime strength level. The Conference also outlined provisions for creation of true reserve forces, which were to consist of different categories and formations suitable for a variety of purposes, including immediate augmentation of the active army, rear service, replenishment of wartime losses, and constitution of whole units in the event of full mobilization. However, the new corps organizations remained subordinate to their respective military district commanders, while changes in reserve and active army organization for the most part corresponded either with Miliutin's earlier proposals or with the recent recommendations of the Organizational Commission. Despite some harrowing moments, Miliutin's reform program emerged relatively intact from its severest peacetime trial. Still, there was only limited cause for exuberance: except for less costly changes, relatively little could be accomplished immediately, either because of trained manpower shortages or because of financial limitations. In the foreseeable future, even those units located along the vulnerable western borders were to remain at reduced peacetime strength, with the Conference somberly conceding that key issues of troop reinforcement and strengthening frontier defenses were dependent upon "the financial means at hand."[60]

Of more immediate importance, Conference deliberations provided added impulse for the adoption of universal military service. The State Council, that body charged with preparing legislation for the Tsar's approval, had begun reviewing the recommendations of the earlier Commission on Terms of Service even as the Special

Conference was meeting. In a series of exhaustive and often stormy sessions extending from 27 January to 14 December 1873, the State Council gradually hammered out the appropriate legislation. Hotly contested issues ranged from the difficulty of implementation among ethnic groups and minority nationalities to the relative length of active and reserve service, and from the criteria for the granting of privilege to the protection of gentry interests. Played out against the backdrop of lively discussion in the open press, debate within the State Council dragged on for nearly twelve months before the Tsar received draft legislation suitable for imperial approval.[61]

The resulting Statute on Universal Military Service, which went into effect on 1 January 1874, solemnly declared that "the defense of throne and homeland from foreign foes is a sacred duty of every Russian subject." The law now required all males, without regard to class or birth, to perform active military service from their twenty-first year. Exemptions from the prescribed fifteen-year service term (six active and nine reserve) were based on physical standards, domestic considerations, and education.[62]

On the basis of various conditions and claims, about 48 percent of eligible males were excused from peacetime service and 24 percent from wartime service. Privileges accorded by the law stemmed not from class but from educational qualifications. Thus, to ensure an adequate supply of commissioned and noncommissioned officers, conscripts with varying levels of educational attainment served shorter periods of active service before relegation to the reserve. Students and businessmen could receive deferments, but not exemptions. Those men with a least two years' formal schooling could volunteer for active service with the assurance that they might receive preferred treatment and a shorter term of active service. These volunteers, together with conscripts of appropriate educational background, might also expect swifter promotion to officer and noncommissioned officer rank. However, as a concession to gentry sentiment, men of education who attained officer rank had to serve an additional three years before attaining noble status.[63]

Still, application of the statute at last gave Miliutin the basis for filling the ranks of a truly expandable cadre and reserve army in the event of war. With variations to account for service among the Cossacks and within specified regions and among minority groups, the statute would endure until 1917 to provide a basic framework governing compulsory military service within the Russian Empire.[64] Miliutin had both answered the growing military needs of the state and taken a large step in the direction of making the army a limited engine of social change. With the onset of universal service, the army's literacy programs might benfit larger segments of society, while the possibility of advancement in rank based in some measure on educational qualifications might open wider the prospect for careers based on talent.[65]

Force Structure and Organization

Changes in army organization and force structure between 1861 and 1876 reflected a combination of influences, including reform imperative, accepted military wisdom on force mix, fiscal constraint, and the growing impulse to balance active and reserve components. Changing circumstances and requirements made for sometimes perplexing shifts in force size and composition, leaving in their wake inconsistencies which would defy resolution until after the Russo-Turkish War of 1877–78.

By Miliutin's initial calculations, the army would have to provide a structure that called for an active force of 730,000 troops expandable to a wartime complement of 1,170,000. In the interests of simplicity and speed of mobilization, the army's peacetime structure would have to correspond more closely with its projected wartime structure, with the number of large units remaining constant in peace and war, while their authorized strength fluctuated with circumstance. Consequently, in 1864, when more complete transition to the military district system called for a new force structure in which the infantry division was the largest peacetime military formation, the Tsar approved a new table of organization providing for forty-seven active infantry divisions (three Guards, four grenadier, and forty line). Each division consisted of four regiments, producing a total of 188 infantry regiments, each of which was carried on the lists at one of four strength levels: cadre (960), normal peacetime (1,500), reinforced peacetime (2,040), and wartime (2,700). Each infantry regiment consisted of three infantry battalions, which in turn were made up of four line companies and one rifle (or "sharpshooter" [*strelkovaia*]) company. In addition, scattered along the northern, southern, and eastern fringes of the Empire were twenty-nine separate rifle battalions. The complement of each battalion stood at one of three force levels, depending upon the army's overall status: wartime (720 soldiers), peacetime (544), and cadre (400).[66]

To facilitate recruit training, Miliutin created eighty reserve infantry battalions, which he distributed throughout the Empire. Except for their primary mission, however, the function of these battalions, which remained stationary and subordinate to their respective military districts, was never clearly delineated. While these battalions served an undeniably useful function, their permanent cadres and transient recruit contingents hardly provided the basis for the large-scale expandable force which Miliutin had envisioned since the mid-1850s.[67]

Ten years after the reorganization of 1864, thanks to changes mandated by the secret Conference of 1873, the number of infantry slowly began to increase. One infantry division was added in the Caucasus, to produce a total of forty-eight. Each division was now divided into two brigades of two-regiment composition. However, in place of the previous three battalions, each regiment now counted four battalions. The heretofore separate rifle battalions were formed into nine rifle brigades. If the constituent elements of these separate brigades were taken into account, the Russian army of 1876 numbered 192 regiments, producing a total active field force of 682 battalions.[68]

The changes of 1873 also brought to life the first genuinely flexible reserve configuration of the Miliutin era. On the eve of the war with Turkey, the newly prescribed reserve structure included one Guards and 164 army cadre battalions. Like their less flexible predecessors, these reserve battalions retained a permanent base, where a peacetime cadre of officers and noncommissioned officers supervised training for a constant flow of new recruits on their way to the active army. Upon

RUSSIAN ARMY CORPS, 1877-78 (TRANSITIONAL)

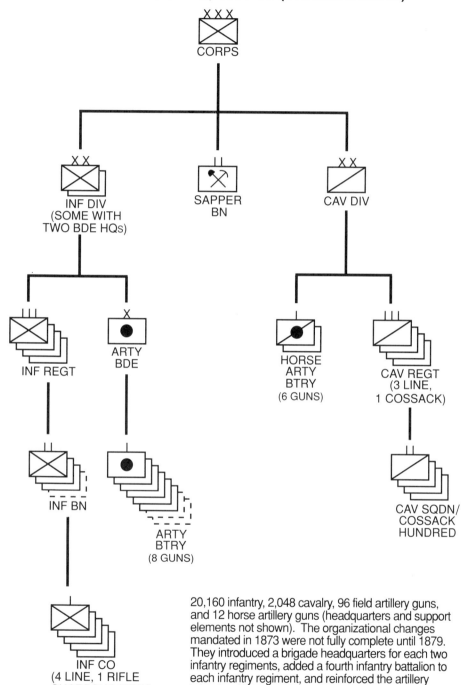

20,160 infantry, 2,048 cavalry, 96 field artillery guns, and 12 horse artillery guns (headquarters and support elements not shown). The organizational changes mandated in 1873 were not fully complete until 1879. They introduced a brigade headquarters for each two infantry regiments, added a fourth infantry battalion to each infantry regiment, and reinforced the artillery brigade with two additional batteries.

mobilization, however, the reserve battalion's four cadre companies each received an influx of reservists sufficient to form another whole battalion, which meant that a reserve battalion could produce a four-battalion reserve regiment, at the same time retaining sufficient cadre to leave an additional battalion in place. In theory, the resulting one Guards reserve and 164 army reserve regiments would nearly double the size of the active field army. But until the new structures were brought to sufficient strength, they could field only thirteen and one-half reserve army infantry divisions with force structures nearly identical to active army divisions. Indeed, the War Ministry's intent was that the two types of divisions be differentiated from each other only by numerical designation. Although the full potential of the new order would not be realized until universal service produced a sufficiently large pool of trained manpower, the foundation was now in place for a system which promised to accommodate the army's growing appetite for infantry in the event of war.[69]

As the cadre and reserve system gradually took root, the War Ministry provided for more thoroughgoing exploitation of its possibilities. In addition to the creation of entire reserve formations, soldiers and junior officers who had completed their active service obligation received mobilization assignments corresponding with their age, experience, and recency of service. In the event of mobilization, those with less than eight years' total service furnished the manpower necessary to raise existing units of the active army to full wartime strength (after 1874, regiments were assigned three strength levels: peacetime [1,500], reinforced peacetime [2,100], and wartime [3,600]). Men with between eight and twelve years' service formed a general pool (*zapas*) from which losses in the field were made good. Finally, those men with between twelve and fifteen years' service were assigned to form the separate reserve formations which might substantially augment the active army.[70] There also existed on paper a militia, or home defense force, called the *opolchenie*, which consisted of those males exempt from conscription and those younger than forty who had already fulfilled their service terms. The *opolchenie* could be called out only by special imperial decree. Full development of these categories came about slowly and their presence was reflected only imperfectly until after the Russo-Turkish War, when universal conscription and the new structures enjoyed full impact. In November 1876, for example, full mobilization could provide an active force of 722,000 and a reserve of 752,000, a combined figure which fell short of organizational requirements by 480,000 troops, or 30 percent.[71]

Artillery organization reflected in broad outline many of the same influences affecting the infantry. In conformity with both infantry and cavalry organization, the field artillery was divided into foot and horse contingents, the primary difference being the degree of mobility required for keeping pace in the field with either infantry or cavalry. After witnessing difficulties with rearmament and reorganization during the early 1860s, the foot artillery by 1869 came to consist of forty-seven artillery brigades to match the number of infantry divisions. Three additional brigades were assigned to Central Asian and Siberian military districts. After changes introduced in 1870, each brigade comprised six eight-gun batteries distinguished by type of armament. As in the case of the infantry, artillery brigades were maintained at varying strength levels, depending upon their level of military readiness. Many of the same circumstances applied to the horse artillery, which consisted of four Guards batteries and seven horse artillery brigades. In addition to conventional foot and horse artillery, various localities within the Empire served as bases for thirty-seven artillery parks, the function of which was to serve as depots for equipment, horses, and ordnance.[72]

Overall organizational changes of the mid-1870s were reflected in the size and expandability of Russian field artillery. On the eve of the Russo-Turkish War, foot artillery in the active army had grown to forty-eight brigades to match the number of active divisions, with an additional three brigades to support separate infantry units. To correspond with the growing need for firepower, each brigade would eventually be allocated eight batteries in place of the previous six. Meanwhile, horse artillery in the active army underwent limited expansion to provide for twenty-seven regular and twenty-one Cossack batteries, all of which were armed with six guns.[73]

At the same time, reserve artillery formations also increased in importance. After 1876, they numbered six reserve foot artillery brigades, each of which counted six batteries. Upon mobilization, each cadre battery expanded to four batteries, which meant that each reserve brigade could field twenty-four batteries in the event of war. After provisions for maintaining a replacement and training pool, the reserve artillery retained sufficient force to allocate ninety-six reserve artillery batteries in support of reserve divisions assigned to the field army, while retaining forty-eight batteries for replacement purposes. By 1876, the total wartime complement of Russian field artillery was 365 batteries and 2,809 guns.[74]

Thanks to changing tactics and missions, the Russian cavalry also underwent considerable alteration before 1877–78. In 1866, the mounted arm consisted of fifty-six regiments (four cuirassier, twenty dragoon, sixteen uhlan, and sixteen hussar), all of which made up ten cavalry divisions. Divisions of the Guards and of the Caucasian Army consisted of four regiments, while regular cavalry divisions consisted of six (two each of hussars, uhlans, and dragoons). Each regiment was composed of four squadrons. In contrast with the expandable infantry regiments, cavalry made little distinction between wartime and peacetime complements (777 and 689 respectively).[75]

After studies of the early 1870s concluded that Russia was falling behind other major European powers in the number of its cavalry formations, the mounted arm enjoyed limited expansion. By 1876, Russian cavalry numbered nineteen divisions (two Guards, fourteen army, two Cossack, and the Caucasian). While the Guards divisions varied in their composition, the normal complement of each regular cavalry division was one regiment each of dragoons, uhlans, hussars, and Cossacks. The usual composition of each cavalry regiment was four squadrons, while Cossack regiments retained their traditional complement of six "hundreds" (*sotnii*). On the eve of the Russo-Turkish War, the field army deployed cavalry forces of 237 squadrons and 130 Cossack hundreds.[76]

Because cavalry proved expensive to maintain and the Russian Empire could call upon other mounted assets, cavalry units did not maintain the same kind of expandable reserve formations as the infantry and artillery. By 1875, the cavalry reserve system had evolved to provide one reserve squadron for each active regiment, for a total of fifty-six reserve squadrons. Their function was the peacetime training of recruits and the wartime dispatch of replacements for combat losses. True wartime expansion of cavalry was a function of the Empire's retention of large Cossack formations available for duty upon mobilization.[77]

One of the major unintended legacies of the Miliutin era was a revitalization of Cossack military service. In 1861, the Cossacks of Imperial Russia formed ten separate hosts (*voiska*) and several lesser formations. These hosts, of which the Don was the largest, were the direct or indirect descendants of frontiersmen and other fugitives who had originally fled to the periphery beyond the reach first of Muscovy, then of Imperial Russia. In the eighteenth century, following subordination to St. Petersburg, the Cossacks had performed useful auxiliary mounted service, while in the nineteenth century they were increasingly called upon to augment the regular cavalry, especially for scouting and reconnaissance purposes, and occasionally as battlefield disrupter and long-range strike force.[78] However, in the post-Crimean period, Miliutin viewed the Cossacks as an expensive anachronism whose populace resisted the imposition of uniform military legislation and whose interests ran counter to social and economic modernization. The War Minister's initial inclination was to limit their service and gradually abolish their special status.

The Franco-Prussian War, however, imposed a mid-course change upon Miliutin. With cavalry in greater need for operations involving border security, screening, and the disruption of enemy mobilization, the curtailment of Cossack service suddenly appeared unwise. Short of spending scarce resources to strengthen the regular cavalry, which was difficult to train and expensive to maintain in peacetime, the best answer for additional cavalry seemed to lie with reaffirming the Cossacks' traditional service and imposing on them a more uniform military obligation.[79] Consequently, an important aspect of cavalry reform after 1874 was not only the gradual incorporation of Cossack units into regular cavalry divisions but also the integration of pure Cossack formations into a rational mobilization regime. As a result, in 1876, the nearly two million Cossack inhabitants of the Russian Empire provided the population basis for the wartime fielding of 887 squadron-size units, 264 guns, nineteen battalions of infantry (*plastuny*), and a number of smaller detachments. Full Cossack mobilization could call to the colors approximately a quarter of a million men, of which the Don Host accounted for about 40 percent.[80]

The composition of the army also included more specialized forces. The 1864 table of organization called for engineer troops numbering eleven sapper battalions and six pontoon half-battalions. Fortress troops provided another six regiments and three separate battalions of infantry and fifty-four artillery companies. In 1876, the number of sapper battalions had grown to fifteen (eleven active and four reserve), while the number of pontoon battalions remained stable. New engineer formations included three railroad battalions, six military telegraph parks, and four field engineer parks. By 1876, the engineers also held responsibility for two siege engineer parks which served as depots for siege artillery.[81]

Finances, changing technology, and new army tables of organization all played important roles in determining the fate of fortresses and their troops. In the beginning of the 1860s, Russia had twenty-nine fortresses, most of which were badly in need of either modernization or complete reconstruction. With memories of the Crimean War still fresh, the initial intent was to allocate ten million rubles to rebuild a substantial portion of them on the model of Sevastopol. When the events of 1870–71 underscored the sharply changing impact of railroads and rifled artillery on siege warfare, the result was consternation and indecision. Only after considerable debate did the secret Conference of 1873 decide to allocate an additional sixty-six million rubles for the reconstruction of fortresses covering the western military frontier, especially Ivangorod, Novogeorgievsk, Warsaw, and Brest-Litovsk. However, this work would require some thirty years to complete, and by then it would be outdated.[82]

With these considerations as background, in 1874 the War Ministry formed a special category of fortress infantry to free regular infantry from fortress duty. Roughly analogous to reserve formations, the peacetime cadres of the twenty-nine fortress battalions (reduced to twenty-four in 1876) could be expanded in wartime to become four-battalion regiments. At the same time, in 1876, fortress artillery troops numbered fifty battalions, with a composition that varied from 400 rank-and-file (peacetime) to 1,200 (wartime). Unlike the fortress infantry, artillery troops could be called upon to discharge detached duty as siege artillery detachments.[83]

A number of local and native contingents rounded out the total manpower of the Russian army. The so-called *inorodtsy*, or native troops, yielded a number of colorful formations, including the Bashkirs and the various Caucasian cavalry detachments. In wartime, their numbers added approximately 23,500 troops to the field army.[84] In addition, each province retained local battalions for guard and convoy activities, but as the army increasingly attempted to distance itself from the internal security function, these troops were gradually supplanted by gendarmes.

In the aggregate, thanks to varying internal and external considerations, the size of Russia's active and reserve military forces fluctuated considerably during the period between the onset of Miliutin's reforms and the conclusion of the Russo-Turkish War. In 1862, the active army numbered 792,761 officers and soldiers. The real—as opposed to paper—reserve backup was slightly more than 100,000. Despite the need to reduce expenditures, the Polish insurrection of 1863 together with final pacification of the Caucasus actually required a temporary increase in the size of the active army. In the spring of 1864 it reached a strength of 1,132,000, then entered a period of reduction, thanks to completion of operations both in Poland and in the south. Initially the Tsar authorized a reduction by 264,000, and by 1867 active army strength dropped to 742,000, a peacetime figure that would remain steady over the next two decades. Thus in 1874 the active army numbered 754,265 officers and soldiers. However, on 1 January 1878, at the height of the Russo-Turkish War, the strength of the active army climbed to slightly more than 1.5 million. Although this figure still fell short of the roughly 1.9 million required for war against an Austro-German coalition, the total available forces mobilized provided eloquent testimony to the growing effectiveness of Miliutin's cadre and reserve system.[85]

Rearmament

While nearly exhausting himself and his political capital on issues of compulsory service and military organization, Miliutin also undertook the rearmament of the Imperial Russian Army. This task was particularly complex because of its urgency, the uncertainty of new technologies, and the enormity of expenditures associated with their procurement. Whatever the complications, changes in armament had momentous organizational and tactical importance in an era of rapidly changing military technology.

Few who witnessed the Crimean War seriously challenged the need to rearm the Russian army with a modern shoulder weapon to replace the .70-caliber smoothbore musket which had been a fixture of infantry existence since the early eighteenth century. In 1857, this weapon was officially superseded by a .60-caliber rifled muzzle loader manufactured in Germany and Belgium. By 1862, 260,106 of these rifles, called the *vintovka* after the term officially adopted for rifled shoulder weapons the previous year, had been issued.[86] However, when the Austro-Prussian War of 1866 seemed to demonstrate conclusively the superiority of breechloaders over muzzle loaders, the Russians were faced with an expensive decision: either adopt a new weapon or modify current weapons to load from the breech. After much debate, experimentation, and comparative testing, the Russians pursued both courses with lamentable results.

Difficulties began in 1867, when the War Ministry initiated purchase of 200,000 muzzle loaders converted to breechloaders by I. Karle, a Swedish inventor. When the Karle system revealed serious defects, the Grand Duke Nicholas Aleksandrovich reached into his own pocket to purchase for limited trials 10,000 rifles of the Baranov model, a Russian conversion system built around a trapdoorlike action that was gaining popularity in both Europe and the United States. No doubt Miliutin was saved considerable political embarrassment when the War Ministry's examination and testing of various alternatives revealed the suitability—chiefly for financial reasons—of a substitute conversion system designed by the Bohemian armorer Sylvester Krnk (also frequently rendered as "Krenk"). In 1869, the Russians abandoned the Karle for the Krenk conversion, purchasing 800,000 of the .60-caliber breechloaders. These weapons fired metallic cartridges and featured a breechblock which the shooter rotated to the left for loading. The Krenk had a maximum range of 2,000 paces, but like other conversion systems of the period, it was fragile and often experienced extraction difficulties under conditions of prolonged fire.[87]

Meanwhile, several Russian officers, including Colonel A. P. Gorlov and Guards Captain K. I. Gunius, had traveled to the United States, where, after testing numerous weapons, they found promise in the rifle of Hiram Berdan, sharpshooting veteran of the American Civil War. The two Russians spent much of 1867 and 1868 at Berdan's side, devising a suitable metallic cartridge system of .42 caliber and suggesting a series of modifications to improve Berdan's basic design, which incorporated a trapdoor action similar to the Baranov. While Gorlov and Gunius continued experimentation and testing, the War Ministry placed an immediate order for several thousand copies of their modified rifle (Model 1868), which came to be called the Berdan No. 1. Early in 1869, even as the Krenk was winning out over the Karle, Berdan himself suddenly appeared in St. Petersburg to tout another variant of his basic design incorporating a sliding bolt action of great strength and simplicity. This rifle eventually was called the Model 1870, or more simply, the Berdan No. 2. In 1870, despite the recent Krenk adoption and thanks to favorable test results, the ballistic superiority of the .42-caliber bullet, Miliutin's advocacy, and other reasons, the Berdan No. 2 won formal adoption as the Russian army's primary shoulder weapon.[88]

A bolt-action, single-shot rifle of rugged design, the "Berdanka" was to win great popularity during the 1870s and 1880s. It fired a lubricated and paper-wrapped bullet of .42 caliber with an effective range of 450 paces and a maximum range of 4,000 paces. Only limited quantities were available from abroad until 1876, when the Tula Arsenal at last completed arrangements for domestic manufacture of the sliding bolt mechanism. By the spring of the following year, even though 232,000 Berdans were on hand, more than half of the Russian Army was armed with the Krenk, a weapon that was obsolete by the army's own standards. Consequently, on the eve of hostilities with Turkey, the Berdan's adoption left the Russian army with two major shoulder weapon systems and several minor ones (the Karle and Berdan No. 1), a turn of events which would cause significant training, supply, and morale problems. As one Soviet historian has described the situation, Russian infantry armament "was not distinguished by uniformity."[89] In addition to weapons disparity, the false starts of the late 1860s created still more expenditures in a state budget already stretched to the breaking point by other military programs.

A minor but complementary program involved the purchase of sidearms for the cavalry and gendarmes. Because the rubles expended were fewer, the impact less dramatic, and the numbers smaller, the War Ministry understandably devoted less systematic attention to the procurement of pistols than rifles. Almost as an afterthought during his stay in the United States, Colonel Gorlov had reviewed several alternative sidearm systems, including primarily those of Colt and Smith and Wesson. Despite some sentiment favoring an English self-cocking (double-action) revolver, three successive small arms contracts went to the American firm of Smith and Wesson, which eventually supplied Russia with 142,333 copies in three variants of the Model 3 revolver, a rugged .44-caliber single-action (non-self-cocking) pistol which featured a break-open design and an automatic cartridge ejection system for quick reloading. The heavy but thoroughly reliable Smith and Wesson remained the standard Russian sidearm for nearly two decades.[90]

More significant armament programs included artillery design and procurement. As in the case of shoulder arms, the Crimean experience seemed to demonstrate the need for rifled artillery weapons. However, unlike the superiority of the new shoulder weapons, that of the new order in artillery was not always immediately evident. Artillerymen needed rifled tubes to enable them to engage infantry at ranges equal to or greater than the latest rifled muskets, but the problem was that rifled technology exacted a number of technical and tactical trade-offs that either cast into doubt the merits of rifled cannon or required changes in employment that taxed the ability of artillerymen to adapt. Additions in range and changes in trajectory deprived artillery of the ability to employ grazing fire effectively. Rifled muzzle loaders were slower to load than their predecessors. Until the 1860s, breechloaders remained an unknown and unproved quantity. Rifled cannon fired smaller projectiles at greater ranges, but until improved rounds and fuzes came into existence, rifled rounds were not nearly as effective as traditional charges of grape, canister, and round shot. Newer metals such as cast steel permitted greater pressures and higher muzzle velocities, but novel casting processes were untried and often enough caused premature failures under conditions of normal use. To worsen the situation, tacticians were unsure of how to employ the new technology. If cannon went forward into the attack with infantry, the infantry could always go to ground under fire, while artillerymen stood naked behind their guns. If cannon remained in the rear, artillerymen lost the ability to afford direct fire support and suffered from infantry gibes. It was against this background of uncertainty that the War Ministry, first under Sukhozanet, then under Miliutin, proceeded with the rearmament of the Russian field artillery.[91]

As was the case with shoulder weapons, artillery rearmament suffered through several false starts before the War Ministry emerged with what seemed to be a workable solution to its problem. Already during the Crimean War the Russians resorted to expedient measures to close the artillery gap. One was simply to begin rifling existing brass smooth-bores. Twelve-pounders underwent immediate conversion, and after the war six-pounders also joined the conversion ranks. However, the product scarcely justified the expenditure, with the result that the War Ministry turned to the Krupp Works in 1857 with an order for several cast steel tubes for experimentation. When the Austro-Sardinian War of 1859 once again called attention to the importance of rifled artillery, the War Ministry, lacking the means and technology to turn fully in the direction of casting rifled cannon, went ahead with producing new model smooth-bores (four- and eight-pounders) of bronze. Between 1858 and 1862 Russian arsenals produced sufficient bronze cannon to reequip completely 129 foot and eighteen horse artillery batteries with 1,104 guns. Meanwhile, experimentation continued under A. V. Gadolin and N. V. Maevskii on the casting of bronze rifled breechloaders, with the result that in 1867 they emerged with an approved system of four- and nine-pounders. With the addition of metal carriages in 1868, these bronze breechloaders became the mainstay of the Russian field artillery until the 1880s.[92]

The pieces designed by Gadolin and Maevskii were the latest word in bronze breechloading technology. The nine-pounder incorporated a sliding wedge breech block system and weighed just over a ton. Its effective range was considered to be 3,200 meters, although it could throw high-explosive shells as far as 4,480 meters. The four-pounder weighed about 1,600 pounds with corresponding ranges of 2,560 and 3,400 meters. In addition to these two main types, Russian arsenals produced a smaller three-pounder mountain howitzer suitable for transport by pack animals to areas otherwise inaccessible to artillery. The mountain gun weighed just over 500 pounds and could hurl its shells 1,423 meters.

Of crucial importance to field artillery employment was the selection of ammunition devised to accompany the transition to rifled weapons. By 1876 the Russians had settled on three different rounds: high-explosive shell with impact fuze, shrapnel with time fuze, and canister. High-explosive shell worked well against wooden and stone structures but proved ineffective against earthworks. Both shrapnel and high-explosive shell were very effective against unprotected personnel. However, the shrapnel round retained its greatest effectiveness at mid-range, after which an appreciable increase in elevation and a corresponding drop off in velocity reduced the area of bursting coverage. The canister round contained fewer projectiles than its smooth-bore predecessor but was considered effective to 400 meters. The pace of change would require time and training for artillerymen to accustom themselves to the strengths and limitations of their new weapons and projectiles.[93]

Although both the Austro-Sardinian War of 1859 and the Austro-Prussian War of 1866 had awakened the Russians to the growing importance of rifled artillery, it was the Franco-Prussian War of 1870–71 that spurred Miliutin to speed the rearmament of tsarist field artillery forces. Since the adoption of bronze breechloaders a few years before, Russian arsenals had been turning out the new pieces at a leisurely pace varying from forty to 250 per year. During the next few years, however, Prussian victories instilled a new sense of urgency, both in time and numbers. On 1 January 1873, Miliutin reported to the Emperor that "the decisive influence which field artillery had on the success of engagements in the recent Franco-Prussian War has elicited everywhere an expansion of field artillery; with us the expansion is all the more necessary because the ratio of our field artillery to other types of weapons at present is far lower than the existing relationship in armies of the chief European states."[94] Miliutin's answer to the challenge was to raise production of bronze nine-pounders to 300 guns annually, so that by the end of 1875 rearmament with rifled breechloaders was substantially complete.[95]

Less successful was a brief attempt to boost firepower through the incorporation of Gatling-type guns into field artillery brigades. In 1870, the Russians had begun experimentation with the Gatling system, which employed ten rifled tubes fed by magazine and rotated by a hand-driven mechanism around a central axis to fire at a rate of approximately 200 rounds per minute. Between 1874 and 1876, after the Russian armorer V. S. Baranovskii had modified the Gatling to reduce its weight and increase the rate of fire, an eight-gun Gatling battery was added to each artillery brigade. However, because of weight, technical difficulties, and limited range, the Gatlings were removed from field service in 1876 and relegated to fortress armament.[96]

The Officer Corps and Entry-Level Education

Miliutin's strong views on the necessity to improve the officer corps meant that officer recruitment and education would not escape reformist scrutiny. Prior to 1861, the officer corps had traditionally come from four sources: graduates of military academies and schools, graduates of the cadet corps and Guards military schools, outstanding noncommissioned officers, and those promoted to officer rank from volunteers, chiefly the half-educated offspring of the Russian nobility.[97] Of the

50,567 officers obtained from these sources between 1825 and 1850, only 14,415 had graduated from military educational institutions. In consequence the officer corps was characterized by great disparities in education, a problem Miliutin could not help noticing even in the 1830s. A second difficulty was a shortage of candidates for officer rank. In 1861, for example, 1,270 new officers entered the army while 4,271 left, a difference which, if sustained, spelled trouble for a military institution on the threshold of significant reform and modernization.[98]

Beginning in 1863, Miliutin implemented a series of measures to alleviate shortages and improve officer education. First he abolished the cadet corps, transforming them into twenty military gymnasia with modified curricula which de-emphasized military drill and pedantry. To replace the corps Miliutin created a system of military schools (*voennye uchilishcha*) for the various service branches. These military schools accepted graduates of the military gymnasia and subjected them to a four-year curriculum emphasizing military subjects. Graduates in the upper half of each class received the rank of second lieutenant, while graduates of the lower half received the rank of ensign. Instructional stress was on tactics at the regimental level, although other courses of study included two foreign languages, chemistry, mechanics, analytic geometry, and Russian literature. The military schools produced between 400 and 500 graduates annually.

The shortfall in qualified junior officers prompted Miliutin to create additional educational institutions, the junker schools, within each of the military districts. Candidates for these schools were young soldiers within the various regiments who qualified for vacancies upon passing rigorous entrance examinations. By background, junkers who aspired to officer rank ranged from graduates of the special secondary schools, to those who had completed a portion of the military gymnasium, to those who had graduated from civilian educational institutions. Within the junker schools the course of instruction lasted for two years. First-year study included general subjects, while the second year included military subjects such as tactics, weapons, topography, and the fundamentals of artillery. The bulk of tactical instruction was at the company level. Upon graduation the junkers returned to their regiments, where their advancement to officer grade depended upon vacancies and the discretion of the regimental commander.[99] By 1877 the junker schools had graduated 11,500 officers, of whom approximately three-fourths were of noble origin. Although graduates of both the military schools and the junker schools assured the peacetime army a sufficient flow of officers, mobilization in 1877 revealed a shortage of some 17,000 junior officers in the wartime table of organization.[100]

Even though the purpose of new institutions was to produce literate officers with specialized military knowledge, Miliutin was aware of the necessity to provide additional resources for extended officer education. One expedient was to provide regiments within the military districts with small libraries stocked with books on tactics, military history, and military theory. Another took the form of a directive of the early 1870s to require junior officers to participate actively in each military district's annual "small maneuvers." Miliutin directed that during these summer exercises senior officers supervise their subordinates in making terrain sketches, maneuver diagrams, and plans for field fortifications. In 1875 Miliutin levied additional requirements on junior officers for participation in tactical exercises on maps and in the field. He wrote that it was "fully desirable that the closest supervisors of young officers must be company and squadron commanders."[101]

Higher-Level Officer Education

Miliutin pinned his ultimate hopes for a reformed army on the younger officers, but in the interim he could not afford to neglect the education of his senior officers. Indeed, no less significant than the shortage of junior officers was a shortage of educated senior officers. At the end of the Crimean War Russia possessed three senior service academies, the Nicholas Academy of the General Staff, the Nicholas Engineering Academy, and the Michael Artillery Academy, the latter two of which had been created only in 1855. Although there was some talk of combining the three into one vast "Imperial Military Academy," each retained a high degree of autonomy until 1862, when Miliutin appointed a trusted colleague, Major General N. V. Isakov, to head the Main Administration of Military Educational Institutions under the War Ministry. Between 1855 and 1862, all three academies had produced an average annual total of only 143 graduates. Under the Miliutin-Isakov regime the situation would grow worse before it became better. Despite chronic shortages of qualified staff officers, the War Ministry tightened entrance requirements to the General Staff Academy. Beginning in 1863 applicants had to pass a stiff entrance examination, have spent no fewer than four years' duty with troops, and have presented a statement from their regimental commanders attesting to their loyalty and devotion to duty. Thanks in part to these requirements, the number of applicants in 1863 dropped to eight, and between 1866 and 1870 only ninety-six officers were to graduate from the Staff Academy.[102]

On the positive side, the quality of graduates steadily improved. While the Artillery and Engineering Academies continued to produce staff specialists in their respective areas, the Academy of the General Staff gradually blossomed into a full-fledged staff college. The Academy provided two years of instruction, with officers in the "theoretical course" spending the initial summer session with troop units, while officers in the "practical course" (topography) took to the field for the same period with their maps and survey instruments. During the regular academic year, the main syllabus consisted of instruction in tactics and strategy, with additional courses in military history, military administration, military statistics, spherical geometry, cartography, sketching, and drawing. Subsidiary subjects included Russian language, artillery and engineering practice, political history, and foreign languages. During the early 1870s, introduction of the Russian equivalent of a "Staff Ride" and the technique of war gaming added six months to the Academy curriculum, bringing the total period of instruction to nearly three years.[103] Curricular changes took place against the background of a lively (but unresolved) debate over whether higher military education ought to emphasize more specialized military subjects at the expense of general education. Miliutin, possibly under Obruchev's influence, apparently believed in an overall emphasis on specialized professional education with the assumption that general education should be left either to the officer himself or to his individual background. In contrast, G. A. Leer (Leyer), Professor of Strategy, objected to the Academy's becoming "a factory for general staff officers" because in any case a staff officer's preparation should emphasize the generalist's approach to complicated, multifaceted problems. What Leer wanted was an institution that would assume the attributes of a "military university."[104]

In no small measure, Leer's approach reflected his own shifting sense of academic priorities. Although he had won early prominence at the Staff Academy with his incisive tactical and engineering studies, Leer became increasingly preoccupied with strategy, especially as it related both to military history and to a larger body of scientific knowledge. In the eyes of Leer and many of his contemporaries, the Crimean War had revealed a yawning gap between theory and practice in military art. To bridge the gap, Leer proposed a systematic study of strategy as it had evolved through the ages, with an emphasis on Napoleon and the modern era. Building on the rich traditions of N. V. Medem and Antoine Henri Jomini and borrowing from contemporary positivist trends, Leer advocated the use of history as a factual data base from which to derive the fundamental and unchanging principles of military art which underlay contemporary strategy. In 1867, he published in serial form in the *Military Collection* an early version of his textbook on strategy. Although he devoted several dozen pages to the importance of railroads, noting that they could speed the ground movement of troops by a factor of six and their supply by a factor of fifteen, Leer stubbornly refused to view contemporary military realities through anything but the theoretical prism of Napoleonic experience.[105] In 1871, while delivering lectures at the Academy on the implications of the Franco-Prussian War, Leer would correctly conclude that steam, electricity, and firepower distinguished the contemporary era from that of Napoleon. However, while admitting the importance of railroads for speedy concentration and timely supply, Leer concluded that the expression "march separately, but fight together" belonged not to Moltke but to Napoleon. Leer further asserted that "if in 1866 and 1870 the Prussians had operated on exterior lines because the situation so required, then it still did not follow from this that interior operational lines and the principle of concentration of forces had outlived their time."[106] Thus Leer's devotion to Napoleon and preoccupation with theory caused him to fix on the wrong trees while the entire forest around him was changing.

Throughout his career at the Academy, Leer would continue to press the limits of theory, thus presenting a figure of sharp contrast with his fellow instructor, M. I. Dragomirov, who remained stubbornly wedded to applied tactics. And much of the Academy's history over the next three decades could be written as a function of their varying methods, focuses, and emphases. At stake was not only the intellectual future of the army but also the age-old question of whether theory or application should receive greater stress at an institution of higher military education. The answer, of course, lay in a balanced approach, but Leer and Dragomirov were to establish legacies which did not easily admit compromise.[107]

Whatever the emphasis, the growing importance of the Academy of the General Staff undeniably contributed to a deepening of Russian military professionalism. As graduates rose to high positions, education came to be viewed as an important key to career success, although academic attainment never completely effaced the advan-

D. A. Miliutin (1816–1912)
Minister of War 1861–1881
Photo: Skalon, *Stoletie Voennogo Ministerstva*

M. I. Dragomirov (1830–1905) Tactician, Commandant of the Nicholas Academy of the General Staff 1878–1889, Commander of the Kiev Military District 1889–1904
Photo: Glinoetskii, *Istoricheskii ocherk*

N. N. Obruchev (1830–1904)
Chief of the Main Staff 1881–1887
Photo: *Russes et Turcs: La Guerre d'Orient*, 2 vols. (Paris, 1877–78)

G. A. Leer (1829–1904) Strategist, Historian, Commandant of the Nicholas Academy of the General Staff 1889–1898
Photo: *Voennaia Entsiklopediia*

tages of birth and privilege. In fact, the most potent guarantee for advancement was probably a combination of education and privilege. Despite the initially small number of graduates, the army's command and staff structure gradually came to reflect the importance of attendance at the Nicholas Academy as a stepping stone to success. As John Erickson has indicated, during the thirty-year span between 1852 and 1882, the Academy produced 1,329 graduates. Of these, 903 received direct posting to the General Staff, while 327 went on to successively higher command assignments, including 197 to regiments, 66 to brigades, 49 to divisions, 8 to corps, and 7 to military districts.[108] In the field, graduates often served with distinction, causing General Obruchev to remark that during the Russo-Turkish War a general officer who gave a command to a line officer wearing the Nicholas Academy badge had more confidence that the order would be successfully executed.[109] Although much remained to be done, Miliutin had begun to preside over a system in which many of his commanders and mid-level staff officers were veterans of both field command and the course of instruction at the Nicholas Academy of the General Staff. Only later would the emergence of something like a corporate mentality among the *genshtabisty*, as those who bore the appellation "of the General Staff" were called, begin to spawn the characteristics of military elitism.[110]

The Dragomirov Ascendancy

While Miliutin could not fail to address issues of training and tactics, the task of communicating the new Russian military vision to the rank and file belonged to trainers and troop commanders. Already in the early 1860s the officer who came both to reflect and represent—even dominate—the new training and tactical order was Mikhail Ivanovich Dragomirov. Born in 1830 in the Ukrainian town of Konotop, Dragomirov was the noble son of a veteran of the Napoleonic campaigns. Like his father, he elected to pursue a military career, successively entering the Noble Cadet Corps (1846) and the Semenovskii Guards Regiment (1849). In 1854 he successfully applied for admission to the Nicholas Military Academy, where he was exposed to the teachings of Astaf'ev and—like many of his contemporaries—was deeply troubled by the Crimean debacle. After graduation with perfect examination scores, he traveled to western Europe, where he studied briefly at the Ecole Militaire at St. Cyr and then served as an observer of training methods in the French and British armies. In 1859 he was designated to serve as official Russian observer in the brief conflict between Austria and Sardinia but arrived too late to witness events at first hand.[111]

Dragomirov emerged from these experiences with deep convictions about tactics and training and the relationship between the two. Like so many of his reform-minded contemporaries, he urged the army to overcome its obsession with parades and petty regulations, and like Suvorov before him, he demanded that troops occupy themselves exclusively with the development of essential combat skills. Indeed, Dragomirov was responsible for popularizing one of Suvorov's maxims, "Train the troops only to do that which is necessary in combat."[112] For Dragomirov, war was not an exercise in the mechanical movement of chess players but a deadly struggle between two living forces in which the unexpected was the rule rather than the exception. To meet the challenges of modern warfare, the objective of training was to develop a versatile warrior, not a marching marionette.

In part, these convictions sprang from the current conception that war embraced two realms, the physical and the moral (or spiritual). In regard to the moral element, Dragomirov's observations and his study of history caused him to conclude that combat subjected men to two conflicting impulses: self-preservation and self-sacrifice. In turn, these impulses were the product respectively of intellect and will. To accommodate the apparent dichotomy between intellect and will in combat, Dragomirov proposed that training for combat be divided into two categories, indoctrination (*vospitanie*) and training (*obrazovanie*). The former involved cultivation of loyalty, courage, and patriotism, while the latter involved marksmanship, bayonet drill, and physical training.[113]

For Dragomirov, intellect and will manifested themselves on the battlefield respectively in firepower and cold steel. Because the very essence of victory lay in imposing one's will on the enemy, Dragomirov held that the bayonet attack was the decisive act in battle. Any exchange of fire was merely a prelude to a climactic struggle of wills as two opposing forces encountered each other in hand-to-hand combat. In accordance with these assumptions, he chose to emphasize the importance of will and the spiritual factor in his training and tactical conceptions. While Dragomirov was not unaware of the physical or material element in battle, his conceptions were more congenial to an emphasis on the moral factor.

His conceptions were also congenial to Russia's new military order, in which the army's rank and file were no longer serfs but free men whose training required an accent on incentive and the positive aspects of combat. Beginning in 1860 as Adjunct Professor of Tactics at the Military Academy, Dragomirov emphasized the need for a unified approach to military training that would direct the whole of the soldier's conscious being to the attainment of military objectives. From 1862, he also served as a member of Miliutin's Special Committee on the Structure and Training of Troops, a position which enabled the young officer to write many of his ideas into doctrine and practice. Dragomirov was instrumental in adding a new dimension to Russian training methods. In accordance with his convictions, he encouraged the War Ministry to divide military training into two categories: indoctrination and training. To develop the will, to indoctrinate, the drillmaster had to instill devotion to sovereign and country, discipline, belief in the sanctity of orders, courage, determination, and esprit de corps. Training involved the development of such skills as marksmanship, bayonet fighting, gymnastics, and tactics. Of the two categories, indoctrination was the more important, for in Dragomirov's scheme an unskilled but dependable soldier was preferable to an accomplished but undependable one. In addition, a motivated soldier would learn skills more diligently than his indifferent counterpart.[114]

Infantry Tactics

Dragomirov's conceptions were more successful in bridging the training gap between an army of serfs and one of free men than they were in bridging the gap between technology and tactics. In addition to larger issues of organizational and social environment, several powerful developments affected the evolution of tactical conceptions and regulations. One was the Russian experience in the Crimean War. Another was the vicarious experience of other armies in wars of the period, including chiefly the Austro-Sardinian War of 1859, the American Civil War, the Danish War of 1864, the Austro-Prussian War of 1866, and the Franco-Prussian War of 1870–71. Dragomirov and his contemporaries searched the evidence of these conflicts to draw important conclusions for the evolution of Russian tactical doctrine. There was considerable divergence of opinion, but the system that emerged was a compromise between the old and the new, a compromise that reflected the prevailing views of Dragomirov over some of his bolder fellow officers.

Combat experience since 1853 indicated that the failure of offensive action stemmed primarily from the heavy losses inflicted upon attackers as they left their preliminary assault positions to close with the enemy. Losses were the result of rifled shoulder weapons, which enabled defenders to engage their attackers at longer ranges with greater accuracy over longer periods of time. In contrast with the smooth-bore muskets of an earlier era, the rifled musket enabled the infantryman to engage targets accurately at distances ranging from 500 to 1,000 paces. This conclusion caused analysts and commentators to seek new ways of conducting offensive action while reducing losses in battle to an undefined but acceptable level. Differing solutions appeared.[115]

Perhaps the most radical solution was that of A. I. Astaf'ev, who in 1856 published a treatise, "On Contemporary Military Art." He acknowledged the impact of modern weaponry on the conduct of battle and concluded that rifled weapons had changed the basic conditions of the battlefield. For Astaf'ev, the chief task for the tactician was to devise a way of utilizing the new power of rifled weapons while simultaneously reducing losses. Experience indicated the necessity of adopting on a massive scale the tactics of the dispersed formation (*razomknutnyi stroi*) and the skirmish line (*strelkovaia tsep'*). Astaf'ev advocated adoption of the open order not only by infantry companies but also by battalions, regiments, and brigades. For him the older column formations retained usefulness only in the approach march. If a bayonet attack proved necessary, then dispersed units would concentrate momentarily to deliver a single massive stroke. Adaptation to the terrain and even entrenchment assumed new significance for troops on Astaf'ev's dispersed battlefield. To go with dispersed tactics Astaf'ev also advocated the adoption of uniforms whose colors would make troops less conspicuous on the battlefield and whose style would facilitate greater freedom of movement. He did not address the important issue of battlefield control of troops.[116]

Following Astaf'ev, Leer and Dragomirov professed more systematic views on the important tactical issues of the day. Each continued to write profusely and to wield enormous influence within the Russian officer corps during the 1860s and 1870s. More than Dragomirov, Leer embodied the new concern with the development of military studies as a science. He defined tactics as the theory and art of the most advantageous application of elements (troops, weapons, terrain) in engagements and battles. Just as in the case of strategy, Leer held that a critical analysis of facts provided by military history formed the basis for the development of tactical theory. In his textbook on tactics, he advocated adherence to the deep order, thereby reflecting a tendency common for the period to underestimate the importance of firepower while overestimating the importance of cold steel. Still, his textbook was useful for the way that it emphasized coordinated action by the combat arms, the importance of dispositions, and the necessity for reconnaissance of the enemy and terrain.[117]

Dragomirov echoed many of these sentiments, although he strongly resisted the attractions of science. He felt that there should be no difference between theoretical and applied tactics, and indeed, his greatest contribution was to insist on the unity of training and tactical practice. In the realm of tactics, he veered away from the radical solution proposed by Astaf'ev, giving far less credence to the importance of firepower in modern battle. Dragomirov saw the object of tactics to be a search for the best combination of firepower and cold steel to achieve battlefield victory. He saw the bullet and bayonet as complementary, and not surprisingly he proposed two basic battlefield formations, open and closed. "Missile weapons require open formation, and cold [steel] closed," he wrote in his textbook for military schools.[118]

From his belief in the efficacy of mind over matter, Dragomirov insisted that the closed formation would remain the primary assault formation. However, he also found the evidence of recent wars difficult to ignore; therefore he conceded that the dispersed formation had earned for itself a prominent place in the attack. It was a place that required a considerable degree of independence. He saw the need to conduct exercises and maneuvers to work out a balance between the two formations, especially in conjunction with cavalry and artillery. In addition, Dragomirov wrote of the necessity to adapt formations to the terrain, labeling the utilization of tactical terrain features either "passive" or "active." The ability to overcome obstacles he called "active" adaptation to the terrain. The utilization of terrain to provide cover he called "passive."[119]

Dragomirov's views on adaptation to the terrain and marksmanship reveal the degree to which his tactical conceptions had become hostages to his larger views on morale and élan. He believed that a preoccupation with aimed rifle fire caused soldiers to become inordinately concerned with marksmanship, a concern that reduced momentum in the attack as soldiers halted to engage targets. Nothing should be permitted to break the flow of concerted energy which culminated in a successful assault. Similarly, a concern with finding cover on the battlefield caused the soldier to become preoccupied with the issue of personal safety. Nothing sapped the strength of the attack faster than a concern for self-preservation. Therefore Dragomirov advocated entrenchment only as a means of creating artificial obstacles to halt an enemy advance.[120]

Thanks probably to Miliutin's insistence, Dragomirov throughout the 1860s occupied the Chair of Tactics at the Nicholas Academy. From that position he exercised considerable influence through his teachings and writings on the evolution of a new generation of tactical regulations. He also participated directly in fashioning those regulations as a member of the Committee on the Structure and Training of Troops which Miliutin created in 1862 to oversee the revision of various field and tactical regulations. It was therefore no accident that the regulations of the 1860s and 1870s bore the stamp of Dragomirov's thought.[121]

Perhaps nowhere was the blending of old and new under Dragomirov's aegis more apparent than in the regulations governing infantry assault tactics. *The Regulation for Infantry Combat Service* (*Ustav stroevoi pekhotnoi sluzhby*), published in 1866, was the first new infantry regulation since 1831 (amended in the 1840s). In general terms, the regulation supported an increase in infantry fire as a prelude to the attack with cold steel. In preparing for and conducting the attack, therefore, the regulation prescribed a combination of the open and closed orders, with a considerable degree of flexibility left to commanders in modifying their formations to suit circumstances. The regulation itself was divided into three parts. The first, "Recruit School," reflected the new attitudes of writers such as Dragomirov who emphasized a positive approach to training in fundamentals. The regulation stated that "if a recruit does not do something or does it poorly, this proceeds with only rare exception not from negligence but from an insufficient understanding of the requirements." In short, the emphasis fell on simplified instruction, the utility of example, the importance of encouragement, and abandonment of all but the most necessary training. This approach corresponded fully with Dragomirov's oft-repeated maxim (from Suvorov) "to teach only that which is necessary in combat."[122]

The second and third sections of the regulation dealt with tactics ranging from company to regimental level. In the Russian system which evolved during the 1860s, the regiment was made up of three battalions, each of which was divided into five companies (four line and one "sharpshooter" [*strelkovaia*]). As applied to this organization, the infantry regulation called for a new emphasis on the two-rank formation and made the battalion the basic tactical entity. A battalion attack formation consisted of one or two lines of closed company columns deployed 150–300 paces behind a skirmish line of sharpshooters. The battalion's frontage extended in width from 200 to 400 paces, and in depth from 500 to 700 paces. In normal deployment for the attack, the battalion's four line infantry companies advanced in close order to deliver a bayonet assault behind a skirmish line established by the fifth company of sharpshooters. As the entire formation moved forward, the skirmishers advanced by leaps and bounds, 50 to 100 paces at a time. The line companies followed in tight formation, usually in nonstop fashion, although in theory they were permitted under fire to halt, lie down, and even disperse. As the line companies approached to within several hundred paces of the enemy position, the sharpshooters moved off to one or the other flank to open the way for the line infantry's uninterrupted bayonet attack.[123]

The prescribed assault formation placed four-fifths emphasis on cold steel and one-fifth on firepower. In practice, an obsession with fire discipline reduced still further the emphasis on fire. Regulations stipulated that the skirmish line open fire at mass targets from 600 to 800 paces and at individual targets from 300 paces. In training and combat the watchword was to conserve ammunition. Dragomirov, for example, demanded that his skirmishers expend not more than one-half their personal load, that is, thirty rounds, in covering the attack.[124] If Dragomirov represented the norm, and if commanders shrank from permitting the line infantry from

REGIMENTAL COMBAT FORMATION (OFFENSIVE), 1876

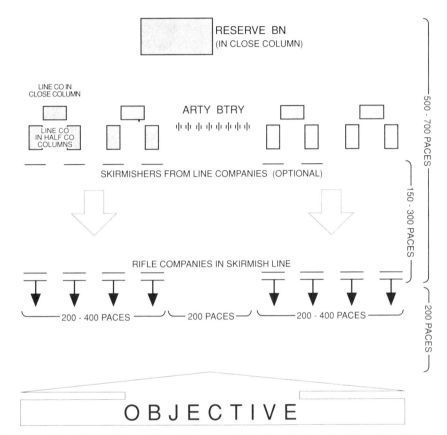

Sources: *Razvitie taktiki russkoi armii*, 1957; and *The Armed Strength of Russia*, 1873

opening fire in the attack, then each battalion limited itself to one-tenth of its firepower potential in the attack. The assumption clearly was that firepower played an auxiliary role to cold steel in the attack.

This sense of priorities was also reflected in organization and attitude. B. N. Grekh, a contemporary infantry officer, would later write that "sharpshooters were strictly differentiated from the line infantry. They trained the first for action by fire in the skirmish line and the latter by hallowed tradition they allocated almost exclusively to the bayonet attack."[125] Although the regulation paid lip service to flexibility, commanders were loathe to diminish their line infantry by using line companies to reinforce skirmishers. And although in theory skirmishers possessed latitude in choice of targets and mode of fire, in practice commanders preferred to keep a tight rein on the skirmish line. Line infantry was taught primarily how to fire by volley from closed formation. Similarly, the regulation, though acknowledging the importance of rifle fire, did not allocate sufficient depth to the formation and thus revealed a basic inability to visualize the battlefield effects of rifle fire. An obsession with orderly formation caused commanders to resist reinforcing the skirmish line and preclude uneven movement from cover to cover. On the contrary, Grekh maintained that the attack was widely practiced "upright with dressed ranks."[126]

The experience of contemporary war seemed to challenge Russian practice, but not conclusively enough to cause alterations in basic infantry tactical doctrine. During the early 1870s, the Franco-Prussian War prompted lively debate about the necessity for further change. One observer, Baron L. L. Zeddeler, wrote an influential article for the *Military Collection*, "Infantry, Artillery and Cavalry in Battle and Outside Battle in the German-French War of 1870–1871." Because of the increased firepower of breechloaders, he observed that "hand-to-hand combat occurred only during the attack of local objectives." Otherwise, "*fire was always preferred to the bayonet.*" The explanation was that "firearms had attained a power so unprecedented that it superseded the bayonet attack, at least in this last war." Therefore the Russians needed to consider changes in the way they conducted fire and the methods they used to close with the enemy. Like Astaf'ev, he advocated reliance in combat on individual fire and the skirmish order because "for battle the *basic formation is dispersed*, and only by way of exception *closed*." Zeddeler held that the Russians should place a new emphasis on individual training, increasing as much as possible the time devoted to marksmanship at various ranges, while teaching the soldier adaptation to the terrain and "strict observance of that formation employed in battle, that is, the open order."[127]

Zeddeler's proposals received scant official support. Miliutin's mechanism for debate and implementation of change, the Committee on the Structure and Training of Troops, rejected an increase in range for marksmanship training to 1,000 paces, citing the urgency of "other requirements for combat training." A proposal for the introduction of quick fire also received short shrift because rapidity of fire was "a question associated with an increase in expenditures." Volley fire by command remained the centerpiece of training because volleys "made a powerful impression on the enemy and in peacetime served as means of training people to give complete attention and subordination of their fire to the will of the commander." Nevertheless, the Committee agreed that long-range fire had its place and that perhaps as many as one-third of the best marksmen might be trained in firing at greater range. The Committee made no move to alter combat formations, although it did concede a certain validity to Zeddeler's observation about the growing importance of the skirmish line.[128]

Cavalry

During the two decades between the Crimean War and the Russo-Turkish War of 1877–78, the role of cavalry in contemporary war was a subject of intense debate. Some military observers believed that the advent of rifled weapons had sounded the death knell for the mounted arm. The Russian and British experience at Balaklava was strong indication that the day of cavalry as battlefield shock weapon had passed. The experience of both the American Civil War and the Franco-Prussian War of 1870–71 seemed to confirm this conclusion. However, the American Civil War also pointed the way to an expanded role for cavalry not as battlefield disrupter but as long-range reconnaissance and strike force. As Russian cavalrymen groped their way from old to new and expanded roles, conservative views and institutional inertia figured prominently in tactical and organizational calculations.

For nearly a decade and a half, the forces of conservatism held the upper hand. Dragomirov, an infantry officer, stubbornly applied his views on the efficacy of shock action to cavalry, with the result that adherents of traditionalism drew strength from one of the leading tactical voices of the period. Considerable reinforcement came from the corps of cavalry officers, a traditional bastion of older and more conservative noble families. The monarchy itself wittingly or unwittingly added to inertia by its preoccupation with pomp and ceremony that detracted from tactical preparation. Of the three combat arms, cavalry revealed the strongest predisposition toward the paradomania of an earlier era.[129]

Only in 1869 did a new drill regulation appear, the *Regulation on Cavalry Combat Service*. Its three parts, individual instruction, squadron instruction, and regimental instruction, corresponded in cavalry terms to the infantry regulation introduced three years earlier. Each cavalry regiment was divided into four squadrons and a reserve squadron. The squadron, broken down into four platoons, was the basic tactical entity. The regulation devoted considerable attention to the uniform training of cavalrymen and their mounts to develop strength, agility, and endurance.

In comparison with its predecessor of 1845, the new regulation devoted considerable space to the mounted attack, an emphasis that corresponded with prevailing ideas on the importance of shock action. At a range of 1,000 to 1,200 paces from the enemy a cavalry unit deployed in several lines for an attack in closed order. As advancing troopers closed the range, their mounts' pace quickened successively from a trot to a controlled gallop to a headlong charge. In addition to closed order, the regulation stipulated the need for open-order attacks to pursue broken enemy formations and to assault enemy batteries. Although the regulation devoted considerable space to formations appropriate to parades and reviews, the writers of doctrine evidently saw less need to emphasize symmetry and smartness on the battlefield. The regulation clearly stated that when the squadron galloped into the attack, "the ranks may not be completely closed, and the dress may not be as good as at other gaits, and these should not be required, so as not to lose speed." The importance of firepower in the attack was denied altogether. No provisions were made for dismounted combat, and the regulation clearly stated that "because of the diminished effectiveness of fire from horseback, the stroke with cold steel comprises the principal, and it is possible to say, the only means which the cavalry employs to bring harm to the enemy."[130]

The views of the traditionalists had scarcely appeared in print when the prophets of change began a campaign which forced gradual adjustments to regulations and training. The War Ministry, itself trying to digest the implications of the American

Civil War, attached growing importance in the early 1870s to the views of N. N. Sukhotin, who emphasized the deep raid and reconnaissance as the proper functions of modern cavalry.[131] In 1871, the War Ministry published a regulation that governed dismounted training for dragoons, and in 1873 there followed another prescribing dismounted drill for hussars and uhlans. The assumption was that cavalry troops of various types would find themselves conducting independent operations to cover crossings, to guard defiles, to attack centers of population, and to conduct long-range raids. The same assumption justified the partial armament of the cavalry with the Berdan No. 2 rifle and the allocation to units of gun cotton for demolition.[132]

Additional missions, when combined with the continuing preoccupation with parade and close-order attack, imposed contradictory requirements on the cavalry. On one hand, parade ground routine and the desire—indeed, the obsession—for precision and polish caused many commanders to save their men and mounts for official functions and inspections. The occasional field requirement to demonstrate capacity for the deep raid to seize limited objectives produced startling but predictable results. The horses survived, but their riders, thanks to garrison life, were all but useless. During field exercises in the summer of 1876 the commander of the Warsaw Military District ordered three squadrons of cavalry to ride approximately 144 kilometers in forty-four hours to seize and hold a crossing on the Vistula. At the end of the ride, one official observer noted that the men were so tired that some had hallucinations while others could not perform even ordinary functions. His conclusion was that "they could not have withstood an attack by even a weak adversary."[133]

In contrast, several other exercises, including one in the Vilnius District in 1876, saw a cavalry division cover about one hundred kilometers and demonstrate readiness for combat at the conclusion of the ride. In 1877, the Grand Duke Nicholas Nikolaevich witnessed an exercise in which twelve squadrons of cavalry rode 170 kilometers in two days without serious impairment of their fighting effectiveness.[134]

The readiness of the cavalry for extended campaign was not conclusive, and inconclusiveness in part reflected larger uncertainties about the employment of cavalry in modern war. Although new views gave rise to optimism, they were only the brighter pieces within a larger mosaic of lassitude, inertia, self-comfort, and institutional stagnation. In this respect, the Russian cavalry probably differed little from its European counterparts.

Field Artillery

Field artillery suffered from some of the same uncertainty which plagued cavalry after 1856, but for different reasons. Although few observers seriously questioned the assumption that artillery retained a role on future battlefields, there was little agreement on the exact nature of that role. In addition, the pace of technological change itself gave rise to a double set of complications. First, it tended to outstrip the treasury's ability to finance new weaponry, especially during an era when other—and equally important—requirements levied incessant demands on the state budget. Second, the pace of change taxed too far the capacity of artillerymen to adjust their views and practices to new technological realities.

The Crimean War rudely awakened Russian artillerymen to the challenges of rifled weaponry. Since the era of Peter the Great, they had ridden onto the battlefield, muscled their smooth-bore weapons into position, then used a variety of shot at close range to blast holes in densely packed enemy formations. Artillerymen had protected the infantry, multiplied its meager firepower, and paved the way for its advance or withdrawal. With the advent of rifled shoulder weapons, however, gun crews suddenly found themselves vulnerable to infantry fire. At Alma and Inkerman, enemy riflemen had mercilessly shot down Russian gun crews who dared accompany infantry into the attack. By 1856 many observers concluded that only the fixed, smooth-bore artillery at Sevastopol had acquitted itself in a way that required the enemy to respect Russian artillery. Field artillery had failed the test of battle.[135]

Technical uncertainties, when coupled with institutional habit, professional prejudice, and conflicting perceptions of past and present campaigns, accounted for lack of strong direction in post-Crimean tactics and training. Because many artillery units initially retained their muzzle-loading smoothbores, crew drill changed little during the late 1850s and early 1860s. The artillery regulation of 1859 prescribed five modes of fire, the majority of which were appropriate to supporting the traditional linear formations of an earlier period. The regulation devoted excessive attention to detail, including an emphasis on uniform crew drill with precise, simultaneous movements for all crews serving the guns of the same battery. While officers paid little heed to the accuracy of fire, they insisted on adherence to standards governing intervals between guns and safe distances between firing batteries and their limbers. To make the situation worse, the War Ministry, hard pressed to meet other financial commitments, allocated precious few funds for the expenditure of ammunition for target practice. It was against this general background in the 1860s that Russian artillery underwent expansion and began to receive rifled cannon in quantity.[136]

As early as the field maneuvers of 1866 Russian artillery displayed sufficient shortcomings to elicit Miliutin's criticism and suggestions for remedial action. In accordance with prevalent opinion on the advantages of rifled cannon, the War Minister outlined a program stressing accuracy at the expense of firepower volume. As batteries took possession of their rifled tubes, crew drill and firing exercises came to reflect Miliutin's mandates and new requirements for firing at longer ranges with fewer rounds. In the new scheme of artillery employment the gun layer emerged to assume an increasingly vital role in crew drill and the conduct of firing exercises. During drill and live-fire exercises the battery commander designated the target and initial range, after which each gun layer assumed actual control of his piece to observe hits, estimate subsequent changes in range, and correct fire. Some experimentation with high-angle fire went on until 1873, after which artillerymen concentrated their full attention on direct fire, which, in accordance with persisting smooth-bore habit, was usually limited to short-range, easily observable targets. In 1873 the War Ministry officially approved fire correction by the bracket method. Meanwhile, artillery officers, occasionally working with mathematicians, began working out range tables and meteorological data and experimenting with such basic concepts as centralized fire direction of an entire artillery battery.[137]

Through the early and mid-1870s a paucity of resources magnified the effects of artillery growing pains. New and more destructive ammunition came into existence, including shell and shrapnel, both with timed and impact fuzes. However, each eight-gun battery received only 128 rounds for practice annually. Of this number only eight were live, the remainder containing reduced explosive charges and smoke for training purposes. After 1876, each battery received four shrapnel rounds per year for practice. Consequently, neither artillerymen nor their officers had any real sense from training of their weapons' potential and limitations.[138]

In field exercises the artillery rode into mock engagements with infantry and cavalry to engage close-range targets by direct fire. In effect, artillerymen of the 1870s employed their new rifled weapons in much the same way their predecessors had employed their smooth-bores during the Crimean era. True, they now enjoyed greater range and accuracy, but few officers either inside or outside artillery had any inkling of what to do with these attributes. They had rifled breechloaders but no coherent method of employing them in support of either infantry or cavalry. Nor were their superiors any better off. In citing reasons for the poor showing of Russian artillery against the Turks in 1877–78, one anonymous author of the Russian official history noted that in part the failure of cooperation with other arms "resulted from an insufficient familiarity of high-ranking officers with the characteristics of rifled artillery."[139]

Field Exercises

The problem of coordinating artillery support for infantry and cavalry was only part of a larger picture of uncertainty associated with field and tactical exercises of the period. The annual summer encampment had long been a fixture of peacetime military life, but Crimean defeat and the subsequent stress on training lent a greater sense of importance and urgency to various kinds of maneuvers. In 1858 the War Ministry published a revised *Military Regulation on Infantry Field Service in Peacetime*, which was intended to replace a regulation of 1846 of the same title. The revised document covered the general situation for field movements and corresponding issues of reconnaissance and security. However, the army did not return to field exercises on a large scale until 1865, when units had returned to Russia after suppression of the Polish insurrection.

Between 1865 and 1877 the army held annual exercises on a scale large enough to require the designation of permanent exercise areas and the construction of camp facilities. Eventually twenty-seven localities were designated "gathering places," in which a large proportion of the army's combat forces could train if sufficient advance planning permitted two shifts. By 1876 ministerial figures placed the number of units taking part in annual exercises at 460 infantry battalions, 321 cavalry squadrons, and 253 artillery batteries.[140]

The camp at Krasnoe selo outside St. Petersburg played a primary role in these summer maneuvers. There the army held title to sufficient territory to conduct tactical exercises involving the three combat arms for periods extending from eight to twelve days. Staff officers came to the imperial capital from outlying military districts to witness these maneuvers, then returned to their own units to duplicate the exercises on home ground. However, the system presented two major difficulties. One was a shortage of maneuver space. Of all the designated areas, only Krasnoe selo and a few others were of sufficient size to exercise large bodies of troops. Many regions, including the crucial Warsaw Military District, possessed sufficient facilities to hold exercises lasting only overnight. A second difficulty was the lack of centralized control over the maneuvers held within the various military districts, a disadvantage of Miliutin's system of decentralized control over Imperial Russia's far-flung military-administrative apparatus. District commanders were subject to varying pressures and wielded control over varying resources, so there was often little assurance that maneuvers held in the outlying districts would duplicate the conditions of the "grand maneuvers" held near the capital.[141]

Inconsistency stemmed not only from decentralization but also from inadequate supervision. Published regulations failed to keep pace with training requirements in a rapidly changing army, and the War Ministry's steady stream of supplementary directives apparently could not compensate fully for the absence of a comprehensive training document. Even worse, members of the royal family dominated the system of military inspection, with the result that compliance with requirements too often remained a function of caprice. Nor was the example of the St. Petersburg Military District always salutary for steady military development. For a number of reasons ranging from organizational inertia to the proximity of the royal family, maneuvers near the capital often assumed the character of well-choreographed stage productions. Despite all the changes, there lurked in St. Petersburg's shadow the ever-present danger of reversion to a more negative past.[142]

A careful reading of after-action reports on the state of training and the results of exercises in various districts reveals much about the readiness of Miliutin's army for war. Even infantry officers often failed to emphasize marksmanship training. Although a basic understanding of Dragomirov's tactical system trickled down to the small-unit level, commanders not infrequently strove for wooden application of a conception which required flexibility. During mock approach marches in unfamiliar circumstances, officers sometimes deployed their skirmish lines before they sighted the enemy. Once friendly forces gained contact and began to develop the situation, officers lost control of their skirmish lines. Coordination disappeared between the skirmish lines and reserves. Worse yet, coordination also disappeared among the combat arms. Battalion and regimental commanders experienced great difficulty maintaining control of their troops over greatly expanded distances. Cavalrymen ignored security and reconnaissance duties while they awaited the right moment to charge against infantry formations. Dutiful artillerymen leapfrogged their batteries forward to support the infantry, but battery commanders demonstrated no particular ability either to mass fire or to draw benefit from the extended ranges of their rifled weapons.[143]

Although staff officers and other reporters often attempted to put the best face on these and other shortcomings, the stream of directives coming from the War Ministry during the early 1870s indicated that Miliutin recognized the need for corrective measures. In addition to prescription, the War Minister consistently underscored the need for additional education to overcome many of the army's worst shortcomings in training and field exercises.[144]

Miliutin's emphasis on field exercises and officer education reflected his concern for several key aspects within the overall military picture. With the assistance of Dragomirov, whose methods and views both shaped and reflected new approaches to training and tactics, Miliutin in less than two decades had engineered a veritable revolution in the way the Russians recruited, trained, equipped, and organized their army. No less important were changes in the way the Russians taught their army how to fight. In many respects, however, the transition from the old to the new remained either imperfect or incomplete. To carry the process to its conclusion required still more time and resources, but the flow of these precious commodities would be cut short by events beyond Miliutin's control. Under these circumstances it is not without justification that many historians have labeled as "transitional" the Russian Army that went to war against Turkey in 1877.

2.

Russo-Turkish War, 1877–1878

> But all other weapons are dwarfed before the breech-loading musket, firing easily 5 to 6 shots a minute and carrying to a range of a mile and a quarter. Therefore the infantry is now more than ever the arm of service upon which all the hard fighting devolves, which inflicts and receives the greatest damage, and to which all other parts of the army are merely subsidiary.
>
> Lieutenant F. V. Greene, Official U. S. Observer, Balkan Theater[1]

ALTHOUGH THE Congress of Paris had made the protection of Balkan Christians a collective responsibility of the European powers, St. Petersburg continued during the post-Crimean era to pursue a policy of maintaining friendly contacts between Russians and the Orthodox and Slavic peoples of the Balkans. This policy both accorded with rising Panslav sentiment in Russia and afforded limited political leverage, especially in Serbia, against Turkey and the great powers, including Great Britain and Austria-Hungary. For a brief period until the assassination in 1868 of the Serbian Prince Michael Obrenovic, rising Slavic sentiment in the Balkans promised the evolution under Serbian auspices and Russian protection of a greater Slavic state embracing not only Serbia but also Montenegro and Bulgaria. Meanwhile, Panslav activism during the early 1870s gravitated increasingly to the cause of Orthodox Slavs in Thrace and Macedonia, where oppressed minorities flocked to the standard of Bulgarian nationalism. A new Balkan crisis erupted in 1875, when Turkish troops in Bosnia-Herzegovina suppressed a revolt of Slavic peasants against their Muslim landlords.[2]

Despite the urgency of events, St. Petersburg was at first understandably ambivalent in its support for various popular movements in the Balkans. On one hand, official opinion within the Ministries of Finance and Foreign Affairs held that Russia could ill afford foreign adventurism and unnecessary wars. On the other hand, public pressure rapidly built to support the aspirations of Balkan Slavs for autonomy and independence. Consequently, Gorchakov, the ailing Foreign Minister, worked for resolving Balkan tensions within the framework of the Three Emperors' League and in concert with the European powers. At the same time, however, Panslavists and Russian consular officials often exceeded their mandates by assuring Balkan Slavs of moral and even material support from Russia. While pro-Slav agitation reached fever pitch in the Russian press, retired and furloughed Russian officers entered Serbian military service. The crisis deepened during the spring of 1876 when Turkish irregulars massacred as many as 30,000 Bulgarians who had belatedly joined the anti-Turkish insurrectionary cause. Finally, in the summer of 1876, Serbia and Montenegro went to war against Turkey.[3]

From this time, events rapidly drew Russia into yet another war with Turkey. Although General Cherniaev, the "Lion of Tashkent," commanded the Serbian army, Serbian resistance quickly collapsed, forcing Russia in October 1876 to impose an armistice on Turkey. After Russia mobilized a portion of its army in November, an ambassadors' conference was convoked at the end of 1876 in Constantinople, where the great powers agreed to support reform and the creation of autonomous Christian provinces in European Turkey. Outright Turkish rejection of this program in January 1877 forced the Russians to confront the real prospect of hostilities. Accordingly, St. Petersburg concluded conventions with Austria-Hungary to carve out possible noncontending spheres of influence in the Balkans and dispatched Ambassador N. P. Ignat'ev to various major European capitals, where he assured the powers that Russia remained amenable to a reasonable solution for the Balkan crisis. In Britain, Ignat'ev's efforts resulted in the London Protocol of March 1877, requesting that Turkey introduce reforms and demobilize. When Turkey refused, Russia could notify the European powers that war was necessary to protect the Balkan interests of Europe as a whole. On 4 April 1877, Russia signed a convention with Romania to permit transit of Russian troops across its territory. On 12 April, Russia declared war on Turkey.[4]

Planning and Early Course

Three basic circumstances—geography, time, and Turkish control of the Black Sea—dominated Russian planning for war with Turkey. Geography and Turkish supremacy on the Black Sea dictated a two-pronged land campaign, one in the Caucasus and the other in the Balkans, with the preponderance of effort in the latter. Moreover, Turkish domination of the sea also meant that any land offensive would have to avoid the coastline to forestall waterborne interdiction of ground operations. Finally, the likelihood of great power interference caused the Russians to anticipate striking quickly to confront the outside world with a fait accompli.

Beginning in the spring of 1876, a series of lectures written by General Obruchev and delivered by Lieutenant Colonel N. D. Artamonov of the Main Staff to officers of the St. Petersburg Military District revealed the broad outlines of Russia's offensive design. To achieve rapid decision in the manner of I. I. Dibich in 1829, Obruchev envisioned a lightning land campaign aimed directly at the Ottoman heart, Constantinople. He proposed mobilizing 250,000 men in Bessarabia, marching directly across Romania to force the Danube somewhere along its middle flow, throwing up defensive cordons east and west to cover a race for the Balkan divide, then pushing forces through the mountains past Adrianople to threaten the Turkish capital. A secondary theater in the Caucasus would tie down additional Turkish forces and prevent incursion into the south Russian steppe. The design was bold, direct, and—at least in lecture form—capable of implementation.[5]

In accordance with plans loosely based on Obruchev's concept, the Balkan campaign against Turkey actually unfolded in three distinct phases. During the first, from 12 April to 3 July 1877, the Russians with the acquiescence of Bucharest shifted troops from Bessarabia to concentrate four corps in Wallachia, conducted a successful forced crossing of the Danube, and established a secure foothold on the southern shore. During the second phase, from 4 July to 10 December, the Russians found themselves bogged down with operations in Bulgaria, including a time- and manpower-consuming siege of Plevna. During the third phase, from 11 December to 19 February 1878 (Treaty of San Stefano), the Russians broke through the Balkans in force to menace Constantinople.

The lightning campaign against a weak opponent which Obruchev and others had calculated to require several months in reality stretched to forty-seven weeks. Disparity between vision and reality was the product of a number of complications and miscalculations, including weather, distance, the availability of manpower, inept planning, Turkish stubbornness in the defense, and the impact of changing technology on operations and tactics. Even before the outbreak of hostilities, the Russians had conducted two partial mobilizations (1 November 1876 and 3 April 1877), neither of which was to provide the decisive manpower edge needed to accomplish the army's trans-Danube mission.[6] Once war had been declared, an extraordinarily wet spring made for heavy going on Romanian roads and kept rivers and streams excessively high for weeks. Despite a relatively smooth mobilization of active and reserve forces, rail movements in both Romania and Russia misfired because of underdeveloped and overloaded rail nets, mixed gauges, and poor management of rolling stock. Distance and lack of access to sea lines of communications magnified these and other difficulties—by direct march from Kishenev via Shipka Pass to San Stefano the Russians would have to cover nearly 1,000 kilometers, the majority of them by putting one foot in front of the other.[7]

Turkish defenses initially favored a bold Russian offensive. Despite naval superiority on the Black Sea and a military modernization program scheduled for completion in 1878, the Sultan's military clung to an outmoded strategy of passive defense. The Defense Council in Constantinople trusted to seapower and retention of a powerful fortress quadrilateral—Silistria, Varna, Ruschuk, and Shumen—to foil Russian designs in the Balkans. Indeed, the reasons for caution were plain: although in the event of war the Turks could mobilize an army of more than 400,000, much of that strength remained untested and unprepared. The standing army was small, support services almost nonexistent, and permanent units larger than battalion size a fiction. Not surprisingly, Turkish tactics stressed the defense of prepared positions. Only several bright spots relieved the larger gloomy picture. One was the recent adoption of a new infantry weapon, the Peabody-Martini rifle, then considered one of the finest breechloading shoulder arms in the world. To go with this novelty, an

otherwise incompetent supply service had the foresight to procure 500 to 1,000 cartridges per weapon. A second cause for optimism was the recent rearmament of the Turkish field artillery with a limited number (forty-eight) of Krupp steel cannon whose power and range (four to five kilometers) outstripped anything in the Russian inventory. Neither of these developments boded well for the Russian infantrymen who would have to confront their Turkish counterparts in the field over open sights.[8]

Dragomirov and Gurko

During the initial months of hostilities, resolute action, good fortune, and Turkish ineptness insulated the Russians from the grimmer realities of ground combat in the 1870s. Major General M. I. Dragomirov's forcing a passage on the Danube in mid-June was a classic example of a successful operation based on extensive reconnaissance, deception, careful preparation, and energetic execution, all

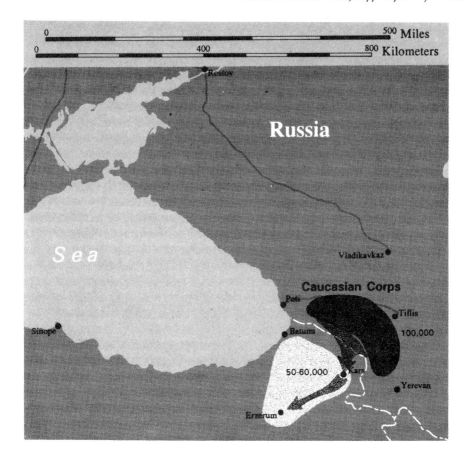

proceeding from bold, confident, and knowledgeable leadership. For nearly a month beginning in mid-May, while Russian forces (260,000 men in four corps: VIII, IX, XII, and XIII) under the overall command of the Tsar's brother, Grand Duke Nicholas Nikolaevich the Elder (Chief of Staff N. N. Nepokoichitskii), concentrated in the region of Bucharest, engineers carefully reconnoitered the Danube before the high command decided to attempt the crossing at Zimnicea, a settlement located on a direct line to Shipka Pass but midway on the Danube between major Turkish garrisons on the opposite bank at Nikopol and Ruschuk. A successful crossing in the face of opposition at Zimnicea would be no small feat, for steep banks lay on the other side of the river, which at that point was more than a kilometer in width and thirty meters in depth. In anticipation of the operation, pontons and other bridging materials were brought overland, in some cases from as far away as St. Petersburg, and assembled on the Olt, a tributary of the Danube. Meanwhile, batteries were erected opposite Nikopol and Ruschuk, and five mine barriers were laid in the Danube to neutralize Turkish ironclads. Several days before the appointed crossing, the Russians put a diversionary force across the river at Galati and feigned preparations for a crossing at Nikopol. Only on 13 June did General Dragomirov begin concentrating his troops opposite the actual crossing site in numbers ostensibly too small to reveal Russian intentions. Finally, small steam-driven launches were brought from Odessa and Sevastopol to assist in the movement of troops and pontons.[9]

Of particular note were General Dragomirov's tactical dispositions for the crossing. At his disposal were troops from his own 14th Infantry Division (the 53rd, 54th, 55th and 56th regiments), the 4th Sharpshooter Brigade, a composite Guards company, two detachments of dismounted Cossacks (*plastuny*), two companies of sappers, two mountain batteries, the Don Cossack 23rd Regiment, a Ural Cossack squadron, four pontoon battalions, and a detachment of Guards sailors (in all, 16 2/3 battalions, six Cossack squadrons and 64 guns). Opposite the Russians at Svishtov were a Turkish battalion and a squadron (770 men) with two guns. Another enemy detachment lay about five kilometers east of Svishtov with five battalions (3,330 men) and a four-gun battery.[10] Dragomirov personally reconnoitered the crossing site with his subordinate commanders, explaining his concept in detail. He planned to divide the assault into three successive waves, incorporating artillery support into each and making provisions for covering artillery fire from the friendly shore. He envisioned obtaining a foothold on the opposite bank with the first wave, then reinforcing and expanding that foothold with the two successive waves to secure Svishtov and destroy or drive off the two local Turkish detachments before major enemy reinforcements could be brought up from either Ruschuk or Nikopol. Until several hours before the crossing, none of the troops knew the actual time or place of attack. Although the Imperial Suite had taken to the field, even the Emperor was not informed of the crossing details until 2000 hours on 14 June.[11]

The crossing began at 0100 hours on 15 June. Turkish sentries sounded the alarm and opened small arms fire on the first wave of attackers, who were themselves under orders not to return fire but to await setting foot on shore before conducting a bayonet assault. Although Turkish bullets sank a ponton with its precious cargo of two mountain howitzers, by 0145 eleven companies of the 53rd Volynian Regiment and several Cossack detachments (about 2,500 men) were safely on the south bank of the Danube. There, despite scattered landings and initial confusion in the darkness, the Russians were able to clamber up the bank and fan out to establish a defensive perimeter along both banks of a small stream feeding into the Danube. Because communications were difficult and because the situation changed by the minute as the alarm spread and frantic Turkish officers fed their troops piecemeal into savage local counterattacks, company commanders acting on their own initiative became the crucial elements in developing the tactical situation. Without waiting for orders they expanded the defensive perimeter, adjusted local positions, launched their own counterattacks—often using improvised formations—and then took measures to secure the crossing of the second wave. In the face of repeated threats from an enemy intent on pushing the Russians into the river, what initially saved Dragomirov's crossing was not his infantry assault tactics but the initiative of small-unit leaders and the hasty employment of ad hoc skirmish formations, which Lieutenant Francis Vinton Greene, the American observer, later noted seemed to come naturally to the troops. In addition, six mountain howitzers (two from the first wave and four from the second) provided close support with canister to blunt enemy counterattacks and sap Turkish resolve.[12]

Perhaps the most critical moment came just after first light when the arrival of local enemy reinforcements and artillery caught the second Russian wave still in the water. Several pontons were swept clean by rifle fire, and one survivor later recalled that "the water around the pontons boiled from ricocheting bullets." However, once again initiative came to the rescue, when one of the junior company commanders, an Ensign Matornyi, sized up the situation, ignored the fusillade, and rallied two companies of the 54th Minsk Regiment to conduct a daring bayonet attack which rolled up the flank of the Turkish riflemen firing at Russians on the water. Although Russian gunners on the northern shore could not support their infantry directly for fear of hitting friendly units, nine-pounders firing across the water effectively suppressed hostile artillery. Meanwhile, Dragomirov himself landed and, after determining that the lodgement area afforded sufficient space to land follow-on forces, ordered a temporary halt in the ground advance to permit consolidation. At 1100, with most of the 14th Division and parts of the 4th Sharpshooter Brigade now on the southern shore of the Danube, Dragomirov ordered an attack on Svishtov heights to secure the local high ground in anticipation of expanding the lodgement to absorb substantial elements of General F. F. Radetskii's VIII Corps. Turkish resistance melted away as the Russians attacked to seize their objectives, and by 1500 Dragomirov's troops had occupied both the heights and the settlement of Svishtov. Russian casualties were thirty officers and 782 soldiers killed and wounded.[13]

On the next day engineers began assembling the first of two pontoon bridges, which were finished respectively on 19 June and 28 July. Meanwhile, various kinds of improvised transport ferried additional troops across the Danube, so that by evening of 20 June all of General Radetskii's infantry and major components of XIII Corps had already completed the crossing. By 1 July elements of four Russian corps (120,000 troops) were already conducting offensive operations south of the Danube. In one daring stroke the Russians had breached the first line of Turkish defenses and were well on the way to realizing Obruchev's bold offensive design of the previous year. Soviet historians still revere actions leading to the forcing of the Danube in 1877 as an excellent example of "a complex strategic operation conducted on a broad front." In addition, they cite as worthy of emulation the initiative of small unit leaders whose actions proved crucial to Dragomirov's success.[14]

Even as the Russians poured into Bulgaria, the Grand Duke issued orders to constitute three separate detachments in the trans-Danube: the Western, the Eastern, and the Advance Guard. Because the width and swiftness of the Danube had precluded swimming cavalry across in advance of the main Russian crossing, the high command lacked information about Turkish numbers and dispositions. When the Advance Guard under Major General I. V. Gurko, popular former commander of the 2d Guards Division, marched to Turnovo on 25 June virtually unopposed, the Grand Duke proposed that the Western and Eastern detachments screen right and left respectively to support a reinforced effort in the center to penetrate the Balkan passes behind Gurko. However, on 28 June the Emperor—apparently on the advice of Miliutin—rejected his brother's audacious design, calling it "too risky" while Ruschuk, Nikopol, Shumen, and Plevna still lay in enemy hands. Moreover, the Emperor asserted that Russian communications in Romania lay exposed to Turkish attack from Vidin. The Grand Duke bowed to the Tsar's will and immediately bade VIII Corps to halt its advance at Turnovo to await the arrival of IV and XI Corps. The Tsar and the high command, rather than throwing caution to the winds, clasped it to their bosoms, thereby failing to profit from the advantage won so handily for them by Dragomirov at Svishtov.[15]

The sudden emphasis on caution in part reflected tensions inherent in the Tsar's earlier decision to join his troops in the field. Although Alexander II had assured the Grand Duke complete freedom of action as nominal Supreme Commander, the influence and presence of the Tsar and his closest advisers figured prominently in key decisions. This arrangement did not sit well with the Grand Duke, who preferred to think of himself as a bold and independent leader, but it did offer justification for his own lapses in command. Meanwhile, the Tsar's personal presence also acted as a damper on the actions of lesser commanders who began to reckon with the practical and personal consequences of failure on the Imperial Suite. More immediately, the seventeen trains or 350–500 wagons needed to transport the imperial entourage imposed a heavy burden on an already overtaxed logistical structure.[16]

Despite the shifting winds of command, Major General Gurko retained orders to seize the Balkan passes, and he now made preparations to penetrate the mountain divide. Because 4,700 Turks occupied Shipka Pass, he reorganized his Advance Guard into a forward detachment (10 1/2 battalions, 31 1/2 squadrons, and 32 guns, in all about 16,000 troops) to discard unnecessary baggage for a rapid passage through the more restrictive Hainkioi Pass. After fighting several minor battles on the Balkan south slope, Gurko arrived at the south end of Shipka Pass in the vicinity of Shipka village on 5 July, a day later than that earlier designated for a joint north-south attack with VIII Corps participation from the north on Turkish defenses within the pass itself. Gurko's inability to communicate with his counterpart in the north over circuitous routes led to two uncoordinated and unsuccessful Russian assaults on Turkish positions in the pass, but on the night of 7 July Turkish forces withdrew on their own accord, leaving their wounded, ammunition, and field artillery (eight steel guns).[17]

Gurko requested reinforcements to continue his drive south, but the changing enemy situation on both sides of the Balkan divide now precluded development of his initiative. Indeed, the Danube crossing and Gurko's sudden appearance on the south Balkan slope had sown panic among the Muslim inhabitants of Rumelia and prompted a realignment of the Sultan's government in Constantinople. As the Prussian Helmuth von Moltke had once observed, the Turks usually started fighting about the time everyone else gave up, and his words were never truer than in 1877. A new commander in chief, Mehemet Ali Pasha, was recalled from Montenegro, and the Sultan rushed reinforcements into the field under Suleiman Pasha to confront Gurko. Thanks to these and other frantic efforts, by mid-July Gurko's 16,000 Russians now faced 50,000 Turks. On 17 July, he attempted to renew his advance from Kazanlik to menace Turkish communications with Adrianople, only to collide

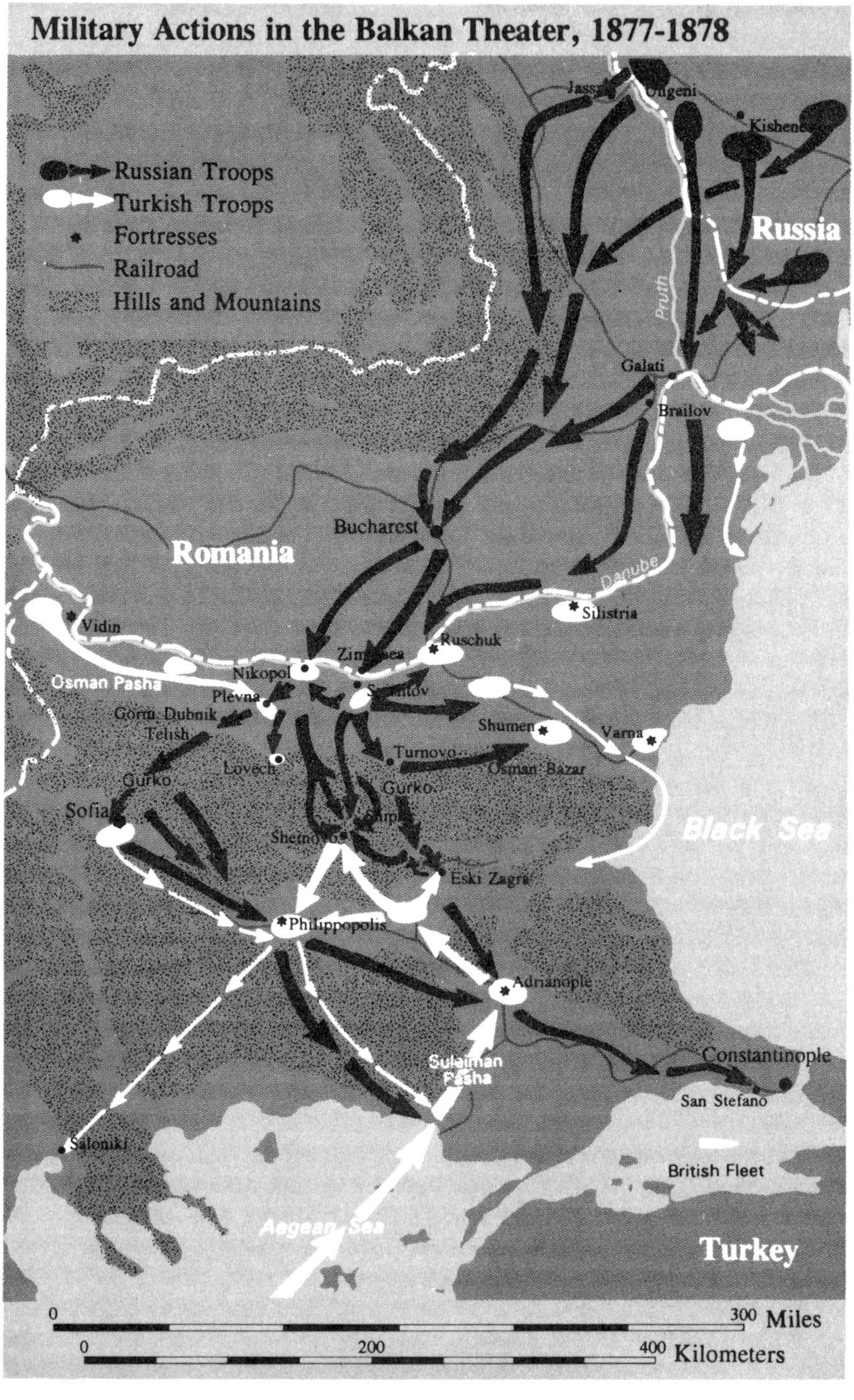
Military Actions in the Balkan Theater, 1877-1878
Russian Troops
Turkish Troops
Fortresses
Railroad
Hills and Mountains
Russia
Pruth
Galati
Brailov
Bucharest
Romania
Danube
Silistria
Vidin
Ruschuk
Nikopol
Osman Pasha
Plevna
Gorni Dubnik
Telish
Lovech
Shumen
Varna
Turnovo
Osman Bazar
Gurko
Sofia
Gurko
Black Sea
Eski Zagra
Philippopolis
Adrianople
Suleiman Pasha
Constantinople
San Stefano
Saloniki
British Fleet
Aegean Sea
Turkey
0
300 Miles
0
200
400 Kilometers

head on with Suleiman's advancing troops at Eski Zagra. After two days' intermittent battle, Gurko withdrew to Shipka Pass, but not before his cavalry managed to disrupt Turkish rail and telegraph communications over a wide area. Major General N. G. Stoletov assumed command of the Shipka defenses, and Gurko retired northward to rejoin his parent VIII Corps.[18]

More than a mere raid, Gurko's action has been remembered in Russian and Soviet military history as a model for the conduct of all-arms operations with a forward detachment. In less than a month he had seized one of the principal Balkan passes, contributed to a panic which nearly toppled the Sultan's government, destroyed parts of two significant rail and telegraph lines, and gathered invaluable information on Turkish dispositions south of the Balkans. What many commentators have failed to recognize is that in accomplishing his mission Gurko also depleted his horses to such an extent that VIII Corps, which had been allocated the lion's share of Russian cavalry, was deprived of reconnaissance for several months.[19]

Radetskii, VIII Corps commander, had been unable to reinforce Gurko in mid-July because rekindled Turkish resolve also affected the operational picture north of the Balkans. After crossing the Danube in late June and early July behind Gurko's forward detachment, Lieutenant General P. S. Vannovskii's Eastern Detachment (XII and XIII corps, one cavalry division and 216 guns) had struck east to the River Lom to begin a leisurely investment of Ruschuk. Mehemet Ali soon heated up what had been a quiet front by reorganizing local troops into a credible fighting force, then conducting a minor offensive which tied down three Russian corps (Vannovskii's and Radetskii's). Meanwhile, General N. P. Kridener's Western Detachment (IX Corps, three cavalry brigades, and 108 guns) moved upstream on the Danube to invest and attack Nikopol, which fell to the Russians on 4 July.[20]

Reduction of Nikopol delayed Kridener's advance to the southwest for several days, thus setting the stage for an epic confrontation at Plevna between Russian and Turk for Balkan hegemony. More than any other set of circumstances, it was Plevna that would reveal the yawning gap between offense and defense opened by changes in technology, including chiefly the widespread adoption of breechloading rifles. Until early July, military operations had retained a high degree of fluidity, with surprise, speed, and maneuver obviating the necessity of costly set-piece battles. When both sides finally settled down to a slugging match at Plevna, the confrontation would subject Russian leadership, tactics, organization, and equipment to their ultimate test.

First and Second Plevna

Russian occupation of Turnovo finally provoked the Turks to action, and on 2 July, Osman Pasha, the best Turkish field commander, left Vidin marching southeast (not into Romania, as the Tsar had feared) with 16,000 seasoned troops and sixty guns. Meanwhile, General Kridener, commanding IX Corps on the Russian right, conducted a successful storm of Nikopol and its lightly defended inner fortress. On 4 July, the day of Nikopol's fall, Russian headquarters learned from

Viennese newspapers that Osman Pasha had left Vidin, while on 5 July Russian cavalry patrols reported a strong Turkish column advancing on Plevna, an important road junction not forty kilometers from Nikopol. That same day Kridener detached Lieutenant General Iu. I. Shil'der-Shul'dner's 5th Infantry Division (9,000 men, forty-six guns) to occupy Plevna "if no special obstacle is encountered." Both Osman Pasha and Shil'der-Shul'dner reached Plevna on 7 July, but the Turks beat the Russians by six or eight hours, and on those few hours hinged the next four months' campaign.[21]

The engagement that followed was subsequently labeled First Plevna, and in broad outline it presaged many of the frustrations and lapses that would come to plague the initial Russian war effort. Shil'der-Shul'dner had designated his route of advance using a map which showed outlying settlements farther from Plevna than was actually the case. In addition, he failed to receive timely reconnaissance assistance from either the Caucasian Cossack Brigade or the 9th Don Cossack Regiment. Consequently, early in the afternoon of 7 July his lead infantry regiment, suffering from the heat, choking on dust, and advancing in road march formation, came unexpectedly under artillery fire when it blundered into the outskirts of Plevna and Osman Pasha's hastily improvised defensive line. The Russians immediately deployed on line, and Russian batteries began a fruitless artillery duel with the Turks. Without bothering to conduct a detailed reconnaissance to determine enemy strength and dispositions, Shil'der-Shul'dner bade his troops to rest in place before going into the attack at dawn on the 8th.

Osman Pasha had deployed his main force in hastily dug field fortifications stretching six to eight kilometers in length along a chain of heights located north-northwest of Plevna. Additional troops he arrayed east of town to block attack along the Ruschuk road. Irregular terrain and forested areas covered approaches from the south and west. Despite the counsel of Shil'der-Shul'dner's chief of staff and artillery commander, both of whom felt the Russians had encountered "a special obstacle," the Russian commander decided to press forward into the attack. Therefore, he split his forces to conduct one assault from the north with the 17th Arkhangel and the 18th Vologda Regiments and another from the southeast with the 19th Kostroma Regiment. The Caucasian Cossack Brigade was to attack the Turkish rear while the 9th Don Regiment screened the extreme right.[22]

The main engagement began at 0530 on 8 July, when the 1st Brigade (17th and 18th Regiments) attacked with skimpy artillery preparation, advancing with five battalions arrayed behind skirmishers in company columns. The Russians came under heavy rifle fire at great distance from the Turkish earthworks, and the attackers were forced to speed their advance. Momentum carried the remnants of the extreme right-hand detachments through Osman Pasha's defenses, and for a short time there was actually fighting in the streets of Plevna. However, heavy losses, the lack of a reserve, and superior Turkish numbers eventually forced the Russians to withdraw. On the Russian left, the 19th Kostroma Regiment had started its attack at 0600, also with very little artillery preparation. It swept through three lines of trenches before heavy losses, lack of a reserve, and an ammunition shortage forced Shil'der-Shul'dner to order withdrawal. The Caucasian Cossack Brigade's role was limited to collecting wounded from the battlefield. Fortunately for the Russians, the Turks themselves had suffered enough losses to preclude effective pursuit.[23]

Russian losses were horrifying. One-third of the officers and one-half of the men were casualties. Two regimental commanders and a brigade commander had fallen mortally wounded. The Russian attack was spirited, so much so that Osman Pasha later recalled that the Russians came closer to defeating him at First Plevna than in either of the two succeeding and more powerful attacks. Unfortunately, the Russians themselves did not know how to evaluate the experience, with the result that, as one Soviet writer has put it, "in a series of subsequent battles and engagements the Russian command repeated the same mistakes which were permitted at First Plevna."[24]

Losses notwithstanding, Shil'der-Shul'dner's reverse did not unduly alarm the Russian high command, whose response was to send Kridener's entire corps (20,000 men, 140 guns) to drive the Turks from Plevna. However, Kridener no longer shared his superiors' enthusiasm for the offensive, in part because he habitually exaggerated Turkish strength and in part because the possibility of failure weighed heavily in his calculations. If defeat opened a hole in the Russian dispositions, the Turks might pour through to capture the Imperial Suite and cut off all Russian forces south of the Danube. Despite his own urging for caution, Kridener soon received orders to attack upon receipt of reinforcements which the high command calculated would raise his corps to parity with the Turks in infantry and superiority in cavalry and artillery. General-Adjutant Nepokoichitskii, the Grand Duke's chief of staff, even ventured the gratuitous opinion that Kridener could make good use of cavalry since "action against the flank and even rear of the enemy can force him to quit even a very strong position."[25]

From the beginning, Kridener seemed more intent on securing his line of retreat than finding and exploiting enemy weaknesses. He repeated Shil'der-Shul'dner's failure to conduct a detailed reconnaissance of the locality and enemy dispositions, although one of his subordinates, Major General M. D. Skobelev, thought he had discerned a vulnerability: lack of strong fortifications in the south and west. However, like Kridener, Skobelev also tended to exaggerate Turkish numbers. Osman Pasha now possessed 22,000 men and fifty-eight guns, and unlike his adversaries he used every available moment to strengthen his dispositions. Although Turkish fortifications were incomplete to the south and nonexistent in the west, several lines of trenches and a number of powerful redoubts now blocked Russian approaches from the north and east. In particular, a series of strongpoints, collectively labeled the "Grivitsa Redoubts," lay across the Russian line of advance from the east along the Ruschuk road. From the south, only hastily dug earthworks defended against a potential Russian advance to the Grivitsa Brook.

Under goading from headquarters Kridener planned an attack for 18 July with thirty-six battalions. His dispositions called for launching twenty-four battalions in a main attack on the Grivitsa Redoubts and twelve in a supporting attack from the south. The latter under Lieutenant General Prince A. I. Shakovskoi would silence Turkish batteries on heights just south of Plevna and then advance to threaten

Osman Pasha's right and rear. The 9th Don Regiment was to secure the Russian right, while the Caucasian Cossack Brigade under General Skobelev prevented the arrival of Turkish reinforcements from Lovech to the south.[26]

The assault began at 0900 on 18 July with a three-hour artillery bombardment from the east that proved ineffective because of distance and failure to mass fires. Shakovskoi's own preparation from the south proved somewhat more effective because opposing fortifications were weaker. However, the Turkish earthworks on the Ianyk Bair heights north of Plevna remained beyond his artillery range, and when Kridener sent a query at 1230 about the effectiveness of Shakovskoi's artillery, Shakovskoi misread the message as a gentle goad to attack in order to employ his artillery to better advantage. When Kridener received a brief message that Shakovskoi was attacking, Kridener incorrectly concluded that his subordinate was responding to a momentary opportunity too promising to ignore and too fleeting to risk losing time while the two commanders exchanged communications. Kridener briefly considered the situation and ordered the entire right to attack in support of what he thought was Shakovskoi's success.

Kridener's attack was initiated by the 121st Penza Regiment, which advanced in company columns from positions in the open 1800 paces from the enemy. In thirty minutes the regiment was stopped one hundred paces from its objective, Redoubt No. 7, with 1,000 casualties. Similarly, a storm of Turkish fire halted the 123rd Kozlov regiment, and to its right, the 122d Tambov regiment before Redoubt No. 8. Kridener attributed the right's terrible losses to insufficient artillery preparation. But, rather than send his reinforcements to Shakovskoi, Kridener committed the reserve to follow his own faltering attack on the right. However, the reserve commanders had been insufficiently informed of Kridener's scheme of attack, with the result that they did not know their objective or routes of advance. After receiving the attack order, their regiments broke through brush and cornfields in company columns before they too were stacked up against Redoubt No. 7. Those soldiers not killed or seriously wounded took cover where they could in the folds of the terrain, but they could neither advance nor withdraw.[27]

The situation was just as perplexing on the left flank. At 1400, Shakovskoi had given the order for his own attack. However, the signal threw his commanders into confusion because they knew nothing of their commander's dispositions and order of attack. When Colonel Rakaza of the 125th Kursk Regiment queried his brigade commander, Major General Gorshkov, about the direction of attack, Gorshkov replied, "Advance straight ahead and kill any swine [*svoloch'*] you encounter on the way." Despite the inexactness of attack orders, several regiments doggedly fought their way to the line of southern redoubts, but strong Turkish counterattacks eventually drove them back, even though Shakovskoi tried to support success with his limited resources. Short of reserves, at 1815 Shakovskoi requested additional troops, writing "six hours of fighting; all cartridges spent; wounded more than one fourth of complement; confusion setting in, can easily grow worse. Request support."[28] However, by this time Kridener had only one battalion left in reserve; therefore, he called off the attack. Luckily the Turks again could not mount an effective pursuit.

The butchery called "Second Plevna" was over. Osman Pasha had lost about 5,000 men, while Russian losses added up to 7,000 killed and wounded. The infantry bore the burden of these losses, with casualties amounting to 25 percent of its officers and 23 percent of its rank and file. Headquarters reported to the Tsar that Russian failure had resulted from "a strong superiority of weapons compared with ours and the Turkish ability to fortify a position well."[29] Unofficial commentators found deeper philosophical implications in the defeat: the physician-writer S. P. Botkin, commenting from the field, wondered, "Who is guilty in all the failures? A lack of culture in my opinion lies at the foundation of everything unfolding before us . . . we have to work hard, have to study, have to have more knowledge and then we won't have to take lessons from the Osmans and Suleimans."[30]

The Defense of Shipka

Observers may have commented variously, but "Second Plevna" had an indisputably dampening effect on the Russian offensive spirit. The Grand Duke immediately went over to the strategic defensive, while the Tsar declared a new mobilization of divisions, including five army, two cavalry, three Guards, and two grenadier, for a total reinforcement of 110,000 men. However, the arrival of these reserves would require time, and time was now working for the Turks. The Porte decided to take the offensive by throwing additional troops into the field, shifting forces to the Balkans from the Caucasus, and launching coordinated operations on both sides of the Balkan divide. Although the Russians were vaguely aware of Turkish intentions, the tsarist high command could not take additional measures to strengthen its hold on the north Balkan slope and Shipka Pass. Without the benefit of offensive momentum and without reinforcements the Russians suddenly found themselves spread perilously thin south of the Danube to face a concerted Turkish counteroffensive. If Osman Pasha attacked from Plevna or, more ominously, if Suleiman Pasha crossed the divide over one of the lesser passes to join Mehemet Ali in the vicinity of Osman Bazar, the Russians lacked sufficient reserves to counter thrusts which might come from a variety of directions. Once the Russian advance slowed, several months would be required to make good the original miscalculations on the number of troops required to accomplish the campaign's strategic objectives.[31]

Nowhere was the situation more precarious than at Shipka Pass, where scarcely 5,000 defenders would soon face the fury of an entire Turkish field army. Located about 1,200 meters above sea level, Shipka Pass represented the best and most direct way through the Balkans to Adrianople and Constantinople. Loss of the pass would be both a setback for Russian offensive plans and a smashing victory for the Turks, who could then push troops through the mountains in mass to threaten the Russians at any point along the north Balkan slope. The pass itself was about sixteen kilometers in length. From the north the road over it ascended along the valley of a stream, then climbed sharply to follow a ridge line dominated by three hills before descending in steep zigzags to Shipka village at the southern entrance of a defile. General Radetskii's VIII Corps bore responsibility for controlling the pass, the defenders of which were under the overall command of Major General N. G. Stoletov. To hold against an enemy many times superior in numbers Stoletov counted a mixed detachment of 4,400 men, including five battalions of Bulgarian militia, the 36th Orel Infantry Regiment, four Cossack squadrons, two sapper detachments, and twenty-seven guns. Opposing Stoletov was Suleiman Pasha with 27,000 infantry and cavalry and forty-eight guns. Stoletov's closest help was VIII Corps general reserve located at Turnovo, sixty-seven kilometers from the northern entrance to the pass.[32]

The principal Russian defensive position was located on three sets of hills which dominated the pass at its highest point. On the highest elevation, Mount St. Nicholas, Stoletov had deployed three Russian batteries and seven of the Krupp guns left there on 7 July by Turkish defenders retiring before General Gurko's advance. On the two elevations north of this position, "Central" and "Northern" Hills, Stoletov arrayed four battalions of Russian infantry and five of Bulgarian militia. He and General V. F. Derozhinskii, the commander of the 36th Regiment, evidently ascribed primary importance to the proper integration of defensive fires. They covered each approach with cross fires and designated three vertical tiers of fire, two lower for small arms and one upper for artillery, to permit both full integration of defensive fires and free artillery fire over the troops' heads. Although the Russians had dug shallow trenches to connect their principal defensive positions and had mined some of the main avenues of approach, they evidently trusted more to the nature of the terrain than to their shovels to halt any enemy attack. Their fortifications lacked depth, and in some cases the troops had only rearranged piles of stones for cover. Stores of all types, whether food, water, or ammunition, were in short supply.[33]

A six-day battle for Shipka Pass exploded on 9 August, when twenty battalions of Turks screaming "Allah" hurled themselves against the "steel" battery on the left spur of Mount St. Nicholas. Exploding mines and savage defensive fires seared through their ranks, but nine times they reformed and struck between noon and 2100, when darkness left them clinging to positions only a hundred meters from their objective. On the next day the Turks contented themselves with infiltrating around the Russian and Bulgarian flanks and emplacing supporting artillery on adjacent heights. By the morning of 11 August, the most critical day of the engagement, the Russians still held the high ground, but they were nearly surrounded and lay exposed to deadly crossfires from ranges of 1,500 to 2,000 paces. At 0600 Suleiman reopened the attack with assaults from three sides, one column against Mount St. Nicholas and two against Russian supporting positions on Central Hill. Lieutenant F. V. Greene later summed up the day's action by writing that the defenders were "engaged to the last man, and trying to hold their own against 25,000 Turks."[34] Toward afternoon the Russian situation finally become desperate: rifle and artillery ammunition were nearly exhausted, there were no reserves, food and water were gone, and the troops lay in the heat constantly exposed to heavy enemy fire. Stoletov grudgingly gave ground, created reserves by stripping them from lesser engaged units, then counterattacked to regain lost ground. About 1630, just as the Russians on Northern Hill began to give way in a movement that threatened to become a rout, troops from the 4th Sharpshooter Brigade of Radetskii's general reserve arrived to shore up Stoletov's sagging defenses. So pressed for time were the Russians that the lead reserve battalion entered the fray mounted on Cossack horses hastily impressed to cover the last few critical kilometers.[35]

Early the next morning (12 August) additional reinforcements arrived in the form of General Dragomirov's 14th Division. About mid-morning, while deploying his troops to relieve pressure on Mount St. Nicholas, Dragomirov was severely wounded in the knee and carried from the field. Shortly afterward Suleiman made his last effort to carry the position by coup de main. In one of the fiercest assaults within an overall engagement known for its ferocity, troops of the Pasha's lead battalion actually gained the trenchline, before a lethal mixture of Russian bullets and bayonets annihilated them, wiping from the face of the earth a unit of 500 men. Overall, the arrival of reinforcements over the next several days gradually raised the number of defenders to about 13,000, but fighting dragged on for nearly two more weeks before Suleiman retired to Kazanlik to reconstitute his army. His losses have been estimated variously between 8,246 and 12,000. The Russians and Bulgarians lost 3,640 killed and wounded. Francis Greene would later write that the first five days of Shipka's defense were notable for tenacious, dogged defense, but he would also note that "there were no skillful maneuvers of the troops on either side."[36] In this the Russians were fortunate, for had Suleiman chosen to use adjoining passes to threaten Stoletov's lines of communication and cut off the possibility of reinforcement, the story might have ended much differently.

Although Radetskii has been criticized for holding his reserves too far north for the timely reinforcement of Stoletov, VIII Corps bore responsibility for defending more than Shipka. Radetskii's chief concerns included Suleiman and the quadrilateral, but it was Osman Pasha who in the end lent VIII Corps' dispositions a certain credibility. On 19 August in a venture that caught the Russian high command by surprise, Osman sallied from Plevna with nineteen battalions against IV Corps. Constantinople had issued orders for him to relieve pressure on Suleiman, and Osman himself evidently wanted to test the strength of Russian dispositions. The Grand Duke had earlier expressed a desire for just such a circumstance that would enable him to take advantage of the superiority of Russian offensive tactics against troops deployed in the open. In the event, there was a breakdown in will, initiative, and coordination. Fourteen battalions of IV Corps defended against the Turks at Pelishat-Sgalovna, while twenty-four other Russian battalions of the Western Detachment were spectators on the flank. Lieutenant General P. D. Zotov, IV Corps commander, who in the absence of the Grand Duke also commanded the Western Detachment, failed to grasp the opportunity which Osman so eagerly thrust into his hands. Evidence indicates that Zotov feared a Russian failure that would expose the Imperial Suite at Gorni Studen to attack. More seriously, Zotov was not confident that General Kridener, IX Corps commander, could execute his part of a coordinated attack. Consequently, after losing perhaps 1,300 men in a fruitless attack on the Russians in defensive positions, Osman retired without pressure to Plevna.[37]

Third Plevna

Although Osman's venture confirmed the worst fears of the more cautious members of the Russian high command, it marked only a rude interruption in the larger planning and organizational effort for yet another assault on Plevna. Even before Osman's sally, the earlier failure of Suleiman's offensive together with the arrival of Russian reinforcements and Romania's entry into active hostilities against the Turks had restored Russian resolve to get on with the reduction of Plevna. This time, however, the Russians were to place much more emphasis on thorough preparation for the attack.

Part of that preparation included a supplementary attack at Lovech to seal off Osman Pasha from a third direction, the south, with its promise of assistance through another of the major Balkan passes, the Troyan. Besides, Lovech controlled an important road junction not fifteen kilometers from Plevna. The settlement itself lay in a three-sided bowl formed by Balkan foothills open to the northeast. Approximately 8,000 Turks with six guns occupied defensive positions on two successive ridge lines, one east and one west of Lovech. The two ridge lines were separated by a valley in which the settlement was located and through which flowed the River Osma. The strongest Turkish position, a redoubt covered by a system of trenches, was located on the western ridge line, on heights just northwest of Lovech. Lovech's reduction was entrusted to Major General Prince A. K. Imeretinskii, who commanded a force of about 27,000 troops in twenty-five battalions and fourteen squadrons with ninety-eight guns. The action, which took place on 22 August, remains instructive for the way it revealed in microcosm some of the possible solutions to tactical dilemmas inherent in the rifle-dominated battlefield.[38]

On 21 August, after a subordinate commander, General Skobelev, had conducted a thorough reconnaissance of Turkish dispositions, Imeretinskii drew up his plan of attack. His orders called for a southeast-to-northwest infantry attack, supported by artillery, on the successive ridge lines. The intention was to sweep over the objectives in two coordinated assaults, a main attack on the left under Skobelev and a supporting attack on the right under Major General M. V. Dobrovol'skii. Cavalry forces, including the Caucasian Cossack Brigade, would guard the main approaches and pursue broken enemy forces.[39]

The next morning the attack began at about 0600 and 0830 respectively with a heavy cannonade on Skobelev's initial objective and a premature attack by Dobrovol'skii's 3rd Sharpshooter Brigade. Because of insufficient time to conduct his own thorough reconnaissance and lack of experience, Dobrovol'skii had deployed the brigade in open positions 1,500 to 2,000 paces from its objective. Because of haste and inadequate preparation of emplacements, only three of his six supporting nine-pounders could put effective fire on the objective. When his troops began to take serious casualties from long-range rifle fire while still in their jumping off positions, Dobrovol'skii ordered an early advance to silence their tormentors. By 1000 his men had seized their first set of objectives, but their losses (600 officers and men) were so great that the brigade was rendered *hors de combat*. Dobrovol'skii immediately halted his attack and requested reinforcements.[40]

Meanwhile, Skobelev enjoyed success on the right. He had begun his attack with a preparation from fifty-six guns, then at noon he had launched his infantry against the first ridge line in a conventional assault. Skirmishers fanned out in advance of half-company columns deployed on line with bands playing and banners unfurled. During the crucial early minutes of the attack, the Russians advanced through vineyards which screened them from hostile observation and fire. Thanks to a combination of circumstances including the screen and the artillery preparation, and perhaps most of all to Dobrovol'skii's disaster on the right, which had diverted Turkish troops and fire from the main attack, Skobelev rolled over his initial objective with light casualties.

The second stage of the attack presented Skobelev with greater difficulty. Although he displaced his guns forward to cover the next phase of the advance, the Turks occupied the second ridge line in sufficient force and with sufficient cover to defy whatever the artillery threw at them. By 1400 Skobelev had led two battalions of the 64th Kazan Regiment to the banks of the Osma, and after some confusion the Russians crossed opposite the redoubt using several fords. They were followed somewhat later by the 5th Kaluga and 7th Revel Regiments to the left and right. As troops of the Kazan Regiment formed up on the flat terrain before their objective they came under heavy direct and indirect rifle fire. Rather than remain vulnerable in the open, small groups of officers and men either sprinted forward 500 paces to a mill surrounded by trees or advanced by leaps and bounds using whatever cover they could find. Those who stopped to await orders or to consider the situation often died where they stood or lay. As soldiers gathered around the mill for cover, their officers rallied them to continue the advance, again by leaps and bounds. In this fashion they crossed the remaining 1,000 paces of open ground and climbed uphill to a fold in the terrain before the first line of Turkish entrenchments. All across the open area soldiers of the Kaluga (on the left), Revel (on the right), and the 6th Libau (slightly to the rear) Regiments followed the unconventional example of the Kazan Regiment. Meanwhile, farther to the rear and left, on the outskirts of Lovech itself, Skobelev had managed to form a conventional battle line with the reserve 11th Pskov and 8th Estonian Regiments. Just as infantry of the Kazan and Kaluga Regiments hit the first Turkish trench line in skirmish formation with shouts of "hurrah," Skobelev launched the two regiments from the rear into a headlong assault against the heights on the left. Again bands played and banners snapped in the breeze. At approximately the same time troops from the Revel Regiment struck the redoubt from the right. The result was a coordinated attack on the redoubt from three sides. Those Turks who did not flee were killed where they stood. Those who fled were either cut down or taken prisoner by pursuing Russian cavalry. Altogether, the Turks lost 2,200 on the field and another 4,000 prisoners and dead in the pursuit. Two days later remnants of only two Turkish battalions of the original seven rejoined Osman Pasha inside the Plevna defenses.[41]

There had been mistakes in planning and execution, but the action at Lovech kept alive a belief in the efficacy of offensive infantry action against prepared positions. Success had turned on a number of circumstances, including careful planning, good fortune, reconnaissance, artillery preparation, the initiative of officers and soldiers, and flexible adaptation to the terrain. The skirmish line had stretched in width to two kilometers, and conventional infantry had joined the sharpshooters to advance as much as 1,200 paces under fire in leaps of approximately 200 paces at a time. Cavalry had discharged its security mission and once again had demonstrated the merits of an effective pursuit. Despite such lapses as Dobrovol'skii's attack, after the war Lovech came to be regarded by some as a textbook example of how to conduct an attack. Among the battle's interesting statistics was the fact that on August 22 alone, the Russian artillery had fired 4,883 shells in support of the attacking infantry. Each rifleman had fired an average of 14.5 rounds, for a total expenditure in the engagement of over a quarter million rounds.[42]

The seizure of Lovech secured the southern flank for a joint Russian-Romanian attempt to crack Plevna's defenses. Earlier the Romanians had consistently refused to cross the Danube until they could reach a suitable arrangement with St. Petersburg "to preserve their individuality." Following agreement, Prince Karol arrived on 16 August at field headquarters to assume command of the Western Detachment. By 24 August 32,000 Romanians and 108 guns joined the Russians at Plevna to raise the attackers' numbers to 84,000 men and 424 guns, against which the Pasha disposed 36,000 men and 72 guns. General P. D. Zotov served as the Prince's chief of staff, but staff coordination between the two allies remained problematic.

The plan of attack envisioned a three-pronged assault on the Turks. On the right, Prince Karol would throw forty-two battalions of Romanians and Russians against the heart of the Turkish defenses at the Grivitsa Redoubts. In the center, that is, to the southeast, the Russians would attack with twenty-four battalions to secure a strong point designated as Omar-bei-Tabia. On the left a secondary attack would come from the south under Skobelev and Imeretinskii through the "Green Hills" to seize the Krishinskie Redoubts. The overall attack was to be preceded by a four-day artillery bombardment to which the allies would devote the fires of twenty siege guns and 152 field guns.[43]

From the beginning the omens were bad. As in the past, artillery fire failed to inflict appreciable damage on Turkish field fortifications. To permit additional preparation, the attack was postponed from 28 August to 30 August. This meant that the element of surprise again failed to figure in Russian planning, and the Turks, aware of impending attack, bent all efforts to strengthening their already formidable defenses. On 29 August a rain storm transformed the battlefield into a muddy black paste, rendering difficult the resupply of artillery shells. Zotov himself seems to have lost confidence in the success of the attack. The tactician P. M. Gudim-Levkovich, who was present during staff deliberations, clearly recalled Zotov's remark, "Nothing will come of this attack: losses will be tremendous, and we will not take Plevna."[44] Although Zotov talked of postponing the attack once more, his arguments failed to sway Prince Karol and the Grand Duke.

Thick ground fog and drizzle greeted the attackers on 30 August, rendering ineffective the last two-and-one-half-hour artillery preparation. At 1500, the Romanian General A. Cernat led forty-eight Romanian and Russian battalions into the heart of Osman's defenses. In a new twist the infantry advanced directly from their assembly areas into the attack without bothering to occupy concealed jumping off positions. The predictable result was heavy losses, and the first wave of fourteen Romanian battalions was beaten back, as were the second and third. At 1800, the 5th Russian infantry division joined the Romanians, and after two hours' hard fighting, allied troops managed to capture a portion of the Grivitsa Redoubts. However, there was no respite, for the attackers now came under heavy fire from adjoining trenches and other nearby redoubts which reconnaissance had failed to locate. Further attacks were fruitless.[45]

Meanwhile, precious little assistance came from the center, where General E. K. Krylov had attacked the redoubt Omar-bei-Tabia at 1500 with thirty battalions of infantry. Troops of the 16th and 39th divisions attacking shoulder to shoulder were cut to pieces. The assault soon flagged, and without reinforcements, Krylov was reduced to demonstrating in support of the other two attacks.[46]

In contrast, the left enjoyed unexpected success. Imeretinskii attacked in three echelons with the dashing General Skobelev leading the first dressed in a white coat and mounted on a white horse. By carefully exploiting the terrain, the Russians were able to occupy jumping off positions within 900 to 1,200 paces of Turkish defenses. As was his habit, Skobelev would launch his first wave into the assault, then at the critical moment, when the attack began to lose momentum, commit additional reserves. By feeding new troops into the assault, Skobelev retained attacking momentum and thus unexpectedly overran the Krishinskie Redoubts, nearly breaking through to the southern outskirts of Plevna. However, Skobelev was now left in a precarious position: he had driven a dangerous wedge into Osman's defenses, but his men could neither halt with fire coming from three sides nor advance without reinforcements. Calls for reinforcement fell on deaf ears, for the high command remained preoccupied with other sectors and subject to confusing and delayed reports describing the action on the far left. When Osman Pasha realized that attacks on his left and center had lost momentum, he shifted fifteen battalions into the battle against Skobelev, who was only barely able to fight his way back to his own defensive lines.[47]

On the morning of 31 August the Russians and Romanians failed utterly to enlarge their success at the Grivitsa Redoubts. With losses numbering 12,700 Russians and more than 3,000 Romanians, the allied high command called off the attack.

The third storm of Plevna demonstrated significant failures in command, control, and tactics. Despite provisions for rudimentary staff coordination, neither the Russians nor the Romanians exercised overall control of the battle. No one had conducted an adequate reconnaissance, and no one had paid heed to the weakness of enemy defenses in the south and west. During the course of the battle itself there was an alarming tendency for subordinate commanders—except for Skobelev—to lose effective control over troops in the attack. Again, as in the first two storms, Plevna demonstrated the devastating effectiveness of rifle fire from prepared positions against infantry attacking in the open. For a variety of reasons artillery once again demonstrated its impotence in the face of what in effect had become a variant of siege warfare. Perhaps most heartbreaking of all, for reasons which remain unclear the Russians failed to employ the field telegraph to ensure coordination across a battlefield which stretched across a sixteen-kilometer semicircle from flank to flank.

The Sofia Road

After the unsuccessful third storm of Plevna, the allied command decided to starve its adversaries into submission. General E. I. Totleben of Sevastopol fame assumed control of engineering operations to complete a close investment of Turkish defenses, while allied forces tightened their grip on approaches from the north, east, and south, leaving only the west and southwest open for Osman to receive supplies and reinforcements along the Sofia road.[48] Russian cavalry soon interdicted that route in strength, but horsemen lacked the offensive power required to subdue well-fortified garrisons, each consisting of six or seven infantry battalions with a few guns stationed at critical points along the road. In its impatience to get on with the reduction of Plevna the Russian high command blamed aging cavalry officers for the failure to seal off the town, and in mid-October summoned Lieutenant General Gurko, the energetic hero of advance guard fame, to lead regiments of the Guards in a spirited offensive against Turkish positions at Gorni Dubnik, Telish, and Dolni Dubnik. Their reduction would complete Plevna's investment.

Totleben and Gurko labored under the illusion that Osman's forces numbered eighty battalions inside Plevna with another 120 strung out in a series of strongpoints between Plevna and Sofia. This assumption affected tactical dispositions and exaggerated Gurko's troop requirements. To reduce the garrisons at Gorni Dubnik and Telish, Gurko mobilized 170 guns and the equivalent of forty-eight battalions, of which more than half were assigned not to the attack but to screening and security operations.[49]

At Gorni Dubnik the disadvantages inherent in the lopsided distribution of forces were magnified by reliance on tactics which the preceding disasters at Plevna had already discredited. The strongpoint in question was located twenty-three kilometers southwest of Plevna, where seven Turkish battalions held a large redoubt and a small redoubt on minor elevations astride the Sofia road with clear fields of fire extending in all directions to distances ranging from 300 to 1,000 paces. At 0900 on 24 October Gurko launched three columns against the redoubts in what was intended to be a coordinated assault against their eastern faces and outer works on their western and southern flanks. When the columns failed to arrive simultaneously at their attacking positions, General L. L. Zeddeler in the center was forced to launch his troops into a premature attack after they had begun to suffer heavy casualties from concentrated enemy rifle fire. Zeddeler, who had earlier warned of the new lethality inherent in breechloading firepower, soon lay gut-shot on the battlefield, but his Guards Grenadier Regiment succeeded in advancing perhaps 1,000 meters under fire to capture the smaller redoubt south of the Sofia road after suffering heavy losses.[50] The other two columns conducted textbook attacks behind deployed skirmishers covering half-company columns in the advance. By noon the troops of these two columns were halted with heavy losses, and they mingled with one another and the grenadiers as they sought cover in ditches and among folds in the ground anywhere from 100 to 800 paces from the objective.

Gurko's solution to the impasse was more of the same. He dispatched reinforcements to the center and issued orders to renew the attack from all sides at 1400. However, in the noise and confusion of battle few troops on the battlefield heard the nine volleys which the artillery delivered to signal a simultaneous advance. Without support to divert attention and fire from newcomers to the fray, infantrymen of the reinforcing Moscow Guards Regiment were forced to go to ground about 650 paces from the Turkish redoubt. When the Guards sappers advanced to assist the grenadiers, hostile fire forced them to dig in on the battlefield, and soon infantrymen across the front followed their example using anything that came to hand—mess utensils, knives, bayonets, even their bare hands. Because some of the Russians had crawled or otherwise worked their way to within 100 paces of the redoubt, supporting artillery began to withhold its fire for fear of hitting its own troops. Although clusters of infantrymen continued to advance in leaps and bounds, by late afternoon Gurko was prepared to call off the attack and withdraw with the onset of darkness.[51]

The survivors among Gurko's troops did not share their commander's intentions. Even after the failure of the afternoon attack, soldiers of the Grenadier Regiment's strongest unit, the fourth battalion, continued to worm their way forward. They advanced by groups of twos and threes until they reached the cover of a small white hut, where they paused to catch their breath and screw up their courage to push ahead. Under lively and accurate covering fire from nearby troops of their own and other regiments, some grenadiers sprinted the short distance from the hut to a dead space in a ditch along the redoubt's southern face. There until just after sunset they waited, their numbers gradually growing until the dead space could hold no more men, at which time they sprang shouting "hurrah" to the rampart and into the redoubt. Their shouts galvanized nearby troops into action, and with the Turks' attention momentarily diverted, the rush to the redoubt became general. As Russians poured into the stronghold to engage their adversaries in hand-to-hand combat, the Turkish commander struck his flag, surrendering 2,000 men. The Turks suffered 1,500 killed and wounded. Russian losses were 3,500 killed and wounded, a sum about equal to the total number of Turkish defenders. The Guards grenadiers alone lost 1,017 officers and men.[52]

Without the victorious conclusion, the same bloody drama was replayed at nearby Telish, where the Guards Jaeger Regiment conducted an attack to prevent Turkish reinforcement of Gorni Dubnik. After conducting a perfunctory reconnaissance and posting Guards cavalry units on his exposed left flank, the Jaeger commander, Colonel Chelishchev, ordered his troops into a textbook attack on a redoubt and several entrenchments located astride the Sofia road. While the 3rd Guards Artillery Battery delivered supporting fire at 1,500-meter range, the jaegers left cover of a ravine to deploy in cornfields facing their objective to the west. According to the unit diary their formation consisted of "two lines of companies with skirmishers in the van." Although cornstalks obscured the attackers' vision, the jaegers immediately came under fire from Turks occupying rifle pits midway between the Russian jumping off point and the redoubt, about 1,200 meters distant. Chelishchev ordered his troops to clear these intermediate positions, and they enthusiastically responded, advancing "as in maneuvers, upright, with officers before their companies."[53] The Russians easily took the trenches, but in the process they exposed themselves to heavy fire from the more distant redoubt. At the same time, Turkish rifle fire put part of the attackers' artillery out of action. While Chelishchev worked to resolve the problem of fire support, the commanders of his lead companies, acting on their own initiative, renewed the advance to silence the fire that was so unmercifully lashing their ranks. Not to be left out, companies of the second line surged forward in what became a spontaneous general infantry assault with sparse artillery support against a strongly fortified position. Momentum and will carried the attack over level ground to within forty paces of the main redoubt, at which point the jaegers either flattened themselves against the ground or used the bodies of their dead comrades for cover. The survivors could neither advance nor retreat, so there they remained, pinned down under the hot sun for four hours, trading shots with their adversaries and dying by degrees. Company officers especially suffered as they attempted without success to rally their men for the final assault. One, an Ensign Perepelitsyn, was struck by nine bullets, five of which individually would have produced fatal wounds. A Turkish sally was driven back, and twice the Turks struck their colors only to recommence firing when the Russians stood up to receive the surrender. Finally, at 1400, Chelishchev was able to disengage his men and withdraw them to a covered position to reorganize for another attack, a course of action he discarded later that afternoon after receiving word that the Turks had managed to slip reinforcements into the beleaguered redoubts. Casualties for the Life Guard Jaegers at Telish stood at more than 900, of which about one-half died.[54]

These losses were horrifying and sobering. Commanders might have been able to rationalize earlier losses, but these last came from the flower of the army, the Guards, on whom the monarchy doted and among whom served the offspring of distinguished noble families. The roll of dead and wounded high-ranking officers included Zeddeler, Rosenbakh, Pritvits, Ol'derogge, and Grippenberg, all members of illustrious Baltic noble families. As Lieutenant Greene later commented, "the breechloader is no respecter of persons, and there was a wail of lamentation throughout St. Petersburg and Moscow."[55] Not surprisingly, Gurko's profligacy nearly cost him his command and earned immediate admonitions from both the Emperor and the Grand Duke.[56]

It was a chastened Gurko who returned to the field for a more circumspect approach to Telish and the reduction of Dolni Dubnik. Three days after the failure of 24 October, he returned to the Sofia road for a thorough reconnaissance of Telish in anticipation of carrying the position by force of artillery fire alone. By now, Gurko's infantry occupied entrenchments on three sides of the Turkish position at distances varying from 1,600 to 2,000 meters. At noon on 28 October ten batteries of Russian artillery opened fire at ranges of 1,400 to 1,900 meters from their targets. Three thousand rounds and nearly three hours later, the Turkish commander surrendered. His troops lacked the extensive cover which more substantial fortifications had afforded the defenders at Gorni Dubnik, and Russian artillery firing a mixture of explosive and shrapnel proved more effective than either at Plevna or Lovech. Gurko's casualties were one killed and five wounded, while Turkish losses were 157 killed, all victims of Russian artillery fire. On hearing of the surrender at Telish, the Turkish commander at Dolni Dubnik quit his position without firing a shot to join forces with Osman Pasha inside Plevna.[57]

At Plevna, close investment accomplished what attack had failed to do. By the end of November, the Turkish garrison had run so low on supplies that Osman Pasha resolved to break through the encirclement before extreme hardship further reduced his troops' combat effectiveness. On the night of 28 November, he concentrated his forces just west of Plevna on the left bank of the River Vid, and at dawn of the 29th, he threw them into a desperate attack against three opposing lines of fortifications. Before Russian reserves could be rushed to the scene, the assault had penetrated two lines. However, the Russians soon counterattacked with fresh troops, hurling the Turks back to their attacking positions, at which point Osman finally asked for terms. He subsequently surrendered his entire command, about 43,000 troops. With the Grand Duke's right flank free of this encumbrance, the Russians might renew their advance to the Balkan ridge and beyond.[58]

Trans-Balkan Campaign

The fall of Plevna dealt a heavy blow to the Turks, depriving them of a veteran field army and their best commander, but the situation was still far from hopeless. The harsh Balkan winter had only begun to wreak devastation on Russian field forces, and by early December ice had cut the Russian pontoon ribbon across the Danube, thus isolating—for a time—the Tsar's corps in Bulgaria. The Turks had only to block the Balkan passes, spend the winter rebuilding and reconstituting their forces for a strategic defense of the south Balkan slope, then let time, distance, and the European balance of power work in their favor. Unfortunately, the Turks were unable to cope with two other circumstances that worked heavily in the Russians' favor: rekindled Russian offensive resolve and poor Turkish generalship. At the urging of Miliutin, the Grand Duke, ever aware of the increasing probability of British intervention as the war wore on, decided to steal a march on Whitehall—and some of his own diplomats—by engaging in winter operations. Lieutenant Greene, ever the acute observer, noted succinctly that "in this war, as in all others, purely military reasons had to be subordinate to the higher political reasons." At the moment, perhaps no one understood this better than the Grand Duke, who determined that time must not be allowed to work in favor of the diplomat and politician at the expense of the soldier.[59] Rather than permit the moment to slip away, the high command was willing to push its armies through dangerous mountain passes into the teeth of Turkish opposition and the Balkan winter. For their part, the Turks unwittingly assisted the Russians both by refusing to believe in the possibility of a winter campaign and by disposing their troops south of the Balkan passes in such a way as to invite defeat in detail. Evidently they shared the convictions of an illustrious commentator, Helmuth von Moltke, who had declared the mountains impenetrable in the winter. As he reviewed the Turkish situation after the fall of Plevna, the Chief of the Prussian General Staff had remarked that any general attempting a winter crossing of the Troyan Pass, one of the chief routes to Sofia, "deserved to be called foolhardy, for two battalions would be sufficient to withstand the attack of an entire corps."[60]

Moltke's opinion notwithstanding, the Russians now felt that they possessed sufficient superiority within the Balkan theater to take the war to the Turkish heart, Constantinople. Russian forces in the Danubian Theater stood at half a million men with 1,343 guns, against which the Turkish command now deployed only 183,000 troops with 441 guns. Except for the quadrilateral, the Turks were arrayed on the south Balkan slope with Suleiman Pasha on the right, Vessil Pasha in the center, and Shakir Pasha on the left. The Russian plan was to hold against Suleiman and attack on their own right and center with three columns: Gurko through the Araba Konak Pass to Sofia, P. P. Kartsov through the Troyan Pass, and Radetskii through Shipka.[61] The offensive began on 13 December, with Gurko maneuvering through several lesser passages to force Shakir Pasha's retirement from blocking positions within the Araba Konak Pass. As the Turkish commander withdrew to a new defensive line, he uncovered Sofia, which Gurko occupied on 23 December. Despite poor weather conditions, Gurko continued to apply pressure, pursuing Shakir Pasha along the road to Philippopolis, thereby threatening the rear of Turkish troops blocking the Troyan Pass. This opened the way for Kartsov, who now crossed the divide with about 5,000 troops to join with Gurko in the march to Philippopolis.[62]

Gurko's success triggered Radetskii's attack against Vessil Pasha whose forces included 35,000 men and 108 guns arrayed in several defensive positions between Sheinovo and Shipka to prevent a Russian incursion through Shipka Pass. Radetskii's plan was to launch a frontal pinning attack against Turkish defenses at the southern entrance to Shipka Pass, then cut loose two of his subordinates, Skobelev and N. I. Sviatopolk-Mirskii, to conduct enveloping attacks over several lesser passes against the main Turkish encampment at Sheinovo. Although the Turks evidently trusted in the weather to provide their first line of defense, Radetskii's concept involved considerable risk. To divide his forces in the face of the enemy invited defeat in detail. Once the columns were actually on the march, there was no way to assure communication among them. Finally, the mountain passages were so restrictive that a winter storm or even minor resistance along any of the three routes might upset the timetable, and with it, the entire operation.

The two enveloping columns set out on 24–25 December. Although marching the longest distance (thirty-five kilometers) through snow four meters deep, the left column under Mirskii (19,000 troops and twenty-four guns) broke through first, and for its trouble was forced to endure some anxious moments before either Radetskii's central column (12,000 troops and twenty-four guns) or Skobelev's right column (16,500 troops and fourteen guns) could effectively join the fray. Skobelev encountered unexpected resistance during his twenty-kilometer march, while Radetskii made little headway in his assault on several successive Turkish defensive lines.

Consequently, neither could directly support Mirskii on 27 December, the day designated for concerted attack against the primary Turkish defensive position at Shipka-Sheinovo.[63]

Nevertheless, an outnumbered Mirskii attacked on the 27th as originally ordered, his left flank dangling in thin air. At about 1000, with the redoubtable 4th Sharpshooter Brigade deployed in the van as skirmishers, Mirskii advanced against outlying Turkish positions with the 33rd Elets, the 36th Orel, and 117th Iaroslav regiments drawn up in two mutually supporting lines. The Russians quickly occupied two villages, but by nightfall their attack had stalled, thanks to increasing Turkish resistance, heavy losses, and a shortage of ammunition. Mirskii now found himself in a precarious position: although his men had carried the first enemy defensive line, his left lay unprotected and he lacked the manpower and ammunition to continue the advance. Indeed, at any moment he expected a powerful Turkish counterattack. During the hours of darkness, he managed to establish contact through messengers with Radetskii, who advised him to maintain his position at least for an additional twenty-four hours.[64]

The next day, 28 December, brought foul weather and the expected attack from Vessil Pasha. In the valley south of Shipka Pass soldiers awoke to a heavy ground fog, which obscured enemy positions but did not deter the Turks from assailing in succession Mirskii's right and left. When both onslaughts failed, Mirskii riposted with his left and enjoyed unexpected success, his troops overrunning Shipka village. Meanwhile, within the pass itself, fog was transformed into a blinding mixture of snow and frozen mist. With visibility limited to perhaps ten meters, at 1200 Radetskii dutifully threw his 55th Podolsk, 56th Zhitomir, and 35th Briansk Regiments into a supporting attack. Two hours and two Turkish trench lines later, Radetskii halted his attack with 1,700 casualties. The issue now remained Skobelev's alone to resolve.

The near-legendary "White General" had delayed his own attack until all his units had completed their tortuous march through Imetli Pass. Although resistance had caused him to arrive in the valley a day late, he carefully fulfilled the letter of Radetskii's orders "positively not to attack before all his men were assembled." Once in the open, Skobelev wasted little time establishing contact with Mirskii and forming his troops for an assault upon Sheinovo redoubts from the west. Under cover of lingering fog, he arrayed his forces in two lines, with the 63rd Uglich Regiment and the Bulgarian militia in the first and the 61st Vladimir and 64th Kazan regiments in the second. With bands playing, Skobelev hurled his regiments into a daring bayonet attack across open ground. Losses were heavy, including many bandsmen, but the Russians gained the redoubts, where for some minutes the adversaries remained locked in hand-to-hand combat. The grim contest ended only when six Turkish battalions laid down their arms. Even as Mirskii and Radetskii advanced to complete the encirclement, Vessil Pasha surrendered his entire force of 33,000 men. Only Suleiman Pasha now lay between the Russians and Constantinople.[65]

Sheinovo probably ranks with Union General George Thomas's triumph in 1864 at Nashville as one of the nineteenth century's "perfect battles." Lieutenant Greene called the approach maneuver "brilliant" and even asserted that in boldness and brilliance the capture of Vessil Pasha's army surpassed Gurko's earlier advance guard action.[66] However, the success was not without immediate and long-term costs. The Russians suffered more than 5,000 casualties, of which 1,103 died. More damaging for the future was the way that the Sheinovo success—like Lovech—kept alive a faith in the efficacy of the textbook-style infantry attack, even against prepared enemy positions. The Russians had been unable to bring their artillery over the wintry mountain passes, with the result that infantry alone carried the Turkish positions at Shipka and Sheinovo. This caused observers such as Lieutenant Greene to note that Skobelev's attack "renders more than doubtful the conclusion which has been hastily drawn from this war (from Plevna particularly), that successful assaults of earthworks defended by modern breechloaders are impossible."[67] If others drew the same conclusion, Sheinovo rather than Plevna would serve as the model which future commanders would seek to emulate, and the costs of emulation would come high.

Overall, in the trans-Balkan campaign the Russians had recouped their fortunes by accomplishing one of the most daring feats in modern military history. They had driven the elements of three large columns (Gurko, Kartsov, Radetskii) through the Balkans under conditions which saw trails covered with ice and snow measuring up

I. V. **Gurko** (1828–1901) Cavalry and Detachment Commander 1877–1878, Commander of Warsaw Military District 1883–1894
Photo: *Russes et Turcs*

M. D. **Skobelev** (1843–1882) Legendary "White General," Commander of 16th Infantry Division at Plevna and Sheinovo
Photo: *Russes et Turcs*

to four meters in depth. They had utilized surprise and turned to their advantage both superiority in numbers and failure of the Turkish field forces to assure one another mutual support. In most cases the Russians had overcome strong defensive positions either by envelopment or by maneuver. In a word, the Russians had convincingly overcome the paralysis of Plevna.[68]

The Russians now poured through the Balkans to press their advantage. Gurko advanced along the valley of the Maritsa to threaten Philippopolis, and Radetskii struck out with several columns for Adrianople. In a three-day battle (3–5 January 1878) at Philippopolis, Gurko shattered Suleiman Pasha's army, the remnants of which fled to the shores of the Aegean Sea after abandoning their stores and artillery (180 guns). On 8 January, after the Turkish garrison had surrendered without opposition, Radetskii entered Adrianople with banners flying and bands playing. While the belligerents at last undertook negotiations, the Russian army eventually halted its advance at the village of San Stefano, about fifteen kilometers from Constantinople.[69]

The Caucasian Theater

The Caucasian Theater was essentially an economy of force operation. Although General Obruchev's plan had originally foreseen employment of 125,000 men and 456 guns south of the Caucasian divide, at the outbreak of hostilities in April 1877, only 50,000 troops and 202 guns were available under the overall command of Grand Duke Michael (Chief of Staff General-Adjutant M. T. Loris-Melikov). Divided into four detachments (the Kobulety, Akhaltsykh, Aleksandropol', and Erevan), the Caucasian forces had as their mission to tie down Turkish troops and, if possible, to reduce Turkish fortresses at Batumi, Ardahan, Kars, Bayazid, and Erzerum. Covering the main approaches against the Russians were two Turkish armies, one of 60,000 under Mukhtiar Pasha and another of 40,000 near Erzerum.[70]

The Russians enjoyed early success, then lapsed into operational torpor. On 18 April, the Erevan Detachment occupied Bayazid without battle. On 6 May, the Akhaltsykh Detachment, enjoying strong artillery support, took the Turkish fortress at Ardahan with light casualties. After these successes, Loris-Melikov attempted to concentrate his forces for a reduction of the Turkish fortress at Kars. However, initial operations against the fortress proved inconclusive, in large part because the Russians lacked sufficient manpower both to undertake a siege and to contend with Mukhtiar Pasha's covering army of 35,000. Meanwhile, the Kobulety Detachment, battling with Turkish reinforcements brought by sea, fared little better in its offensive against Batumi. Soon all detachments were reduced to conducting defensive operations, and by midsummer a thoroughly frustrated Loris-Melikov had even withdrawn his main column to the Russian border.[71]

The high command, however, was not about to allow the Caucasian Theater to wither so easily. To renew the offensive the Tsar in July decreed a mobilization of reinforcements, eventually dispatching from the interior of Russia by rail and on foot the 1st Grenadier Division and the 40th Infantry Division. Equally important, on the

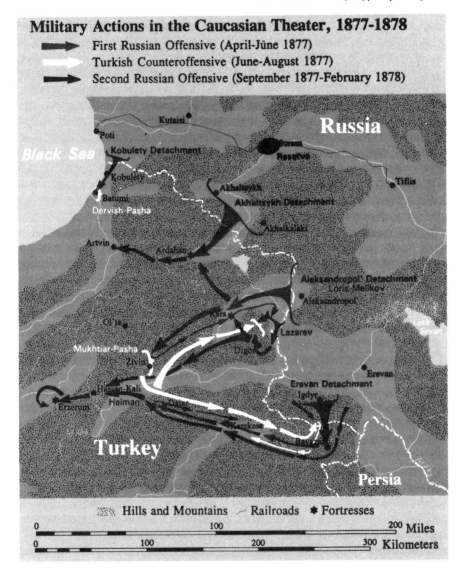

advice of Miliutin the Tsar posted General Obruchev to the Caucasus to assist in planning a new offensive.[72]

The arrival of fresh units in September raised the detachment operating against Kars to a strength of 50,000 troops, 184 field guns, and twenty siege guns. With these forces Obruchev proposed attacking Mukhtiar Pasha in the field to defeat the Turkish covering army, thus isolating Kars. However, a three-day battle (20-22 September) produced inconclusive results and 3,000 Russian casualties. Momentarily, the situation seemed worse than ever for the Russians. Then, for reasons which remain unknown, Mukhtiar withdrew his main forces to static defensive positions on heights at Avliar, Aladja Dagh, and Vizinkioi, thus unwittingly offering Obruchev an opportunity to atone for earlier losses.

Despite his scholarly appearance and record, Obruchev was not one to waste time pondering alternatives. He immediately prepared a new plan that called for the annihilation of Mukhtiar's army. In brief, his concept called for the Russian main force under Loris-Melikov (thirty-two battalions, four squadrons, and 112 guns) to conduct a demonstration against the Turkish front. Meanwhile, a second force under General I. D. Lazarev (twenty-three and one-half battalions, twenty-nine squadrons, and seventy-two guns) would pass completely around the Turkish right flank to fall on Mukhtiar's rear. Obruchev trusted the field telegraph to assist in coordinating the demonstration and envelopment.

Lazarev left the main camp during the night of 28–29 September, stringing telegraph wire behind his advancing troops. By moving at night in great secrecy, his flanking force was able to reach the Turkish left rear on 2 October, at which time he drove off a 6,000-man Turkish detachment and consolidated his position. After an exchange of information via telegraph during the early morning hours of 3 October, Grand Duke Michael ordered his main force to attack at daylight. Lazarev joined in the attack at noon, after which joint efforts of the pinning and attacking force produced the expected results. With all escape routes blocked and Russians advancing from the front and flank, Mukhtiar's army suffered a crushing defeat, losing 16,000 casualties and 8,500 prisoners. During the general confusion of battle, the Pasha himself managed to escape to Kars.[73]

Despite the strength of its fortress, Kars was not long to remain a refuge. To be sure, without additional troops and guns, Loris-Melikov had earlier discarded the possibility of taking Kars—thanks to alleged prewar Prussian assistance the Turks had modernized the fortress so that it presented a formidable obstacle to Russian conquest of Armenia. The fortress consisted both of a masonry citadel built on a rock overhanging a gorge just north of Kars and twelve detached forts, all of which added up to a powerful defensive line more than seventeen kilometers in length. Altogether, the Pasha counted a defense force of 25,000 troops and 300 guns, of which 192 were steel.[74]

To attack this powerful complex, the Grand Duke approached Kars at the end of October with 50,000 men in forty battalions and fifty-six squadrons and 126 field guns. To the latter he was gradually able to add sixty-three siege guns. When the fortress commandant, Hussein Khalil Pasha, refused terms of surrender, a Russian council of war agreed to take the stronghold by storm on the night of 5–6 November. The way to the actual attack was paved by an elaborate deception plan. For several consecutive nights before the one designated for the attack the Russians launched demonstration attacks. On the night of the real attack, Grand Duke Michael divided his forces into five attacking columns and two demonstration columns. The latter were to jump off first against four of the forts. When they had diverted the Pasha's reserves, the five genuine attacking columns were to assail various forts and strongpoints in the heart of the defenses.

Thanks to elaborate preparations, the Russian assault caught the Turks by surprise. The demonstration columns did their work by attacking from the southwest, diverting the Pasha and his reserve from the real attacks of the remaining five columns. Russian resolve and bayonets did the rest. Lieutenant General Ia. K. Alkhazov's column captured forts Haziz and Karadej. Behind him Major General Shatilov's column took Fort Arab Tabia. The remaining three columns fought a furious see-saw battle for Fort Kanli, the key to the fortress defenses. However, the attackers made such steady progress that by morning they had taken the town and were engaged in mopping up the citadel. The Turks lost 7,000 casualties and another 17,805 prisoners. Russian losses were 2,270 killed and wounded.[75]

Following the capitulation of Kars, the final act in the Caucasian drama was the taking of Erzerum. Even before the Russian blockade, Mukhtiar Pasha had left the fortress to organize the defense of Erzerum. On 23 October, while the Grand Duke and General Lazarev were planning to storm Kars, Mukhtiar had arrayed thirty-five battalions across the Erzerum road to block General V. A. Heiman's advancing Erevan Detachment. During the course of a day-long battle, Heiman turned the Pasha out of his position. However, Heiman was slow to pursue, with the result that Mukhtiar Pasha again escaped the Russian net, this time taking refuge for the winter in Erzerum. When Heiman moved forward to invest the town, his troops began to suffer heavily from typhoid and cholera. The Turks managed to hold out over the winter until they were forced to hand over Erzerum in accordance with the San Stefano agreement of 19 February 1878.[76]

By terms of the agreement reached at San Stefano and later modified by the European powers at the Congress of Berlin in mid-1878, the Turks were also forced to acknowledge Romanian independence, to grant autonomy to Bulgaria, Montenegro, and Serbia, and to permit the Russians once again to deploy substantial naval units on the Black Sea. Although great power interference at Berlin prevented the Tsar from enjoying all of the fruits of his Balkan victories, the might of the Russian army had been instrumental in bringing about a major shift in the balance of power in southeastern Europe. However, the price was steep. Total Russian casualties exceeded 100,000, and the treasury lay exhausted. When Russian diplomats departed for Berlin in the spring of 1878, Miliutin advised them to accept whatever terms were possible, because there could be no thought of continuing the struggle.[77]

Rear Services

The outbreak of pestilence among Heiman's troops was reminiscent of one of the saddest chapters in the overall sad epic of the Crimean War of the prereform era.[78] One of Miliutin's reform objectives had been the improvement of rear services, including support and medical care. The question now was whether Miliutin had made any headway against the traditional scourges of Russia's field armies, chronic supply shortages, and heavy losses from wounds and sickness.

The answer to this question was only partially affirmative, with important differences among localities and categories of support. By far the most positive change occurred in the realm of ammunition resupply, where staff officers working within the framework of regulation and organization were able to impose priorities over other types of support and assure troops and guns of a generally adequate flow of artillery and small arms munitions. Whatever shortages occurred, including those experienced in the prolonged fighting for Shipka Pass, appear to have been local phenomena and primarily a function of difficult access and scarce immediate resources. Fortunately for the Russians, a combination of circumstances, including strict conservation measures and a reduced rate of usage, prevented the trans-Balkan campaign from taxing the ammunition resupply system beyond its modest capacity.[79]

Medical care was another area in which the 1877–78 record seems to have improved markedly over 1854–55. Although Vsevolod Garshin's short story "Four Days" dramatized the trauma of wounded soldiers left to their own devices in the field, the overall impression was quite different.[80] Thanks to an influx of trained medical personnel, the advantages of improved evacuation by rail, the designation of unit stretcher bearers, and the creation of division-level field hospitals, wounded and sick soldiers now stood a far better chance of survival than twenty years before. For wounded soldiers, the mortality rate dropped to between 10.8 and 14.1 percent respectively for the Danubian and Caucasian theaters. In general, the mortality rate for sick and wounded personnel fell to approximately one-third of the Crimean War rate.[81]

Still, there were important lapses in hygiene and medical care. Responsible staff personnel were not always kept informed, and medical facilities were often poorly located, understaffed, and given short shrift in resource allocation. Thus there were instances in which wounded were transported in freight cars, in which hospital linens went unchanged for a month at a time, and in which sufficient medical personnel were not available to satisfy short-term demand caused by episodic surges in wounded and sick troops. Nonetheless, the overall picture had changed for the better since the Crimean War, with progress stemming not only from systemic improvements but also from contributions made by nurses, the Russian Sisters of Mercy, and by the Society of the Red Cross.[82]

Because of tainted drinking water and recurring difficulties with bad food and field hygiene, troops were still far more likely to fall ill than to fall victim to an enemy's bullet. Statistics indicate that Russian combatants in the war of 1877–78 were apt to become seriously ill at a rate several times that of incapacitation resulting from battle casualties. However, again in comparison with the Crimean War, the chances of recovery had improved dramatically, primarily because of greater stress on field hygiene and more timely and better medical care.[83]

In contrast, the overall picture for supply services remained grim. As the war unfolded, the concept was that the military districts adjacent to the theaters of operations were to push their support staffs and structures forward to serve the supply needs of the field armies. In reality, money was short, supply stores and facilities were insufficient, and the transportation net and means were grossly inadequate. Added to these shortcomings was a supply staff lacking the necessary authority and integration with the field forces to make any kind of rudimentary support system capable of satisfactory performance.[84] Consequently, the army intendance service quickly turned to civilian contractors to compensate for shortfalls endemic to the structure. The most notorious of these contractors was the partnership of Greger, Gorvits, and Kogan, which accepted government credits in return for an agreement to monopolize army contracting services in the Danubian theater.[85] From the beginning, the supplies provided by these contractors were either late, inadequate, spoiled, or nonexistent. The farther the army marched into the Balkans, the worse the situation became, until much of the already spare and corrupt structure degenerated into chaos, causing troops south of the Balkan divide to experience severe shortages in rations and forage.[86] Although staff officers were cashiered, depots pressed forward, and temporary changes written into the regulations on field administration, the supply snarl remained a stubborn fact of military life until war's end. Without too much exaggeration, one contemporary observer expressed the general sense of frustration when he noted that "the supply functionaries by their actions both in the Crimean and in the latest Turkish campaign accorded much greater service to our enemies than to our armies."[87] Meanwhile, contractors continued to make enormous profits, but investigations never uncovered the deeper sources of corruption, even though recurring rumors ascribed contractor immunity to well-placed bribes with the Grand Duke Nicholas Nikolaevich and his chief of staff, General-Adjutant Nepokoichitskii.[88] Understandably, the whole story remained obscure even after the war, when the War Ministry moved to prevent a recurrence by means of improved regulations for field administration.

For a variety of reasons, then, troops in the field were often ill fed, especially when combat or the pace of advance took them away from their own regimental kitchens. Soldiers sometimes made good shortages from captured stores and warehouses, and the local Slavic population was often generous in sharing its own meager resources. Whenever possible, the troops reverted to tradition and lived off the land, but this was far from a satisfactory solution to a systemic problem. The same kind of difficulties held true for boots and clothing, especially winter clothing, with the result that the Danubian corps were literally only half-outfitted for the coming of an unusually severe Balkan winter.[89]

As was the case in other major areas of concern, including tactics and armament, Miliutin's army presented a support picture that was transitional. The system no doubt represented in many aspects an improvement over what had previously been the case, but much work remained to be done. If the reforms of the 1860s and 1870s could put a mass cadre and reserve army into the field, then the infrastructure was still inadequate to the task of keeping that army fed and clothed within the theater of operations.

Lessons Learned and Unlearned

A historical survey of the operational and tactical aspects of the Russo-Turkish War of 1877–78 leads to a number of conclusions, some evident, others less obvious but nonetheless significant. One major conclusion was that plans for and conduct of the war were heavily and even decisively influenced by the lack of Russian naval support on the Black Sea. The absence of a genuine Black Sea fleet caused the Russians to limit their planning to a land campaign with all its advantages and disadvantages. Popular pressures notwithstanding, the failure of diplomacy and the decision to declare war revealed a basic misunderstanding of the necessity to pursue political objectives with reference to actual military and naval means. Or, to characterize the situation another way, the Russians permitted a gap to open between military and naval capabilities on one hand and policy and aspiration on the other. What might have happened if the Russian army had broken through to the Bosporus to confront the heavy guns of the British Royal Navy remained sobering conjecture, although figures such as Miliutin seem to have retained a healthy sense of Russia's inadequacies.[90]

The War Minister's presence aside, the Russians demonstrated a curious inconsistency in exercising effective command and control at higher levels. Despite occasional difficulties with mobilization, the Russians deployed their forces with a fair degree of efficiency and got them into the field in a timely manner. Commanders and staff officers employed the railroad and telegraph to operational advantage but not always with the required degree of consistency. For example, there was no practical reason why the Russians could not have utilized the telegraph more intelligently to speed communications that would have permitted the conduct of economy of force operations on the north slope of the Balkans. This in turn might have saved the high command from making decisions which in the end prolonged the war by affording the Turks breathing space.

At the same time, the scale of operations and the size of battlefields begged for effective means of control beyond the traditional combination of messenger, sound, and visual communications. Tactical use of the telegraph before Kars aptly demonstrated the utility of modern technology on the battlefield, and a similar application at Plevna might have saved much time and many lives. Beyond the limits of the battlefield, a failure to maintain instantaneous communication would necessitate mobilizing additional manpower which otherwise might have been conserved in economy of force operations informed by telegraph.

Commanders themselves were inconsistent in their convictions and subsequent actions. Officers in important positions too often vacillated between underestimating and overestimating enemy strength and capabilities. Vacillation stemmed in part from a failure of reconnaissance and in part from a cautiousness that Turkish actions usually did not justify. For various reasons the Russians failed to employ effective reconnaissance at the theater and tactical levels. Shortage of horses and the necessity of undertaking screening missions dissipated cavalry resources and diverted them from reconnaissance and deep raiding missions.

At the purely tactical level, command and control suffered from frequent failures of perception and coordination. Given available technology, linkages were unnecessarily weak among main elements of the army at Plevna. In battle the infantry would often find itself bogged down, and the only way to reinforce an attack was to have the commanding officer on the scene in the manner of Skobelev feeding in reinforcements as he perceived a loss of momentum. Commanders also tended to lose coordination between skirmish lines and follow-on infantry formations. One of the consequences was a heavy loss of both senior and junior officers who exposed themselves to enemy fire while exercising command well forward to coordinate assaults and conduct personal reconnaissance. These and other difficulties had already been evident in prewar maneuvers.

Failure of coordination extended to the use of combat arms in combination with one another. Cavalry did not always provide an adequate screen, and commanders in general did not seem to employ cavalry consistently in conjunction with infantry, the result perhaps of the changing role of cavalry on a more lethal battlefield. More significantly, commanders did not seem to understand how to employ artillery in support of an infantry attack. In general, this shortcoming corresponded with another—the failure to appreciate the importance of rifle fire in support of the infantry attack. Commanders did not distinguish between destruction and suppression. If artillery shell fire lacked sufficient power to destroy enemy earthworks, then perhaps it might have been employed more effectively in conjunction with rifle fire to suppress enemy fire during an attack. The rifle had become king of the battlefield in part because commanders failed to challenge it effectively with other assets at their disposal.

While there was not enough emphasis on suppression, there was too much on the speed of advance and the élan needed to sustain that speed. The emphasis on speed, while not completely incorrect, obscured the need to recognize and profit from partial success. Attacks became all-or-nothing propositions, with troops often either advancing to victory or withdrawing to ignominy. If, for example, troops possessed entrenching instruments, and if an attacked bogged down, they might have entrenched themselves in place—as at Gorni Dubnik—thus preserving at least some of their hard-won gain and opening the way for renewed advance at a later time. As it was, in too many cases only the chance terrain feature offered much hope for cover on the way to victory and shield on the way to defeat and reorganization.

In the fight for life and victory, astute commanders and troops readily caught on to the advantages of even incremental technological advances. Troops picked up the discarded weapons and entrenching tools of their defeated enemies. Wiser commanders such as Skobelev evidently encouraged limited scavenging, provided that it did not lead to irreconcilable supply problems. Consequently, by the time Skobelev's troops reached the outskirts of Constantinople, many of them carried entrenching tools and whole regiments had discarded their Krenks for captured Peabody-Martinis.[91]

If battlefield experience demonstrated the value of new technology, the same experience also underscored the inadequacy of tactical conceptions and doctrine in dealing with the increased lethality of modern weaponry. In the offensive, there was too much reliance on speed and élan and not enough on firepower to carry attacking troops through the extended danger zone of both grazing and plunging rifle fire. At the same time, however, the ability of troops to persist in the attack after sustaining heavy losses demonstrated the basic soundness of Dragomirov's training concepts. Only highly motivated and well-trained troops would have exposed themselves to such peril at such risk. Of course, the danger inherent in partial success was the possibility that Dragomirov's methods might be confused with his tactical doctrine, with the former unjustifiably reinforcing the latter.

Finally, the historian cannot escape the conclusion that sound theater-level perspective and leadership practices might both obscure shortcomings in tactical doctrine and enable the perceptive commander to overcome those shortcomings. The old Suvorov triad of "speed, assessment, and attack" might serve the commander well if he interpreted it to mean the desirability of using speed and offensive maneuver to reach key terrain first, then holding it defensively against even superior attacking forces. One might conclude this was how Osman Pasha had turned the tables on the Russians at Plevna. One might also conclude that reference to the triad could help account for the way that the Russians had forced the Danube, the way that Gurko had conducted his forward detachment operation, and the way that various Russian commanders had employed surprise, maneuver, and economy of force to turn the Turks out of their positions on the south Balkan slope. At critical junctures, old-fashioned leadership had also assisted in the development of Russian success. The danger was that observers would interpret Russian victories, including especially Lovech and Sheinovo, to demonstrate the essential validity of Dragomirov's tactical vision, when in reality these and other battles needed to be viewed within their full context to yield meaningful conclusions. Without systematic and ruthlessly honest analysis, essential flaws within the larger organizational and tactical system might easily escape detection and correction.

3.

The Army of P. S. Vannovskii and A. N. Kuropatkin

> Political reaction, which resulted in stagnation and routine, could not but be reflected in the military sphere, where it affected not only the naming of commanders, but also the character of troop training, the technical equipping of the army, and other issues.
>
> P. A. Zaionchkovskii[1]

As was the case during the previous era of reform, the fate of the Imperial Army between 1881 and 1904 turned on individuals and their relationship with the Tsar. Of the four military men who figured prominently in army affairs during the interwar era, two were tacticians turned military educators and two were chiefs of staff turned military administrators. One of the quartet, M. I. Dragomirov, had also enjoyed prominence during the reform period. He attained still greater prestige and importance after the Russo-Turkish War of 1877–78, serving as Commandant of the Nicholas Academy of the General Staff between 1878 and 1889, then commanding the Kiev Military District until his retirement in 1904. Even after his departure from St. Petersburg, Dragomirov's views continued to dominate and reflect prevailing instruction on tactics within the Imperial Russian Army.

Dragomirov's ascendancy in tactics was matched in the realm of strategy by G. A. Leer, his successor (1889–98) as Commandant of the General Staff Academy. An avid student of the Napoleonic campaigns, Leer completed his pilgrimage from tactics to strategy, then proceeded to dominate Russian military thinking about theater-level war in much the same way that Dragomirov dominated thinking about battlefield tactics. Each was a strong-willed man of ideas, and each for his own reasons chose to ignore the impact of continuing changes in means and method on the conduct of modern war. In the end, each would bear untold responsibility for sowing the intellectual seeds of Russian failure on the battlefields of Manchuria.

They were not alone. If Dragomirov and Leer preached a less than perfect gospel, then two other major military figures of the period, P. S. Vannovskii and A. N. Kuropatkin, presided over a less than perfect congregation. They inherited leadership of an army which in reality had been only partially reconstructed to confront war in the industrial age. In their successive capacities as war ministers, Vannovskii (1881–97) and Kuropatkin (1897–1904) orchestrated a continuing program of expansion and modernization which occurred during a period of shifting threat and constant financial stringency. Unfortunately, neither officer shared D. A. Miliutin's vision and energies, and more importantly, neither possessed their sovereigns' trust and confidence in the same way that Miliutin had enjoyed those of Alexander II. Consequently, they were to build selectively and uncertainly on Miliutin's achievements to produce an army which possessed many of the outward attributes of a modern military colossus. Yet, for all the accomplishment, they failed to support internal professional development and organizational integration in ways which might have compensated on future battlefields for imperfect assimilation of technology and faulty tactical and strategic conceptions.[2]

Changing Personalities and Priorities

Hindsight usually sharpens perceptions, and in retrospect the lessons of the Russo-Turkish War seem neither so obscure nor so controversial that they could not have been systematically studied to foster additional productive change within the tsarist army. However, the complexities of modern armies are such that perception of need is only one of a series of preconditions necessary for constructive change and adaptation. To begin, some kind of permanent mechanism must exist to study the past systematically for whatever lessons it might hold for future combat. Then, another mechanism must exist to translate lessons learned into changed regulation and organization. Next, the political and military leadership has to display sufficient awareness of the need for constructive evolution in order to support change. Further, the leadership must exhibit the will and determination to implement and supervise change. A final important factor would be the existence of a disciplined and educated corps of military professionals, officers and noncommissioned officers, to oversee at all levels the actual implementation of change.

At the close of the Russo-Turkish War, these preconditions were present in varying degrees within Imperial Russia and its army, but the situation changed markedly in 1881. In the spring of 1879, Miliutin created the Military Historical Commission, the purpose of which was to gather and publish documents from the war, then use them as the basis for a thorough study of the conflict with an eye to profiting from the lessons of combat. Constituted under the Main Staff with a general officer as chairman, the Commission was ordered, in Miliutin's words, "to compile a complete systematic description of the war without engaging in inopportune criticism but delineating with complete veracity the factual aspect."[3] In addition to the Historical Commission, there continued in existence the Committee on the Structure and Training of Troops, a mechanism which Miliutin had created within the War Ministry during the early 1860s to implement doctrinal and organizational change within the army. In large part, the capacity of these two mechanisms to effect meaningful, coordinated change hinged on productive linkages and exchanges between the two. Such a relationship would remain the function of a resolute, integrating leadership; however, scarcely three years after San Stefano, the War Ministry was shorn of Miliutin, the official most likely to provide such leadership.

Miliutin's resignation came as a direct result of the assassin's bomb which on 1 March 1881 struck down Alexander II on the bank of the Catherine Canal in St. Petersburg. The new Tsar, Alexander III, had never shared his father's reformist convictions and was openly known to sympathize with some of Miliutin's more prominent political and military enemies, including at earlier times Prince A. I. Bariatinskii and General R. A. Fadeev. On 5 May 1881, less than a week after Alexander III had proclaimed a return to autocracy in the wake of wholesale changes in government, Miliutin requested permission to retire, thus ending an era for the army.[4]

The next quarter of a century aptly demonstrated the degree to which the Russian army remained both a hostage and a beneficiary of an autocracy which harbored deeply militaristic inclinations. Different rulers brought altered priorities and perspectives. Born in 1845, Alexander III (1881–94) was a physically imposing man whose convictions were conservative and nationalist and whose sympathies were Panslavist. Unlike his father, he refrained from direct involvement in military affairs, even though he had once proclaimed Russia's only true allies to be her army and navy. He ignored military minutiae and absented himself from military parades, but he remained devoted to the army and bore responsibility for sanctioning Vannovskii's essentially conservative approach to military matters. More telling, the Tsar proceeded to ensconce less than competent members of the royal family in vital command and inspectorate positions, a practice reaffirmed by his successor with lamentable consequences.[5]

His successor, Nicholas II (1894–1917), was an interventionist in military affairs but lacked the temperament and solid military background to ensure that his interference did more good than harm. A contemporary once faintly praised him as a man with the education of a Guards colonel of good family.[6] Although Nicholas could design uniforms and prescribe buttons, larger military realities often lay beyond his comprehension.[7] Like many of his ancestors, Nicholas took solace in military order. However, also like many of them, his sense of serenity derived not from the realistic resolution of free-play military exercises but from the imposition of parade-ground exactitude. Not surprisingly, by 1900, field exercises, especially those held in the environs of Krasnoe selo, came to resemble military parades. For the wrong reasons, nearby Gatchina, site of an imperial residence and scene of Paul I's ill-fated military experiments a century before, cast a long shadow over the army of Alexander III and Nicholas II.[8]

The advent of the new order also marked a change of governing style which had important implications for the War Ministry and the army. Heretofore, the army as an institution had enjoyed ready access to the sovereign through Miliutin, while the War Minister himself was left relatively free to exercise sway over his own jurisdiction. Although Alexander II might vacillate, Miliutin usually retained sufficient initiative to hold potential opposition at bay. Under both Alexander III and Nicholas

II, however, the relationship between the military and the throne was different, with both emperors playing factions at court and within the ministries off against one another to retain the throne's position as final arbiter. This left additional maneuver room both for ad hoc advisers and for other ministries such as Finance and Interior to garner support for their programs, including railroad construction and educational reform, at the expense of the War Ministry and the Russian army. It also permitted favorites and members of the imperial family to insinuate themselves into the military policymaking process to a degree which had been impossible under Alexander II. Thus neither Vannovskii nor Kuropatkin wielded Miliutin-like power over the War Ministry.[9]

The new politics were immediately evident in Alexander III's replacement for Miliutin. General Petr Semenovich Vannovskii was a fifty-nine-year-old officer of the old school who had learned his trade on the job and who possessed sufficiently conservative leanings to make him congenial to the Emperor. The two owed the origins of their working relationship to the Russo-Turkish War, when Alexander as Tsarevich had nominally commanded the Eastern Detachment with Vannovskii as his chief of staff. The scion of a noble family from Minsk province and a graduate of the 1st Moscow Cadet Corps, Vannovskii had seen combat in the Crimean War and had risen during the 1870s to division and corps command.[10] Despite higher unit experience, Vannovskii's vision apparently never rose much above the level of a company or squadron commander.[11]

At the same time, his contemporaries depict him as a fundamentally honest, honorable, and sometimes crude man of limited intellectual capacity. He preferred to keep officers around him in a state of constant uncertainty, proudly asserting, "You know, I'm a dog, I bite everyone."[12] Yet one of his virtues was that he recognized his own limitations and succeeded in persuading the Emperor to appoint General N. N. Obruchev, one-time political radical and more recently Miliutin's right-hand man, to become Chief of the Main Staff. The justification was that Vannovskii needed a subordinate of education and intellect to enable him to deal with problems of military science and theory. This left Obruchev in a difficult situation: as one of the leading military thinkers of the era, he had access to information and high persons without the full trust and authority necessary to implement measures he deemed appropriate for the army's continued welfare and development.[13]

To be fair with Alexander III and his new war minister, the problem of continued military reform was more than a question of opinion and politics. More importantly, change was also a question of money, and throughout the last two decades of the nineteenth century, financial considerations determined the shape of Russia's military future perhaps more than at any time since the post-Napoleonic era. Already in September 1881, at the dawn of his retirement, Miliutin noted the appearance in the military press of all sorts of orders related to fiscal stringency.[14] Ten years later, Vannovskii submitted a lengthy report to the Emperor tracing a decade of careful stewardship over the land forces of Imperial Russia. The War Minister prefaced his review with a commentary on the necessity for financial stringency, noting that in 1881 the new Emperor "had deigned to emphasize the necessity in all measures related to military affairs observation of the strictest economy, and if possible, even the reduction of expenditures in the military sector."[15] In effect, for ten years Vannovskii worked within a budgetary framework in which with only several exceptions he had to fund new projects from existing revenues.[16]

Linked to the question of finances was the growing scale of Russia's real and potential military commitments. During the 1870s, when Russia experienced Austrian and German resistance to Panslavist aspirations in the Balkans, Alexander II initially confronted the possibility of simultaneous war with Austria-Hungary and Germany. However, even after conclusion of the Austro-German Dual Alliance in 1879, the German Chancellor Otto von Bismarck was successful in convincing St. Petersburg that the arrangement was purely defensive. In June 1881, Russia adhered once again to the Three Emperors' League, thus acknowledging for a time that a certain community of ideological interests continued to unite St. Petersburg, Vienna, and Berlin. Throughout most of the 1880s, Bismarck was able to balance various combinations to assure an uneasy peace in which no state, including Russia, felt compelled to challenge the status quo.

Consequently, only two serious incidents rippled the otherwise seemingly placid surface of Russian foreign affairs during the first half-dozen years of Alexander III's reign. The first came in 1885, when N. K. Girs, Gorchakov's replacement as Foreign Minister, anxiously confronted the prospect of war with Britain in consequence of Russia's emergence to confront British imperial interests across a common border with Afghanistan. Following the Russo-Turkish War, the Russians had continued to press their military advantage in Trans-Caspia, leading in 1881 to General M. D. Skobelev's successful assault on the Turkmen oasis at Geok Tepe. Three years later, the capitulation of Merv to Russian authority led to a serious incident in the spring of 1885 at Panjdeh, where Russian troops routed Great Britain's Afghan surrogates. Before diplomacy could heal the rift, there were serious mutterings of war in both London and St. Petersburg.[17]

St. Petersburg weathered a second war scare in 1887, when Russia threatened military intervention in Bulgaria to secure predominant influence in Sofia following the abdication of Prince Alexander of Battenberg. After Bulgarian nationalists offered the crown to Ferdinand of Saxe-Coburg, a candidate enjoying the support of Vienna, the nationalist press in Russia was filled with talk of war and vituperation against Girs. Although Russian intelligence sources reported extensive Austrian war preparations, the scare gradually subsided with the dawning realization that neither major outside claimant for influence in Bulgaria was actually ready for war. Meanwhile, a proud Russian Emperor swallowed his frustration over a failed Balkan policy, and the more bellicose Russian nationalists drank deeply from the cup of anti-Austrian bitterness.[18]

When the Three Emperors' League came up for renewal in 1887, Russia not surprisingly refused to take part in any diplomatic arrangement which included Austria-Hungary. That same year St. Petersburg and Berlin concluded a special Reinsurance Treaty, the purpose of which was to preclude war between the two in the event that either entered conflict with a third party. Despite failure in the Balkans, the Reinsurance Treaty bought time and a measure of security, thus ensuring that the first decade of Alexander III's reign would close on a placid note.[19]

The situation changed markedly in the 1890s. The very loose community of interests which had appeared to unite the Russian and German inheritors of the Three Emperors' League collapsed. Diplomacy failed either to redress the effects of 1870–71 or to provide lasting proof against resurgent nationalism. Post-1871 France, isolated on the continent, thirsted for revenge and restoration of lost patrimony. Germany desperately needed Austria-Hungary to serve as a counter against France in central Europe. Meanwhile, Austria-Hungary, once again supported by Germany, acted to frustrate Russian designs in the Balkans. Once Bismarck had disappeared from the scene, the heretofore unthinkable happened: in 1890, Germany refused to renew the Reinsurance Treaty, and, in 1894, republican France joined forces with autocratic Russia in a defensive alliance. To complicate matters, Russia grew more active in the Far East, pressing its commercial and political interests to the shore of the Yellow Sea and into direct competition and potential confrontation with China, Great Britain, and Japan.[20]

These developments drastically altered Imperial Russia's requirements for military planning and preparedness. However unrealistic their projections, Russian military planners throughout the 1880s could hope to localize any future conflict with Austria-Hungary and Turkey, neither of which were first-rate military powers. By the mid-1890s, however, the advent of contending alliance systems aroused apprehensions over the possibility of a general European war and inexorably drew renewed Russian attention to Germany, a nation whose military forces and industrialized infrastructure immensely complicated the military threat and planning calculus. The logic of the Franco-Russian alliance suddenly rendered more probable that which until now had been only a nightmare scenario—the likelihood of simultaneous war with Austria-Hungary and Germany.[21] Meanwhile, the Russian army expanded— even during a period of financial stringency—to meet what St. Petersburg perceived as a threat of the first magnitude on the Empire's western borders. Problems of western defense also resurrected the expensive issue of border fortifications. In addition, as the Russians expanded their influence in the Balkans, their boundaries in Central Asia, and their sphere of influence in the Far East, the requirements for military security seemed to grow nearly exponentially. Unfortunately for the Russians, neither the army nor the treasury could keep pace with the combined growth of boundaries, influence, and interest, and once again a dangerous gap opened between state policy and military capability. Rapid technological change contributed to the increasing political, financial, and military complexities of the situation.

It was this fundamentally altered situation which General Aleksei Nikolaevich Kuropatkin inherited on the last day of 1897, when an aging Vannovskii finally handed over the reins of the War Ministry. Kuropatkin also inherited a Tsar of mercurial temperament and near-legendary indecisiveness. Characteristically, Nicholas II kept the issue of ministerial succession in doubt, having intimated that he was inclined to replace Vannovskii with either Kuropatkin or Obruchev, then finally giving Kuropatkin the nod—without directly informing Obruchev.[22]

The Army of P. S. Vannovskii and A. N. Kuropatkin / 93

P. S. Vannovskii (1822–1904)
Minister of War 1881–1897
Photo: Skalon, *Stoletie Voennogo Ministerstva*

A. N. Kuropatkin (1848–1925) M. D.
Skobelev's Chief of Staff 1877–1878 and
1880–1881, War Minister 1897–1904
Photo: Skalon, *Stoletie Voennogo Ministerstva*

On paper, Kuropatkin seemed well suited to the job. Born in 1848 into a noble family from Pskov province, he was a graduate of both the 1st Cadet Corps and the Paul Junker Academy. Endowed with intellectual talent and personal charm, he completed the Nicholas Academy first in the class of 1874. He was also a decorated combat veteran, having served as Skobelev's chief of staff both in Turkestan and in 1877–78. Kuropatkin returned to Central Asia in 1880–81, where he served with distinction in the assault against Geok Tepe. A general at the age of thirty-six, he was a soldier-scholar, and his writings marked him as a serious student of military art. Also something of a diplomat, Kuropatkin handled himself with ease before various audiences, always displaying confidence and self-assuredness.[23]

Yet Kuropatkin's undeniably positive attributes and meteoric rise apparently obscured more fundamental personality faults. He found it difficult to tolerate views contradicting his own. His apparent firmness and decisiveness in minor matters at least partially concealed a deeper indecisiveness and a reticence to tackle larger problems. Count S. Iu. Witte, Nicholas II's Minister of Finance, tellingly recounted the words of Skobelev's sister, Princess Belosel'skaia-Belozerskaia, who once said that her brother had loved Kuropatkin and considered him "a good executor and an extraordinarily brave officer." However, she also declared that "as a military leader he was completely unsuited for wartime." Kuropatkin's difficulty was that "he could only implement orders, but lacked the capacity to give them." In her brother's words, she said that Kuropatkin was "brave in the sense that he was not afraid of death, but a coward in the sense that he was not able to make a decision and bear responsibility."[24] These were harsh words for a man who professed to fear "only his God and his Tsar." Were later events not to bear them out, they could be dismissed as mere gossip.

Between 1881 and 1904, then, the Imperial Russian Army would confront the challenge of retaining evolutionary momentum under one war minister of limited vision who would fight only grudgingly to retain the status quo and a second of great vision who would fight for relatively little. Meanwhile, at the hands of Dragomirov and Leer, tactical and strategic thought first blossomed, then ossified. At the same time, the accelerating pace of technological change and the expanding tasks of the army would impose greater competition for fewer resources. Together, these ingredients added up to a recipe for intellectual and organizational stagnation—even in a period of ostensible growth—which could easily lead to military disaster.

Changing Institutional Priorities

The measured approach to change inherent in the Miliutin order was quickly overwhelmed by changing political and financial circumstance. In 1879, Miliutin had pinned his hopes on the creation of the Military Historical Commission, which he designed to produce a calculated study of war experience both to illustrate the need for change and to point the way to constructive alteration. However, the Commission soon became mired in detail and, lacking resources and energetic leadership, failed to produce any concrete results until the last years of the century. This was in part a function of inexperience and institutional drift and in part a function of design. Too many careers and reputations might not withstand the intense light of close historical scrutiny.[25] Consequently, in 1897, at the end of nearly two decades of the Commission's work, Vannovskii sent its members back to the drafting tables with instructions to excise from their analysis any kind of criticism directed at individuals while at the same time emphasizing the heroic achievements of the troops.[26] Thus the complete results of the Commission's work lay buried for nearly another decade, and along with them, the last hopes for any of its deliberations to generate systematic change.[27]

Vannovskii's decision simply reflected in writing what had already been a disastrous story for official military history between 1881 and 1897. With the Commission hamstrung almost from the start, there had been no central clearing house for the 1877–78 war experience, and consequently, the impulse for military change came not from the Commission but from a variety of sources, including would-be reformers with parochial axes to grind, agitation in the pages of the military press, and the scattered impressions of senior officers who now populated influential positions.[28] During the Vannovskii years, all these disparate sources supplied grist for the mill of the Committee on the Structure and Training of Troops, which ground out a steady stream of revised regulations and tables of organization.[29] To the Committee's credit, it proved capable of producing the required documents, but too often revision came in response to only the most immediate, acceptable pressures, with little thought given to the integration of novel conclusions into a harmonious whole. To paraphrase the German philosopher Schopenhauer, every writer of official regulations could take the measure of his own horizons to be the measure of the world.

Yet this sad state of affairs had its paradoxically positive side for the further development and application of military history. A genuine thirst for knowledge and truth, when coupled with selective official inertia, added fresh impetus to historical study among a variety of audiences. The 1880s and 1890s witnessed a burst of original research and publication from a myriad of students and agencies, ranging from individual officers to departments of the Main Staff and the Nicholas Academy. Study groups sprang up in the military districts to spur officer professional development, and the pages of the military press overflowed with historical accounts.[30]

A missing key ingredient, however, was consistent official sanction, and in this respect Vannovskii's attitude was telling. In the early years of Vannovskii's tenure as War Minister, M. V. Grulev remembered an oppressive intellectual atmosphere and noted that his unit's library was purged of "harmful books" for burning.[31] In 1898, on learning of Kuropatkin's approval for the Society of the Zealots of Military Knowledge, Vannovskii remarked that "not for anything would I have permitted such a society."[32]

Missing—also thanks in no small part to Vannovskii—from the burgeoning interest in military history was a coherent picture of the most recent Russian combat experience. After 1881, the 1877–78 war was simply not on the official agenda, with the result that its lessons were largely lost to a generation of officers who taught and studied at the Nicholas Academy. With few experiences of the last war to guide instruction and research, military theorists either turned more certainly to the wars of Napoleon or less certainly to the wars of German unification for their instructional examples and materials.[33] Too late, in 1902, War Minister Kuropatkin reviewed the list of wars between 1871 and 1902 studied at the Academy and concluded that it "did not yield the required benefit" because the course of instruction "did not aid in the auditors' familiarization with recent Russian military history."[34]

Unofficial studies could not completely fill the void. It was only in the late 1880s and early 1890s that serious monographs on 1877–78 began to appear, and these were not consistently supported and promulgated to the best institutional effect. Good though some of them were, it was difficult to draw from them a comprehensive picture of the war and profit from the experience of mistakes from the more immediate Russian military past. Too late, on the eve of the war with Japan, the Military Historical Commission was pressed into rushing its analytical work into print—and then for the wrong reason. War Minister Kuropatkin directed that the volumes in which he figured as Skobelev's chief of staff be published out of sequence![35]

Structural Change

Even if Vannovskii had chosen to pursue Miliutin's methods, the flow of events quickly deprived him of the initiative. While the Military Historical Commission was still gathering materials, various committees commenced to rewrite Russian field regulations. Meanwhile, the War Minister found himself confronted with an organizational crisis of the first magnitude. No sooner had Miliutin submitted his resignation than General R. A. Fadeev began agitating through well-placed connections at court for a thorough review of the War Ministry's organization. Fadeev was a holdover from the 1860s and 1870s who, like the late Prince Bariatinskii, had never made his peace with the substance of Miliutin's reforms. Thanks to influence at court, Fadeev's connivance bore fruit on 21 June 1881, when Vannovskii, only recently appointed, was forced to announce the naming of Count P. E. Kotsebu, Commander of the Odessa Military District, as chairman of a special commission to consider a reorganization of Russian military administration. Kotsebu's mission was "to conduct a review of the fundamental bases of our existing system of higher military administration—central, local, and field—after comparing the cited bases with the system used in Germany and also with the order which had existed with us in the beginning of this century under Emperor Alexander I."[36]

Kotsebu's Special Commission, the distinguished membership of which numbered some thirty senior officers, including three grand dukes, Vannovskii, Obruchev, and a "who's who" list of veteran commanders from 1877–78, met nine times between early October and early November 1881 to consider nine discrete questions related to Russian military administration.[37] The issues at stake can be reduced to two cardinal concerns which had bedeviled Russian thinking about military organization since at least 1802. The first was what the relationship should be at the center between the war planning and war fighting function and the administrative and support function. Under the Miliutin regime, the former had been subsumed into the latter, with the Main Staff remaining an integral, subservient component of the War Ministry. This arrangement inevitably presented problems, especially in easing the transition from peace to war, but suited Miliutin, who wished to retain unchallenged control over a unified military-administrative apparatus. The alternative was to adopt the Prussian model, in which the Chief of the General Staff reported independently to the throne and retained absolute control over war plans and war conduct. In such a case, the War Ministry assumed only a housekeeping function, supporting the requirements and designs of the general staff. Fadeev's ability to resurrect this question some two decades after Bariatinskii had first raised it represented a direct threat to Vannovskii's power and the organizational integrity of the Miliutin legacy.

The Kotsebu Commission's second large concern had to do with the way in which central administration was reflected in the functioning of the military districts.[38] The essence of the Miliutin design had been to duplicate the unified central organization in the districts, thus handing unified control over each district's war fighting and war supporting structures to the district commander and his staff. Here again, the problem was one of facilitating the transition from peacetime to wartime without disrupting support while simultaneously retaining maximum combat capacity. As in the case of central military administration, a good argument could be made for separating the war fighting (combat) structure from its support (housekeeping) structure. And here again, the disciples of Bariatinskii thought that the Prussian territorial model, in which corps commanders and inspectorates held sway over support structures, offered distinct advantages over the Russian system.[39]

As the debate over these central concerns unfolded within the framework of the Kotsebu Commission's mandate, proponents of the Russian and Prussian systems advanced their arguments and cast their votes. In the end, the partisans of the Miliutin system triumphed, thanks in no small measure to the presence both of Vannovskii, who saw his own situation threatened, and of Obruchev, who saw his former master's edifice endangered. Even the Tsar, who formerly had been known as a supporter of Bariatinskii and Fadeev, agreed with the Commission's majority to sustain the status quo, although not before he asked members to reaffirm their positions. However, the Commission did succeed in prompting the Tsar to order a much-needed review of the regulations for field administration.[40]

The Main Staff

Thus, even before the Commission had completed its work, Obruchev could report to Miliutin that "the Palazzo will continue to stand, and all will know what to call it," but victory did not come without serious long-term costs.[41] Vannovskii and Obruchev had indeed preserved their positions, but they had also put off a badly needed review of general staff functions within the War Ministry. Obruchev's Main Staff remained embedded in the bureaucratic apparatus of the War Ministry on an equal footing with other directorates. Now comprising six sections and two subsections, the Main Staff since 1869 had borne responsibility for maintaining data on forces, managing personnel assignments, orchestrating the complement of forces, and discharging matters on the structure, service, quartering, training, and supply of troops. Related assignments included planning for troop movement, conducting geodesical work, compiling military-statistical data, and collecting intelligence. Throughout the 1870s and 1880s, the Main Staff's functions grew more complex, thanks to the increasingly sophisticated requirements of planning for mobilization, deployment, and strategic dispositions. In effect, the Main Staff was being called upon to fulfill the functions of a European-style general staff without corresponding mandate, stature, and complement.[42]

As Chief of the Main Staff, Obruchev struggled without success to improve the situation, submitting in 1894 a comprehensive plan for reorganizing the Main Staff "to discharge work of a high strategic order relative to the assignment of troops to theaters of war, the formation of armies, the composition of plans for their concentration and initial actions, the preparation of field administration in border districts, [and] organizing the collection of information on the enemy."[43] When Vannovskii refused to consider a comprehensive reorganization of the Main Staff in isolation from other directorates, Obruchev waited three years, then submitted a second plan for reorganization, again without result.[44]

Obruchev's successor, General V. V. Sakharov, raised the issue once more in 1900, proposing both extensive reorganization and expansion and a scheme whereby the Chief of the Main Staff would in effect become deputy war minister with the right to report personally to the Tsar. Neither Kuropatkin nor the Military Council approved elevated status for the Chief of the Main Staff, with the result that reform was put off until 1903, when the substance of Sakharov's proposal finally took effect without any change in the Chief's stature. Accordingly, the Main Staff was reorganized into five directorates: First Quartermaster General, Second Quartermaster General, Adjutant General, Military Communications, and Military Topography. The Directorate of the First Quartermaster General was divided into two sections, one for training and general staff-type affairs, the other for military and civil administration of the Asiatic military districts and the Caucasus. The Directorate of the Second Quartermaster General was also divided into two sections, one for military statistics (especially data on potential theaters of war) and intelligence, and the second for mobilization. The Directorate of the Adjutant General assumed the same two-section organization as the first two, with one section for line unit postings and inspectorate matters (personnel accounting, awards, promotions, and assignments), and a second section for administrative-economic matters.[45]

Sakharov's reorganization represented a step forward, but came too late to make a decisive difference in preparation for war in the Far East. The measure did facilitate better control by the Chief of the Main Staff over the myriad of activities and offices for which he bore responsibility. It also afforded enhanced recognition of the Chief's status as primary assistant to the War Minister. However, the Sakharov scheme did little to resolve organizational dilemmas associated with the absence of a thoroughly integrated staff and planning process, which according to Obruchev's design of 1894, would have provided a single institutional focus for the planning and conduct of military activities from mobilization through dislocation and movement by rail to concentration, deployment, and initial march-maneuver within theater. As a result, these various planning and oversight activities remained fragmented both within the Main Staff and War Ministry and between the directorates and staffs of the central organs and their counterparts in the military districts. As the entire mobilization process grew still more complex, the risks inherent in fragmented planning and execution would become still greater.

Field Administration

If evolving higher staff organization reflected difficulties inherent in making the transition between preparation for war and its actual conduct, so also did practices governing field administration. The experience of 1877–78 had clearly demonstrated the requirement for more specific regulations regarding wartime administration, especially in providing surer linkages between fighting front and supporting rear. Russian forces had begun the war against Turkey operating from supply bases within the Odessa Military District, then had laboriously pushed the whole support structure forward, first to Bucharest, and finally to northern Bulgaria. The whole process was accompanied by delays, corruption, and significant losses to pilferage. No sooner had Count Kotsebu's Special Commission completed its work than General P. L. Lobko, Assistant to the Chief of the War Ministry Chancery, received a mandate to rewrite the regulation governing field administration.

Lobko's draft regulation, completed in late 1885 and reviewed extensively between 1886 and 1889, was finally published in 1890 as the *Regulation on the Field Administration of Forces in Wartime*. The new regulation placed an army or several armies operating within a theater of war respectively under an army commander or commander in chief of armies. The commander in chief, "representing the person of the Emperor," retained control of military forces and civil administration for the theater, both inside and outside imperial territory. The commander in chief also retained the right to conclude an armistice with the enemy and possessed absolute authority over all forces and ranks, including members of the imperial family serving with troops.

Perhaps more significantly, the regulation on field administration specified the composition of staffs operating within a theater of war and how those staffs were to be derived from the peacetime structure of the military districts. In the event of war, the commander of a border military district became an army commander, and his section chiefs became chiefs of corresponding sections within the army's field staff. After the district had spun off these staff members, their deputies remained behind to become the chiefs of the supporting sections within the remaining structure of the military district. The whole district remained subordinate to the army commander, with its structure providing supplies and equipment to the army in the field.[46]

The administration of a field army consisted of a field staff and five directorates. The army commander's field staff consisted of three sections: quartermaster general, adjutant general, and military communications. The quartermaster general section exercised supervision over operational, intelligence, and topographical matters. The adjutant general section managed personnel and medical support matters, while the military communications section oversaw roads, communications, and local security and administration. Other key parts of army field administration included five directorates (supply, artillery, engineers, finance, and control). Overall responsibility for coordination between the directorates and the field staff lay with the commander's chief of staff. In the event that more than one army took the field, the staff and directorates of the overall commander in chief bore an organizational structure analogous to that of the subordinate armies.[47]

The regulation on field administration thus laid out a rational scheme for coordinating staff and administrative functions under wartime conditions. However, the regulation retained at least one undesirable feature of its predecessor: it left the field army commander a free hand to devise ad hoc detachments "conforming with circumstances," something that would return to haunt the army in Manchuria.[48] In addition, the regulation failed to treat two important issues which, in no small way, would determine how well the entire system actually worked. The first had to do with the problem of accommodating change over time as new relationships developed among the military districts. This was especially serious for the western border districts, which would confront additional requirements for planning, readiness, and mobilization, while their counterparts in the interior gradually assumed more of the burden of providing trained military manpower and remote rear support. Related to this was a larger problem of coordination. Except for planning prior to hostilities, no arrangement was made for the organs of the center to serve as coordinators between fighting front and state rear. Indeed, right up to World War I, the practice was for the organs at the center to atrophy at the onset of hostilities, while key staff officers left St. Petersburg to join higher headquarters in the field.

Structural details aside, the functioning of the system was predicated upon the availability of professionally trained officers "of the General Staff" who possessed a common knowledge of staff procedures and practices and a common understanding of the nature and requirements of modern war. Yet a review of organizational tables for 1 January 1904 throughout the army and its central organs reveals that only about 45 percent of those officers holding responsible staff positions within the system (588 of 1,241) displayed the necessary credentials.[49] Why this was so and what attributes officers bearing the coveted designation "of the General Staff" brought to their assignments form crucial parts of the explanation for the army's poor performance in the field during 1904–5.

The Officer Corps: Composition and Competence

Within the context of the post-Miliutin military order, the Russian officer corps faced two significant challenges: expansion and the growing complexity of the profession of arms. As the army of Vannovskii and Kuropatkin expanded in size from the early 1880s to the early 1900s, it gradually came to resemble Miliutin's ideal of a genuine cadre- and reserve-based force. In raw manpower, the peacetime composition of the army grew from 30,768 officers and 844,396 troops in 1881 to 41,079 officers and 1,066,894 troops in 1904.[50] By 1903, according to Mobilization Schedule 18, the fully deployed regular and reserve forces of Russia (not counting the Priamur Military District and the Kwantung Military Region) amounted to slightly more than three million men, of which the reserve accounted for nearly two-thirds.[51]

Simultaneously, the complex interplay between evolving organization and changing technologies, means, and methods dramatically underscored the importance of the officer corps' technical competence. However, the pace and impact of change now decisively called into question the traditional path to professional competence: long-term apprenticeship and on-the-job training at successively higher levels of the hierarchy. Needed now was a balanced combination of experience *and* formal education, not only at the entry level but also at periodic stages in the ascent to high command and staff positions. The kind of military knowledge required for professional mastery was now expanding at a rate that eluded traditional emphasis on each individual's first-hand familiarity with a well-defined trade.

In the aggregate, the implications of expansion and the changing parameters of technical competence placed a premium on institutional flexibility and educational adaptation within the officer corps. However, the throne's stubbornly conservative approach to key issues of officer recruitment, retention, and preparation acted as a brake on institutional adaptation. Just when expansion and change underscored the importance of flexibility, the prevailing emphasis on seniority rule and partially retrogressive educational policies worked against coherence and flexibility. Consequently, the officer corps continued to display a dangerous heterogeneity of outlook and competence. Indeed, on the basis of recent studies of the officer corps, the historian might reasonably identify four or five distinct segments within the officer corps, none of which necessarily communicated well with the others.

A rigid emphasis on seniority endowed Russia's expanding army with superannuated general officers and an increasingly aging and disgruntled corps of senior commanders. Men who had attained general officer rank during the 1870s often surrendered their status only in the event of death or outright senility. P. A. Zaionchkovskii has aptly characterized the situation when he noted the ages of senior generals slated in 1903 to take important field commands in the event of war in the west. Without counting four members of the imperial family who would hold titular command, the eight senior officers nominated for corps command and above included three between the ages of fifty-five and sixty, three between sixty and seventy, and two between seventy-three and seventy-six.[52]

Worse than age was the problem of senior officer education. Despite the impact of Miliutin's reforms, there was no requirement at the higher reaches of the personnel system for education beyond senior command experience, with the result that by 1902 only about one-half of the army's general officers had completed institutions of high military education. The remainder constituted a mixed bag, with many since their cadet days having failed to keep pace with expanding military knowledge. Or, as A. I. Denikin, who rose to division command in World War I, put it, "the commander, starting with the regimental commander, could get by serenely with the same scientific baggage which once upon a time he had carried away from the military or junker academy, could get by without following the progress of military science, and it never occurred to anyone to take an interest in what he knew."[53] Perhaps the classic example of senior command ignorance was General N. P. Linevich, commander of the Priamur District, who, upon seeing a howitzer for the first time, asked, "What kind of gun is that?" Worse, Linevich, the general who would command a field army in Manchuria, could neither read military maps nor understand rail movement schedules.[54]

Those officers who could read maps and devise schedules were more often than not graduates of the Academy of the General Staff who found their special qualifications alternately advancing and hindering their careers. Despite the anti-intellectual overtones of the Vannovskii years, entrance to the Staff Academy remained incredibly competitive, with roughly 600 officers annually vying for eighty vacancies in the basic two-year course. After the Academy reform of 1893, officers in the first, or "junior," year received instruction in tactics, topography, topographical drawing, administration, military history, military statistics, fortification, artillery, philosophy and psychology, Russian, physical geography, political history, and either French or German. Officers of the second, or "senior," year received instruction in strategy (including the tactics of large armies), administration, military statistics and geography, cartography (including geodesy and astronomy), telegraphy, Russian, and either French or German. During the winter months, seniors engaged in tactical map exercises and wrote two compulsory themes, one on the strategic dispositions of army corps and the second on marches. After 1893, those officers who scored "very good" (an average of ten out of twelve units) in their primary subjects were permitted to matriculate into a third-year "supplementary course," the successful graduates of which had their names entered on the list of candidates for appointment to the Officer Corps of the General Staff. Those officers who failed to enter the third-year course were returned to their units for reassignment under something of a cloud. Nevertheless, all graduates of the two-year course earned the right to wear on their right breast pocket the coveted academic order, a silver medallion engraved with branches of oak and laurel and bearing the initials of the Academy.[55]

For graduates of the third-year course, the prestigious designation "of the General Staff" was far from a guarantee of unimpeded success. The uninitiated often looked upon this newly emerging elite, or *genshtabisty* as they were nicknamed, with disdain, calling them "moments" or "pheasants."[56] Those General Staff officers who were less successful in advancing their careers often found themselves relegated to a series of bureaucratic assignments, gradually becoming high-ranking clerical personnel.[57] More successful *genshtabisty* received brief command assignments, then proceeded forward with a succession of mixed command and high-level staff assignments. By 1903, nearly 30 percent of infantry regimental commanders and approximately one-half of infantry division commanders were graduates of higher military educational institutions, especially the General Staff Academy. However, officers with a higher military education were never numerous enough to fill all responsible command and staff positions, thanks in no small part to the Staff Academy's highly restrictive entrance standards.[58]

A recurring theme related to the emergence of a corps of general staff officers was conflict between new and old military elites. The stronghold of the hereditary Russian nobility was the Guards regiments, entry to which came via completion of the Corps of Pages or a few select military academies. Once a position with the Guards was attained, little premium was placed on educational advancement. An officer could remain with the Guards, where he participated in the social diversions of the capital, then possibly elect transfer to a line unit, where his right to automatic advancement in rank facilitated pursuit of competitive command assignment. This placed the older elite in direct competition with the emerging general staff elite for scarce senior command positions. Thus older, ascribed attributes entered into conflict with newer, earned attributes.[59]

There were also more traditional rivalries which interfered with the development of a more cohesive officer corps. William C. Fuller has eloquently summarized the effects of parochial sentiment on the outlook of various groups:

> The Guards believed themselves to be the true patrician elite, the officers of the General Staff deemed themselves the intellectual cream of the service, the cavalry despised the infantry, and the technical services such as engineers and artillery reviled cavalry and infantry alike.[60]

The presence of older and newer rivalries and factionalism made life still more difficult for officers of the line who spent long years rising through the ranks, only to find access to high command increasingly elusive. With scarce funds allocated increasingly to weapons acquisition programs, pay stagnated, thus adding to the line officer's precarious financial status. With advancement more problematic than ever, many senior line officers served out their time in dull garrison assignments where they took solace in drink and aged quietly, while becoming increasingly ingrown and deaf to changes in technology and structure inherent in the army's slow transition to military modernization.[61] Except perhaps in scale, the picture was probably not much different from that of any other peacetime army of the era.

Below the colonels and lieutenant colonels were scores of company-grade officers who possessed less than sterling social and educational qualifications. Alexander III reversed many of Miliutin's educational reform priorities, with the result that junker academies declined in quality, while the military academies were shorn of their civilian instructors and nonmilitary curricula.[62] The result was that the military academies tended to remain a preserve of the nobility, but turned out qualified graduates in a narrow military-technical sense. For a time, the junker academies steadily worsened, so that their graduates, men of diverse noble and nonnoble background, lacked the tools and incentives for advancement.[63] Many senior commanders, including M. I. Dragomirov, simply dismissed the average junker academy graduate as being unsuitable material for command.[64]

Although institutional mechanisms existed to span the gap between entry-level and advanced educational requirements, they were too few and too small. The Officers' Rifle School enrolled 160 captains annually, and the purpose of its one-year course was to train future infantry battalion commanders. The Officers' Cavalry and Artillery Schools, which served an analogous purpose for their arms, were even more exclusive, respectively enrolling sixty-four officers for two years and sixty officers for seven months.[65] Neither the size of the student contingents nor the length of their prescribed courses of instruction corresponded with expanding requirements for competent midcareer officers.

Thanks to varying educational and recruitment patterns, the pre-1904 Russian officer corps presented an extremely variegated picture. At the very upper reaches of the command structure, aging senior officers with the most diverse educational backgrounds predominated. Below them and at the middle levels, a mixture of old and new elites contended for position and influence. Exclusion from either group usually meant stagnation for line commanders and staff officers. At the lower reaches, many junior commissioned officers lacked general education, while others were simply incompetent. Meanwhile, nearly all officers who did not possess independent means suffered from low pay and perceived declining social prestige.

The government proceeded to make this sad state of affairs still worse with a misguided emphasis on ritual and status to compensate for what many saw as debilitating outside influences. During the 1890s, courts of honor and dueling were reinstated. While these were no doubt intended to reinforce a sense of social and corporate identity, they were poor substitutes for a rigorous system of schooling which would have contributed to genuine officer professionalism.[66] Consequently, at the turn of the century, the officer corps failed to display a uniform sense of either consciousness or competence. This was a dangerous state of affairs in an army which would be called upon to assimilate new technologies over a comparatively short span of time.

New Technologies, New Armaments

The Russian Army had fought the Turks in 1877–78 with a mixture of rifled weapons ranging from the primitive Karle to the more advanced Berdan No. 2. Despite wartime efforts to put better weapons into the hands of the fighting troops, 200,000 outmoded Karle and Krenk rifles remained in service three years after the war. However, both active and cadre reserve units, including the dragoons and the Cossacks, were completely outfitted with Berdan rifles by the beginning of the 1880s. By 1884, rearmament with the Berdan was complete throughout the army.[67]

Unfortunately for the army and the treasury, the pace of technological change over the next decade was sufficiently great to render the Berdan obsolete. Two advances in particular—the adoption of smokeless powder and the invention of a satisfactory system of magazine feed—lay at the heart of new small arms systems which began to appear just about the time that the Russians had completed their transition to the Berdan. Smokeless powder enabled designers to attain greater range and muzzle velocity to drive smaller bullets at higher speeds over greater distances with more accuracy and greater penetrating power. Because smokeless powder burned more efficiently and cleanly, its adoption resolved the problem of fouling, one of the main drawbacks of black powder technology. In turn, alleviation of this drawback removed the last major practical obstacle to the development of reliable repeating rifles. Inventors had been experimenting for years with various kinds of actions and feeding systems. By the late 1880s, along with the transition to smokeless powder, most innovators seem to have settled for military purposes on the bolt-action rifle fed by spring action from a staggered column of cartridges enclosed wholly or in part by the receiver mechanism of the rifle. A soldier equipped with such a weapon could deliver a high rate of lethal fire to effective ranges of 500 meters and maximum ranges to a kilometer or more. Unlike the situation with black powder weapons, higher pressure and greater muzzle velocity meant that bullet trajectories remained relatively flat to ranges of 500 meters. The designers now had a weapon which met military requirements for qualitatively greater range, accuracy, and rate of fire.[68]

Despite recent rearmament with the Berdan system, the Russians could not long ignore developments elsewhere in Europe, and in 1889 the Tsar ordered work on "rifles of reduced caliber and cartridges with smokeless powder." He named two commissions, one to direct and allocate resources and the other to conduct actual experimentation, to oversee what would become the foundation for a crash rearmament project. Based on the work of the experimental commission headed by Lieutenant General N. I. Chagin, the War Ministry eventually considered adoption of two basic models of .30-caliber rifles. One was the work of the Belgian armorer Leon Nagant, and the other the work of the Russian officer, Captain S. M. Mosin. For reasons of simplicity, ruggedness of design, and cheapness of manufacture, Mosin's design won the competition. The fact that Mosin was Russian probably did not harm his cause. His design was designated the .30-caliber rifle model 1891, and it would remain more than fifty years the standard infantry weapon of two successive armies, the Imperial Russian and the Soviet. At 4.5 kilograms (eleven pounds) with bayonet, the Mosin was two-thirds of a kilogram lighter and five inches shorter than its predecessor. Fed from an eight- round clip-loaded magazine, its maximum range was 2,700 paces, or about two kilometers. Because of resource and monetary constraints, the Mosin was introduced in two phases. The first phase, which was completed in 1897, saw the infantry of the regular army and cadre reserve forces completely rearmed with two million weapons. The second stage, completed in 1903, put 1,700,000 Mosins into the hands of the cavalry and the remainder of the reserves. When the Tsar went to war with Japan in 1904, his forces had been completely equipped with the latest shoulder weapons that scarce funds could buy.[69]

Although the Belgians had lost out to Mosin in the rifle competition, the 1890s brought Nagant a Russian contract of another sort. Since the 1870s, the standard Russian sidearm had been the clumsy but reliable Smith and Wesson revolver of .44 caliber. After the War Ministry had adopted the small-caliber rifle, the next step was to adopt a small-caliber pistol for standard issue. In 1895 the War Ministry recommended adoption of the seven-shot Nagant revolver, caliber 7.62 millimeters. In the agreement reached with the Belgian armorer, the Russians purchased manufacturing rights to the revolver, and in 1898 the Tula arsenal began production at the rate of 20,000 weapons per year. A very reliable weapon, the Nagant was manufactured in two versions, double-action and single-action, with the former for officers and the latter for the rank and file. The prevailing wisdom was that only officers would be able to use double-action weapons rationally to avoid wasting ammunition.[70]

Artillery rearmament came much more painfully. Although the Russians had begun showing interest in steel guns during the 1860s, technological problems and limited financial resources meant that in 1877–78 the Russians had fought the Turks with bronze breechloaders. When early combat experience seemed to demonstrate the superiority of Krupp rifled steel guns, Alexander II named a special commission to consider various armament alternatives. It recommended immediate adoption of the Maevskii steel system. Thanks to a hasty mobilization of resources, the War Ministry was able to complete rearmament of the regular field artillery with Maevskii guns by 1881. Nine more years dragged by before rearmament was complete in the reserves and within fortress artillery.[71]

The experience of the Russo-Turkish War also prompted a serious reevaluation of the need for weapons having a high angle of fire. Because field guns possessed an essentially flat trajectory, they demonstrated limited utility in the attack of fortified positions. During the 1880s, the Russians began serious experimentation with a six-inch (152-millimeter) mortar, but officers of the field artillery objected to its incorporation into their units on the grounds of its limited mobility. This problem was overcome in 1885, when General A. P. Engel'gardt devised a wheeled carriage with a system of buffers to absorb the mortar's recoil. Mortars of the Engel'gardt design weighed 1,300 kilograms, fired a twenty-five-kilogram high explosive round to a range of 3,200 meters, and possessed a rate of fire of four rounds per minute. Between 1887 and 1895, the Russians incorporated seventy-two of these mobile six-inch mortars into their field artillery. Although these weapons could keep pace with field guns, their limited reliability and range were to impose severe limitations on their employment in 1904–5.[72]

As was the case with shoulder weapons, the Russians had scarcely completed rearmament of their field artillery when the pace of technological change rendered it obsolete. By the mid-1890s, two advances, the adoption of smokeless powder and the appearance of workable recoil mechanisms, made possible the next technological jump, the move to quick-firing field guns. Previously, the absence of mechanisms to absorb the effects of recoil meant that field guns had to be laboriously relaid after each shot. Delays associated with relaying, together with black powder fouling, accounted for a low rate of fire by even the latest steel guns. Once further development incorporated recent advances into designs based on steel gun technology, a series of additional advances were possible. These included mounting gun tubes on sliding mechanisms which absorbed recoil by means of various buffering techniques, then returned the tube rapidly and accurately to its original position for quick reloading. Improved breech blocks of either sliding wedge or interrupted thread design further facilitated the loading process. Together, these innovations spelled greater lethality: smokeless powder meant higher muzzle velocities with less fouling and longer range, while improved recoil mechanisms and breechloading meant greater accuracy and rapidity of fire.[73]

Neither the Tsar nor the War Ministry was favorably disposed to additional arms expenditures.[74] Yet the Russians could not afford to fall behind. Indeed, the War Ministry's assessment of the problem eloquently summarized the dilemma in which the Russians found themselves:

> Our artillery must, no doubt, follow the example of the artillery of western armies in order not to fall behind them in armament and effect of fire; but rearming the artillery will cause us great difficulty both because of significance of expenditure and because of limited production resources and consequent inability to fill orders for a large number of guns, carriages, and shells in any kind of short time period.[75]

The inevitable came in 1897, when German adoption of a quick-firing field gun forced the Tsar's Main Artillery Directorate to press development of a native Russian version of the latest quick-firing technology. In 1899, the directorate tested a quick-firing three-inch (seventy-six-millimeter) gun designed by N. A. Zabudksii, and in 1900 it was officially adopted for general use. Its muzzle velocity was close to 650 meters per second, and it could do the same damage to targets at distances of two kilometers that the earlier steel guns could do at 500 meters. The new gun possessed a maximum range of nearly ten kilometers, and it could hurl a shrapnel round to distances of five kilometers, a two-kilometer increase over previous designs. Although the rate of fire varied from five to twenty rounds per minute, some observers, including War Minister Kuropatkin, were troubled by carriage instability, a problem they felt made the gun inferior to the French model 1897, the gun later popularly known as the "French 75."

Despite Kuropatkin's reservations, in 1900 the gun was rushed into production with an initial order of 1,500 from the Putilov iron works to be followed in 1902 by another order for 900. Because government arsenals had produced a competing model, there was talk that hasty adoption had been more the result of private influence than military imperative. However, modification to the basic model in 1902 silenced most critics by alleviating the problem of carriage instability, and after further modification, including the addition of a shield for the crew, the gun went on to remain the standard light field piece for Russian and Soviet forces for nearly forty years.[76]

Almost simultaneously with the adoption of the quick-firing field gun the machine gun made its debut to render the battlefield still more lethal. Since the 1860s inventors had produced various types of weapons capable of firing multiple rounds either simultaneously or sequentially. Various designs appeared, ranging from the American Gatling gun to the French Mitrailleuse, but armies, including the Imperial Russian, tended to reject these weapons as heavy, clumsy, and difficult to supply. However, with the advent of smokeless powder, inventors began to experiment with true automatic small-caliber weapons, that is, weapons which were capable of loading and firing themselves by utilizing the force of either recoil or expanding gas. Fed with cartridges from a magazine or a belt, the new weapons were relatively light and capable of continuous automatic firing until either the ammunition supply was exhausted, the barrel overheated, or the gunner took his finger from the trigger.

Like most others, the Imperial Russian Army approached the machine gun with great circumspection. The Russians had toyed with the Gatling system in the 1870s, and in 1891–92 they field-tested the multibarreled Nordenfeldt in the Turkestan and Siberian Military Districts. The initial findings of the Artillery Committee of the War Ministry were that "under present armament of the infantry and field artillery, machine guns in general and earlier systems in particular [*mitrailleuse*-type weapons] have little significance for the battlefield."[77] Furthermore, once the infantry was completely rearmed with magazine-fed rifles and the field artillery received even better weapons, the committee speculated that machine guns would become relatively less important. In contrast, the Main Artillery Directorate concluded that they might be useful for fortress defense, and placed an order for 250 of the *mitrailleuse* variety.[78]

This state of affairs did not long persist. In 1896, the War Ministry placed two orders for experimental weapons, one in London and the other in Berlin, for a total of 379 machine guns of the Maxim design. These weapons were subjected to a series of field tests, not only in the environs of the imperial capital but also in the Turkestan, Siberian, and Orenburg Military Districts. Tests were conducted at ranges varying from 400 to 2,200 paces, and results indicated the need for a number of modifications. Faults notwithstanding, the report of the War Ministry concluded that machine guns would be a useful addition to the armament of the army because "they could significantly increase [enemy] casualties." In 1899 the Russians purchased fifty-eight Maxims for 170,056 rubles, or nearly 3,000 rubles each. In 1902, the War Ministry concluded a contract with the British firm Vickers which permitted the Russians to manufacture the Maxim under license in Russia, in return for which the Tsarist government paid fifty pounds sterling per weapon produced, bringing the cost of each gun down to 942 rubles. While the Tula arsenal prepared to begin production in 1904, the War Ministry placed two rush orders abroad to supply needs in the Far East and along the Vistula fortress line. These orders respectively secured 200 guns of the Madsen type and an additional 250 of an unspecified type.[79]

Again, as in the case of the Mosin rifle and the three-inch field gun, the Russians had made an excellent choice of a weapon system. The Vickers-Maxim was a reliable heavy machine gun available with two mounts, wheeled and tripod. With a barrel jacket to permit water cooling, the weapon weighed sixty-six kilograms, and its rate of fire was 250 rounds per minute. With some modification, it would remain the standard heavy machine gun for the Russians until World War II. The Russians estimated that the gun could replace the firepower of fifty infantrymen in a defensive position.[80]

The last quarter of the nineteenth century also drew the Russian army slowly into the air age. In 1885, two officers completed a balloon flight from St. Petersburg to Novgorod, and further experiments at maneuvers between 1886 and 1889 appeared to justify the creation of several provisional balloon detachments. In 1891, after balloonists successfully photographed the imperial capital from altitudes of more than a kilometer, the War Ministry approved the purchase of thirty-three balloons to assist in western fortress defense and another thirty-two for "free flight." For the time being, their chief mission would remain observation and reconnaissance, since in 1892 the tactician N. A. Orlov had determined that "bombardment from balloons is an enterprise which would simply be funny to anyone having an understanding of war."[81]

Army Strength and Organization

Between 1878 and 1904, army organization changed by increments to reflect various influences, including lessons of the Russo-Turkish War, the adoption of new armaments, the appearance of new threats, and perceptions of changing tactical realities. Each of these factors played a role in determining not only the size of the army but also the way that it was configured for combat. After San Stefano, the active army was returned to its normal peacetime strength, which between 1881 and 1891 amounted to approximately 31,000 officers and 850,000 troops. The availability of reserves in the event of mobilization could roughly triple these figures. The army's peacetime strength rose slightly during the Afghan crisis of 1885, then levelled off until 1891, when the full implications of the changing international situation reinforced Russian apprehensions over the possibility of future simultaneous conflict with Austria and Germany. As the Russians sought agreement with France for mutual assistance in the event of war, especially with Germany, manpower requirements grew, so that by 1894 the army's peacetime strength had risen to 35,000 officers and 940,000 men. After 1895, Russia's growing interests in the Far East required additional commitments, so that by 1902 active forces in the Priamur Military District and the Kwantung Region numbered nearly 81,000. Indeed, by 1904, the peacetime strength of the Russian army had climbed to 41,000 officers and slightly more than one million men. Full mobilization would approximately triple these figures.[82]

The experience of 1877–78 coupled with varying training and control requirements also figured in the changing nature of peacetime military organization. During the 1860s, Miliutin had envisioned the division as the largest peacetime entity within the military district. However, by the mid-1870s, this conception underwent limited revision. Because the absence of large-scale formations failed to reflect projected wartime higher command arrangements and inhibited exercise experience with large units, the Russians slowly resurrected their peacetime corps organization, starting in 1874 with the Guards. After the conclusion of hostilities with Turkey, the peacetime Russian army reverted by 1881 almost entirely to a corps-based organization, with the establishment of nineteen: one Guards, one grenadier, fifteen army, and the 1st and 2d Caucasian. The corps was a combined arms organization, with each consisting of two or three infantry divisions, two or three artillery brigades, one cavalry division, and a sapper battalion.[83]

Just as before 1877–78, reserve troops consisted primarily of infantry and artillery. Following a reorganization of the early 1880s, the reserve infantry comprised ninety-six army battalions of five-company composition, one four-company Guards battalion and one fortress battalion. The reserve artillery consisted of six artillery brigades of six-battery composition. Upon mobilization, each reserve infantry battalion deployed four of its companies as battalions, leaving behind an additional company-based battalion for replacement and training purposes. Without counting the stationary battalions, the resulting 384 battalions provided sufficient force structure to generate twenty-four reserve infantry divisions. Similarly, upon mobilization, the six reserve artillery brigades (thirty-six batteries) expanded fourfold to deploy as 144 batteries. Ninety-six of these were combined into twenty-four artillery brigades to serve as support for the twenty-four reserve infantry divisions, while the remainder were utilized for replacement and training purposes.[84]

In the event of war, the Russian army retained the capacity to generate additional reserve forces largely for replacement purposes. Thus there existed seven reserve cavalry brigades, the function of which was to send replacements to designated units in a theater of war. Also, the capacity existed to create 199 reserve replacement battalions, one for each infantry regiment and rifle brigade. Upon mobilization, their training cadres were drawn from a mixture of Guards units, rifle brigades, and reserve battalions, while their rank-and-file came from the *opolchenie*, or home-defense forces.[85]

Infantry

On the eve of the Russo-Turkish War, the Russian infantry had just begun transition to a "four-sided," or "rectangular," organization. The transition was complete in 1879, and by 1881 the regular army counted in its complement forty-eight infantry divisions, including three Guards, four grenadier, and forty-one army (or "line") divisions. Each division was divided into two brigades of two-regiment

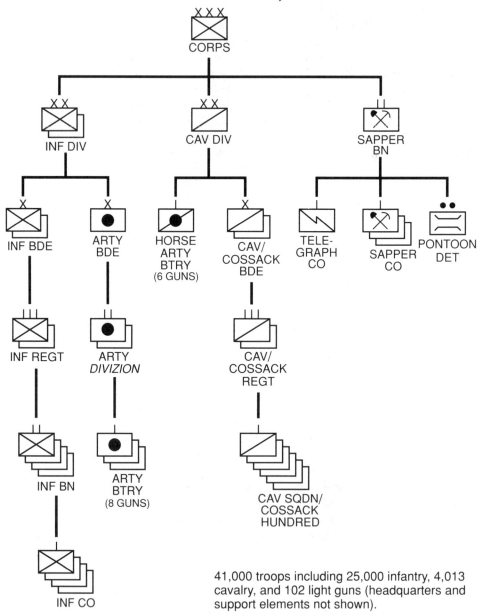

composition. The army numbered 192 regiments, and each was composed of four battalions of four-company composition. In addition, there were ten rifle (or "sharpshooter") brigades, each of which consisted of four rifle battalions. Eight separate Finnish rifle battalions (created in 1882) and thirty-four separate line battalions (two Orenburg, seventeen Turkestan, four West Siberian, four East Siberian, and seven Caucasian) rounded out the active army infantry organization. In all, the active army and unit reserve in 1882 consisted of 948 battalions, without taking into account miscellaneous local and dismounted Cossack formations.[86]

Within the regiments themselves, subtle changes occurred before and during the Russo-Turkish War to reflect the organizational impact of tactical change and a corresponding impulse for decentralization. Prior to 1877, all infantry companies had been divided into two platoons (*vzvody*), each of which comprised two half-platoons. The half-platoons were further broken down into two sections (*otdeleniia*). To facilitate control in the field, companies on the eve of the Russo-Turkish War were broken down into half-companies, each of which comprised two platoons. Platoons were further subdivided into four sections. Greater organizational articulation even before the onset of hostilities thus had acknowledged the growing importance of the infantry company as the lowest unit of independent tactical action. Subsequent combat experience, with its emphasis on cover and dispersion over greater areas, only reinforced this trend. In addition, as the requirement for coordinated fire and movement came to be understood during the 1880s as a commmon function of all infantry formations, the difference between sharpshooters and line infantry gradually faded. Adoption of the Mosin rifle with its extended range accelerated this process.[87]

Less subtle and more alarming to potential adversaries was a pronounced trend in the Russian army between 1881 and 1903 to multiply infantry force to meet the growing requirements imposed by territorial expansion and alliance obligation. Beginning in 1882, the War Ministry gradually increased the minimum troop strength of regiments and rifle brigades and battalions deployed in border areas. In addition, the War Ministry ordained the creation of four additional active infantry divisions, so that by 1903 their total number had risen to fifty-two. Finally, the number of rifle brigades more than doubled, producing a 1903 total of twenty-four, including eight in Turkestan and six in East Siberia. Added to these field units were fifty-nine battalions of fortress infantry and, after 1901, five machine-gun companies with eight guns each. In 1903, the active field army was composed of twenty-nine corps, including one Guards, one grenadier, twenty-one army, two Caucasian, two Turkestan, and two Siberian.[88]

Corresponding with these changes, the War Ministry took measures to increase the quality and size of reserve forces, cadre and replacement. In 1884, a new regulation prescribed that each active regiment designate a small training detachment to supervise its own reserves. In 1888, the individual service obligation was extended to eighteen years, including five active and thirteen reserve. Also in 1888, the War Ministry reinforced and restructured infantry reserve formations, creating four new reserve regiments and transforming twenty-eight reserve battalions located within the western border districts into two-battalion regiments. The majority of reserve battalions were gradually combined into reserve brigades, the number of which totaled fifteen in 1896. In 1890, cadre reserve units began drawing their reserve complement solely from the general pool of trained manpower, while active units began drawing their augmentation from those young males eligible for first call to active duty. The cumulative effect of these changes was gradually to produce a greatly enlarged infantry reserve, which by 1903 amounted to twenty-six brigades and a number of smaller formations. Upon mobilization, these brigades possessed the capability to deploy fifteen first-echelon and between fifteen and twenty second-echelon infantry divisions.[89]

Various augmentations to active and reserve forces meant that the infantry arm grew considerably in strength between 1881 and 1903. Figures revealed that in 1881 the active infantry numbered slightly more than 612,000 troops, and in 1899 approximately 754,000.[90] If all types of reserve infantry (excluding the *opolchenie*) were taken into account, the same figure for 1899 rose to 1,969,000.[91] In 1903, the number of active infantry battalions stood at 1,041, including 832 within various regiments, 150 within the rifle brigades, and 59 designated for fortress defense. The number of reserve battalions was 126, including 122 within the reserve brigades and four of independent designation. Thus in 1903 the total number of active and reserve infantry battalions bearing peacetime unit designations was 1,167.[92] Because reserve units were capable of fourfold expansion, the actual wartime number of battalions could exceed this figure by more than 400.

Cavalry

In 1881, Russian cavalry in the active army numbered two Guards and eighteen army divisions. On the basis of experience during the Russo-Turkish War, the War Ministry reached three major conclusions about Russian cavalry. First, the mounted arm's size had fallen disproportionately below the infantry; second, Russia no longer matched the cavalry power of its neighbors; and third, Russian cavalry had no unified tactical organization. Because the changing nature of military operations no longer justified the existence of uhlans and hussars, in 1882 both types were transformed into dragoons. Over the next decade, the War Ministry added six cavalry regiments to the active army. In addition, a new table of organization for 1883 raised the complement of each regiment from four to six squadrons. In accordance with a new emphasis on organizational homogeneity, Cossack regiments were withdrawn from the organization of regular cavalry formations and organized into composite Cossack divisions.[93]

The question of numbers also figured in a new emphasis on cavalry reserves. Prior to 1877, the few existing cavalry reserves were designated in the event of war to provide replacements for active units. In 1883, a revised regulation governing the organization of cavalry reserves drew cadre forces from active regiments to create the foundations for a genuine cadre reserve. Two cadre elements formed a reserve cavalry brigade, of which there were eight. Three of these brigades formed the nucleus for a cadre reserve cavalry division.[94]

By 1895 these measures, when combined with a reorganization of the regular cavalry, produced an active Russian cavalry force of twenty-two divisions composed of ninety-one regiments. In addition, the active army included in its cavalry complement two Cossack brigades, sixteen Cossack regiments, eleven Cossack squadrons, and four Cossack divisions, while the cadre reserves added eight more brigades. Thanks to reinforcement and reorganization, the Russian cavalry, which had numbered about 70,000 in 1881, rose to 80,621 in 1900.[95] By 1903, the active Russian army possessed twenty-five cavalry divisions, including two Guards, seventeen army, and six Cossack.[96]

Artillery

The rearmament of the field artillery with steel guns during the early 1880s prompted few immediate changes in artillery organization and strength. During the early 1890s, however, the expansion of the other arms began to affect the size and composition of the field artillery by calling attention to various organizational discrepancies, especially the declining ratio of supporting guns to field troops. Until the advent of the quick-firing gun, the principal organizational accommodation to changing technology was the formation after 1889 of mortar regiments to accommodate General Engel'gardt's six-inch mortars. Eventually the army added to its composition five mortar regiments, each of which consisted of either three or four four-gun batteries. Meanwhile, reorganization of the reserve infantry in 1888 prompted reinforcements for the reserve artillery. As the fifteen first-echelon reserve divisions slowly evolved, they would require thirty additional cadre reserve and two general reserve batteries. Moreover, the five rifle brigades formed in European Russia in 1890 required the creation of ten light batteries. By 1895, the field and reserve artillery numbered 428 batteries with the possibility of augmentation from fifty-four "flying parks," depots from which equipment and guns could be drawn in the event of hostilities.[97]

In 1895, possibly as a result of the need for greater flexibility to facilitate control and concentration of fires, the field artillery adopted a new organization, the *divizion*. Not to be confused with the Russian *diviziia*, a designation for infantry and cavalry formations, the artillery *divizion* was a grouping within the artillery brigade which brought two or more batteries under a unified system of command and control. The *divizion* enabled artillery commanders both to mass fires from fewer guns and to support units smaller than a division with only a portion of the artillery brigade's firepower assets.[98]

The most dramatic change in Russian artillery formations occurred during the second half of the 1890s, when the number of batteries was markedly increased to match the growing strength and numbers of infantry units. Throughout the 1880s, the rule of thumb had been to allocate three guns in support of each infantry battalion. During the 1890s, however, the ideal became four guns in support of each thousand infantrymen. Beginning in 1895, a haphazard effort to apply the new standard occasioned the creation of an additional fifty-two batteries, which were distributed unevenly throughout the active artillery. By 1898, a more measured approach spawned an additional 108 batteries.[99]

By 1903, these increases resulted in an active field artillery force of fifty-six brigades ranging in composition from four to nine eight-gun batteries. Other active formations included eight artillery regiments, four separate *diviziony*, five mobile fortress artillery batteries, and ten Cossack artillery batteries. Reserve foot artillery accounted for an additional seven brigades and several miscellaneous formations. By 1903, the total number of foot batteries had risen to 498. Horse artillery contributed another 48 batteries for a total of 546, an increase of 40 percent over the same number for 1881. Between 1881 and 1903, the total manpower strength of active and reserve artillery grew from 107,601 to 154,925. Although by the turn of the century the total number of artillery pieces exceeded 5,500, the Russians, unlike the Germans and the Japanese, possessed no modern heavy field artillery.[100]

Auxiliary Troops and Cossacks

On the eve of the Russo-Turkish War, engineering troops had made up fifteen sapper battalions, six pontoon half-battalions, three railroad battalions, six telegraph parks, and four field engineering parks. During the war, the engineering forces rose in composition to about 38,000 officers and men, the majority of whom served in five sapper brigades. After 1877–78, engineering forces entered a period of declining strength, which was arrested only in 1883 when requirements in Siberia led to the creation of an additional sapper brigade and two separate field companies. In addition, during the following year the War Ministry created a sapper reserve force by pulling cadre for two reserve sapper companies from each of the active seventeen sapper battalions.

Expansion continued in 1886, when the War Ministry created a special railroad brigade by detaching the railroad battalions from their parent sapper brigades, unifying them under a single headquarters, and supplementing each with a fifth company. The purpose of the additional company was to serve as the base for an additional cadre battalion in the event of mobilization. At the same time, an additional battalion was formed to run the Trans-Caspian Railroad. A sprinkling of other units, including two river mining companies, nine fortress sapper companies, six fortress telegraph parks, and one balloon park with four sections, rounded out the organizational structure of the engineer troops for the 1880s.[101]

Just as in the case with artillery, infantry expansion during the 1890s triggered corresponding changes among engineer troops. The rationale for engineer augmentation was that each army corps should be supported by a sapper battalion, with each constituent sapper company supporting an infantry division. As the number of engineering battalions multiplied, they became more uniform in composition, usually consisting of three sapper companies, one field telegraph company, and several pontoon detachments. As a result of expansion, the number of sapper battalions grew from twenty-nine and one-fourth in 1881 to forty-six and one-half in 1894. In 1903, the total number of engineer battalions was fifty-four, including seven railroad battalions and forty-seven sapper battalions.[102]

Because there was never sufficient regular cavalry to discharge the myriad of missions imposed on the mobile arm, Cossacks continued to play an important role in the overall force structure of the Russian army. Although the Miliutin reforms in Cossack service had little time to take effect before the Russo-Turkish War, the Cossacks successfully mobilized 3,672 officers and 140,882 men for service in the Balkan and Caucasian theaters. These figures represented approximately 70 percent of the cavalry forces of the Russian Empire. However, the mobilization and war experiences revealed sufficient shortcomings to elicit a series of postwar measures aimed at improving the training, equipment, welfare, and rotational service system among the Cossacks. The War Ministry also made provision to render more efficient the Cossack mobilization process. Although each host continued to experience a variety of internal difficulties, ranging from insufficient arable land to shortages in horses and modern armaments, the War Ministry managed to maintain among the Cossacks an immense pool of trained manpower suited to the peculiarities of mounted service. Figures for 1900 revealed that 67,000 Cossacks served in the active army. In the event of war, their number could have been expanded to slightly more than 190,000. Official records indicated that in 1903 the fully mobilized unit strength of the Cossacks stood at 149 regiments, forty-two separate hundreds, twenty-two dismounted battalions, and fifty-nine horse artillery batteries.[103]

Mobilization Schedules and Planning

In the event of war, the establishment of a true cadre and reserve system simultaneously permitted the fielding of more formations and presented greater planning challenges. Each peacetime cadre reserve brigade, which consisted of four battalions, was to be deployed in wartime as one or two infantry divisions. At the outbreak of hostilities, twenty-one of the reserve brigades would become thirty-five infantry divisions (46th through 81st). In analogous fashion, each reserve battery would deploy as four batteries, a phenomenon which explained why peacetime batteries were equipped with thirty-two guns. The number of cavalry formations would be doubled by calling Cossack regiments to active duty from home leave status. As Colonel F. Chenevix-Trench, a former British military attache to St. Petersburg, noted in 1886, "the Russian Army must always be a very formidable foe from its great numerical strength."[104]

Yet numbers meant little without mass and concentration, and it was military planners who bore responsibility for drawing up detailed mobilization schedules to provide for the dislocation, transit, and concentration of these units along with the active army at the appropriate time and place. As a British handbook on the Russian army noted,

> In Russia, owing to the bulk of the army being quartered in the western portion of the Empire, and the bulk of the reservists residing in the center, the mobilization of the army is much less in the hands of the commanders of the active troops, and more in those of the head-quarters staff than in any other of the great Continental States.[105]

Given this situation, the object of mobilization schedules, then, was to match men, equipment, and transportation with missions across vast distances in the shortest possible time.

From the late 1860s, the Main Staff of the War Ministry developed mobilization schedules at regular intervals in anticipation of a probable war against either Germany, Austria-Hungary, or both. Turkey and Romania figured in planning as additional and potentially hostile variables. By the early twentieth century, the Main Staff also had to contend with the possibility of war in the Far East.

In the west, three factors dominated Russian planning: the lack of a satisfactory rail infrastructure to permit rapid mobilization and concentration, the likelihood of Austro-German numerical superiority, and the necessity to retain a strong presence on the right bank of the Vistula with secure crossings to the left bank. Sketchy plans initially dating to the 1870s called for concentration in the central area of the western frontier and retained a strongly defensive character. Only after 1883 did the Russians dare plan taking the offensive, and then only against Austria-Hungary, "the real political enemy of Russia." If circumstances required offensive action farther south, the Russians would also attack Turkey, with elements of the Black Sea Fleet and troops from the Odessa Military District scheduled on mobilization day plus eleven (M + 11) to begin landings aimed at denying enemy access to the Black Sea by controlling its exit to the Bosporus.[106] During the 1880s, there was a good deal of fear in Britain and India that the Russians might strike south through Afghanistan to invade the subcontinent, but sober analysis indicated that St. Petersburg lacked the ability to project meaningful combat power across such vast distances.[107]

At the same time, as the possibility of war with Germany and Austria-Hungary became more probable, mobilization problems closer to the center assumed greater importance. Throughout the period between 1881 and 1904, the central difficulty plaguing planners was the Russian Empire's underdeveloped railroad network. As E. Willis Brooks has pointed out, Miliutin was largely unsuccessful in pressing for railroad construction to serve mobilization needs. Priority went to developing a rail net that would serve the general economic infrastructure, and economic and military requirements often did not coincide. This was especially true in areas along the western frontier, which at least initially benefited from less railroad construction than either the Russian interior or, more alarmingly, the eastern frontier areas of Russia's potential adversaries.[108] Miliutin's immediate successors were only marginally more successful in arguing for the strategic development of the Russian rail net.[109] During the 1890s, Count Witte's Trans-Siberian Railroad siphoned off still more resources from European Russia, although it served the economic *and* military interests of Far Eastern expansionism.[110]

In consequence, Russia's comparative rail disadvantage against possible western adversaries grew still greater during the late 1890s. At the end of 1899, European Russia counted 41,515 versts of track, while Russian Central Asia and the Far East counted another 7,409.[111] In contrast, Germany by 1897 had 43,067, while Austria-Hungary had 30,221. Because these figures do not convey a sense of density and location, they represent only rough comparative indices. More telling was information related to hard military realities, and the overall picture was not good. Along the immense span of the western military frontier, Russian military planners envisioned four potential theaters of military action: northwestern, western, southwestern, and southern. The northwestern had three double-track lines linking the Baltic provinces with the center; the western counted ten lines, of which only three were double-track. The southwestern had three, two of which were single-track. The southern had five lines, of which only two were double-track. Worse, the system of lateral feeder lines joining all these "was extremely inadequate," in the words of a Soviet military historian.[112]

Russia's comparative rail disadvantage translated into a strategic disadvantage which increasingly became a source of alarm for the War Ministry. In 1900, Kuropatkin reported that "it is urgently necessary for us to increase the carrying capacity of railroads which retain strategic significance, and no less necessary is the further development of rail nets leading to the western frontier."[113] He went on to address the implications of the situation:

> For the concentration of our army to the west, we have only seven principal rail lines (eleven tracks), which permit us the possibility of transiting up to 167 trains per day, and that from the tenth day of mobilization. In Germany, seventeen through-lines (twenty-three tracks) lead to our frontier, and in Austria-Hungary, eight (ten tracks). Already from the third day of mobilization, these lines will permit up to 812 trains to reach the frontier daily.[114]

It was this calculus plus the geographical peculiarities of the western frontier which dictated a cautious approach to planning for war in the west. Retention of the Polish salient posed a special problem. If too many forces were concentrated there, they presented an invitation to entrapment by enveloping north-south pincers. If too few were left there, the Russians could not guarantee the security either of Russian Poland or of crossings to the left bank of the Vistula, the retention of which was a prerequisite for future offensive operations against either Austria or Germany. For these reasons, the most prudent approach to planning for war in the west seemed to involve initially retaining enough field forces at wartime strength and echeloned in sufficient depth to cover the western fortress system, thereby securing the Polish salient while mobilization wound its slow pace. Under these circumstances, the Russians chose to shift a greater proportion of their peacetime deployments into the Kiev, Vilnius, and Warsaw Military Districts, especially after 1887. As William C. Fuller has pointed out, the number of troops stationed in these districts rose from 227,000 in 1883 to 610,000 a decade later, with the latter figure representing about 45 percent of the Russian army's peacetime strength. Until the Russians completed their concentration to support the Polish salient, initial strategic operations would assume a defensive character. Only in the event that Germany became preoccupied in the West did the Russians anticipate shifting to the offensive, and then only against Austria-Hungary. Whether defensive or defensive-offensive, various plans incorporated provisions for an initial mass Russian cavalry raid of over 120 squadrons to disrupt Austrian and German mobilization.[115]

Within this context, the issue of fixed frontier defenses attained growing impor-

tance. Although the Crimean and Franco-Prussian wars had rekindled interest in fortress systems, until the late 1880s engineers and theorists remained divided on their shape and utility in future war. After 1877–78, several considerations, including the changing international situation, the appearance of new munitions, and increased vulnerability around the periphery, caused planners to reassess the importance of various fortress complexes, especially those on the Empire's northwestern and western frontiers. Successive commissions outlined the problem, recommending a series of incremental improvements to coastal defenses and fortress systems controlling key Baltic approaches to Finland, St. Petersburg, and northwest Russia. Between 1884 and 1887, an inspection of Russian fortresses opposite Germany and Austria-Hungary revealed that these systems were inadequate for modern war. Worse, tests conducted in 1886–87 to determine the effects and implications of the newest large-caliber, high-explosive projectiles forced the War Ministry into a dramatic recalculation of its fortress requirements. The impact of the new technology was so devastating that War Minister Vannovskii reported to the Tsar that "the disturbed balance between attack and defense in favor of the former everywhere forces a transition to the enlargement and reconstruction of fortresses." By 1891, Vannovskii was writing to the Tsar that "forts have already lost their primary significance as artillery positions and must serve chiefly as strong points for the support of separate defensive positions."[116]

Russian engineers rose to the challenge by proposing greatly enlarged fortress complexes of earth and concrete that would embrace an entire fortified region. Rather than rely on static positions and heavily armored gun emplacements, the Russians sought a harmonious combination of "living force and dead mass," that is, of a flexible defense in depth built around concealed and mobile guns and requiring elaborate communications, well-defended approaches, and effective covering forces. During the 1890s, this solution governed the modernization of fortresses on the western frontier, especially in Poland, where Russian military engineers either rebuilt or constructed elements of two separate defensive systems. The first was a network of border fortifications laid along a line extending from the Veprezh to the middle Vistula to the Narew. The second was a central fortified region built around three immense fortress complexes (Warsaw, Novogeorgievsk, and Ivangorod), with flank protection from fortresses at Kovno and Osovets (also covering the approaches to St. Petersburg), and defense in depth assured by additional fortresses at Zergrzh, Brest-Litovsk, and Dubno (also covering the central and southern approaches to the Russian interior). Total cost to the treasury was about 53.5 million rubles, and this expenditure did not include acquisition of modern fortress artillery to replace outmoded ordnance dating to the 1870s.[117]

From 1882 until 1897, while the fortress issue remained unresolved and while the mobilization calculus remained weighted against Russia, the Main Staff clung tenaciously to its initial defensive designs—at least against Germany—even in the face of growing French pressure to assume an increasingly anti-German mobilization posture. Beginning in 1892, the French pressed for integrated military planning, while Obruchev as Chief of the Main Staff successfully counseled that Russia retain maximum freedom of action by choosing to concentrate in the Polish salient for operations against either Austria or Germany. Obruchev's stand reflected two considerations. The first was that Austria remained the real political enemy of Russia; therefore the Russians ought to retain within the framework of agreement with France the ability to strike Austria first while assuming the defensive against Germany. The second was a growing realization that changing military realities had altered the very nature of the political significance of mobilization. Given the new emphasis on speed and mass in seeking rapid military decision, "mobilization nowadays could no longer be regarded as a fearful step, but on the contrary it was a very decisive act of war."[118] For this reason, Obruchev remained very circumspect about entering into collaborative mobilization planning with the French General Staff, which was very intent on honing its own anti-German ax.

Only in the beginning of the twentieth century did the Russians show much inclination to stray from Obruchev's essentially conservative approach. About 1900,

General M. I. Dragomirov, Commander of the Kiev Military District, proposed that St. Petersburg coordinate its operations in the event of war more closely with those of the French with the object of directing the main blow against Germany.[119] For the time being, this served only as a harbinger of the future, although in 1901, V. V. Sakharov, Obruchev's replacement as Chief of the Main Staff, allowed the French to introduce into the joint military convention the offer of financial assistance for the construction of strategic rail lines. This was the first step down the slippery path to greater military cooperation in favor of French military interests to the detriment of long-term Russian foreign policy objectives.[120]

Against this background, between 1883 and the advent of war in the Far East, Russian planners worked out a series of mobilization schedules, the last of which prior to war in the Far East was Schedule No. 18, which went into effect on 1 January 1903. In the event of war against both Germany and Austria, Schedule 18 called for concentrating against them 1,472 infantry battalions, 1,035 cavalry squadrons, 4,558 guns, 32 machine guns, 175 engineer companies, 64 companies of border gendarmes, and 47.5 fortress artillery batteries. If Turkey remained neutral, Russian mobilization planners could count on an additional 64 infantry battalions, 44 cavalry squadrons, 220 guns, and 4 engineer companies. If Germany turned to deliver its primary blow against France, the Russians could count on numerical superiority, varying from roughly 1.4:1 in infantry to 2:1 in cavalry and 1.1:1 in artillery.[121]

As the Soviet military historian P. A. Zaionchkovskii has noted, however, "the fundamental issue assuring the success of military actions, especially in the first months of a war, lay not in an enumeration of the correlation of forces overall, but in their concentration in the theater of war."[122] And it was in this sense that Russia confronted its gravest difficulties. As early as 1870, the Germans had shown themselves able to transit for mobilization purposes ten pairs of trains daily on a single-track line and twenty pairs daily on a double-track line![123] In 1902, Kuropatkin reported that the Germans and the Austrians could put their railway systems on a full mobilization footing by the third day of mobilization, while Russia could do the same only by the ninth or tenth day. Germany could initiate offensive operations on M+12 and Austria on M+16 while Russia was still in the process of concentrating its field armies and the reserves and Cossacks of the second echelon had not even begun mobilization transit. Kuropatkin concluded that "this backwardness in the development of rail means unavoidably emphasizes [in] the initial form of the strategic deployment of our forces the idea of defense."[124]

With the sword of backwardness hanging over its head, the Main Staff reallocated resources within the overall plan to compensate for shortcomings in infrastructure and speed. What the Russians lacked in speed they made up in mass. Schedule No. 18 squeezed more from the reserves, drawing from them thirty-one infantry divisions, ten army corps (XXII-XXX and the Finland), and one cavalry corps consisting of three Cossack divisions. When these forces were combined with regular formations, the Main Staff could throw seven field armies against the Austrians and Germans, and, in the event of simultaneous war with Turkey, an eighth (Caucasian). The initial seven armies would make up two fronts, the German and

Austro-Hungarian, each with its own commander in chief, both of whom would be subordinate to the supreme commander.[125]

In war planning, the mobilization schedule was only part of a larger operational picture. Once the troops were concentrated according to schedule, field staffs at front, army, and corps levels had to deal with actual deployments and initial march-maneuver to contact. P. A. Zaionchkovskii, writing about primary and subsidiary staff activities at the turn of the century, simply asserts, "There was no plan of military actions."[126] He cites F. K. Gershel'man, who had been named chief of staff of the Warsaw Military District in the spring of 1901, as an important source on this matter. Like any good staff officer before his departure to the provinces, Gershel'man spent ten days in St. Petersburg, snooping around the "holy of holies" (Main Staff) to familiarize himself with plans involving the Warsaw District in the event of war with Germany and Austria. He found not a few notations, proposals, and ideas, but little else. Scraps did not constitute a whole picture, and Gershel'man left for Warsaw, muttering in his memoirs that he "found nothing finished, found no work carried to completion."[127]

With special reference to lower-level staffs, Gershel'man remembered after a visit of the War Minister to Warsaw in 1902 that

> In matters on the defense of the western frontier, nothing was definite, nothing finished. We dissipated ourselves in a mass of frequently contradictory assumptions, touching on everything, stopping for nothing, and eliciting in ourselves only complete distrust for our decisions, which changed constantly. . . . Under these conditions we were completely unprepared for war on the Western front, and besides, in order to progress we would have distanced ourselves from the desired objective.[128]

War Minister Kuropatkin reinforced Gershel'man's impressions, with special reference to the Kiev Military District. On 4 March 1903, Kuropatkin met for the first time with the commanders and chiefs of staff of the three armies (3rd, 4th, and 5th) which would fall under him as commander of the Southwest Front in the event of war against Austria. He asserted that "once more at the meeting was affirmed our unpreparedness for the offensive." Further, he commented bitingly:

> No concept about offensives had been put together in any of the armies. As for the defensive, the 4th Army was the best organized of all, but in it our positions at Rovno, Lutsk and Dubno were more designated [on the map] than fortified. . . . In the 5th Army nothing had been done regarding defense, and incidentally very large Austrian forces could collapse this army and force the 3rd Army to withdraw toward Brest.[129]

V. A. Sukhomlinov, who was Dragomirov's chief of staff in Kiev, later took exception to this indictment, but could provide no detailed rebuttal for these "fantastic conversations," saying that he "had no notes from this time at hand."[130] Kuropatkin no doubt exaggerated the situation, but in light of later events his assertions probably ring close to the truth.

The situation was little better in the Far East, which, because of its remoteness from the center, operated under its own mobilization and planning regime. There the Siberian and Priamur districts and the Kwantung Special Fortified Region took their planning guidance from Mobilization Schedule No. 8. Aggregate peacetime strength at the beginning of the twentieth century stood at 77 battalions and seven companies of infantry, 33 squadrons of cavalry, and 156 guns. Additional local forces included 20–28,000 frontier guards organized in four brigades to secure the Chinese Eastern Railway. Yet Schedule No. 8 called for wartime forces of 127 infantry battalions, 81 cavalry squadrons, and 212 guns. The difference would have to be made up by mobilizing cadre and noncadre reserve units from European Russia, then laboriously transferring them across the single-track Trans-Siberian rail line (with a break at Lake Baikal) to the Far East.[131] In 1904, the carrying capacity of this system was eight to ten pairs of trains per day.[132] In the event of war with Japan, the skeletal Far Eastern rail net would find itself taxed to the utmost.

The gap between mobilization schedules and war planning revealed in clear outline much of the achievement and failure of the Vannovskii-Kuropatkin years. The War Ministry, working within the tsarist political framework, was able to tap significant manpower and financial resources to recruit and arm a mass army in the industrial age. The Russians were also able to spawn the planning and directing mechanisms necessary to mobilize and command these masses. Without a suitable rail infrastructure, however, the Main Staff's ability to concentrate that army in a timely manner for decisive operations remained problematic.

In the end, the issue was one of linkages, in this instance, of linkages between means and objectives. Under the pressure of wartime requirements, there was every possibility that insufficient rail carrying capacity would remain only one of a number of serious deficiencies in linkages between supporting base and fighting front. Another deficiency would be imperfect intellectual linkages, a function of varying military and educational experiences. Another would be the appearance of inevitable gaps between theory and practice during two and one-half decades of peace. With all its strengths and imperfections, the army that went to war against the Japanese in 1904 was largely the creature of Vannovskii and Kuropatkin working within the larger tsarist context.

4.

The Legacy of G. A. Leer and M. I. Dragomirov

> Leer had his court complement of facts borrowed primarily from the wars of Napoleon and strictly dressed in the uniform of his Leerist principles.
>
> A. A. Svechin [1]

> I consider machine guns an absurdity in a field army of normal composition.
>
> M. I. Dragomirov [2]

A NUMBER OF important influences shaped the evolution of military thought and doctrine within the Imperial Russian Army between 1878 and 1904. These included not only the considerable force of intellectual and institutional habit but also the experience of the Russo-Turkish War, the evidence offered by other conflicts, new technological and organizational developments and their consequences, and perceptions of the need for change flowing from requirements of organizational evolution. These influences were filtered separately or in combination through the perceptions and experiences of various observers, commentators, and functionaries, some of whom occupied positions which enabled them to translate their views and preoccupations into instructional and doctrinal reality. Despite the pace and impact of technological change, continuities in military thought and tactics remained striking. As in the previous period, the pattern of change was incremental, not revolutionary or discontinuous.

Dilemmas of Military Theory and Application

Memories of 1877–78 joined with perceptions of 1870–71 to prompt practitioners and commentators alike to reexamine traditional military verities. New technologies and methods enabled commanders to assemble masses of men, equipment, and horses, then project them more quickly than ever before into potential theaters of conflict. Issues of time and space became still more vital as theorists envisioned the outcome of future war to be determined largely by which side would win the race for deployment to and concentration within theater. The railroad and telegraph fundamentally altered traditional conceptions of assembly, deployment, and concentration, with the result that military men were forced to accept as conditional rather than absolute long-cherished convictions about the importance of such fundamentals as interior lines and mass. Given the lethality of modern weaponry and the problem of extended frontages, the solution lay, as Gunther Rothenberg has written, in "outflanking the enemy in a single, continuous strategic-operational sequence combining mobilization, concentration, movement, and fighting."[3] Helmuth von Moltke's oft-quoted maxim to "march separately and fight together" ("getrennt marschieren, zusammen schlagen") perhaps most aptly summarized the theoretical and practical challenges confronting military thinkers of the 1880s and 1890s: the notion was at once Napoleonic and contemporary. The emphasis on mass and concentration remained traditional, but the underlying assumption was that changing technologies and methods were busily reshaping the conditions and recasting the means.[4]

Nowhere was change more evident than on the battlefield itself, where the evidence offered by Plevna and Gorni Dubnik of the growing lethality of modern technology was too startling either to ignore or to dismiss out of hand. Various Russian commentators acknowledged the growing power of the tactical defense, but, as in the period after the Crimean War, no one seemed prepared to abandon his faith in the need for offensive action to produce battlefield decision. What was needed was a combination of resolve and improved technique to restore confidence in the tactical offensive by reducing casualties to an unspecified but acceptable level. Under these conditions, one of the principal developmental tendencies within tactical thought and doctrine involved an understandable search for a more perfect version of the familiar.

In theoretical perspective, the challenge was how to understand the impact of mass armies and changing technologies and methods on the complex interplay of offense and defense within both theater and the narrower confines of the battlefield. Again, Moltke thought he had the answer when he stressed the importance of assuming the strategic offensive within theater, then going over to the defensive, thus forcing his adversary to spend manpower and energies in a series of disastrous tactical confrontations against powerful defensive dispositions. If circumstance required offensive decision, the assailant turned to the new technologies and methods at his disposal to conduct a frontal pinning attack, then to envelop the enemy's comparatively weaker flanks to seek a classic victory of annihilation by means of encirclement. The events of 1870–71 and 1877–78 offered two powerful examples: Sedan and Sheinovo.[5]

For Russian students of war, the primary task was one of building an effective intellectual context within which to view the complexities of Sheinovo and Sedan in all their dimensions. For reasons of misunderstood science, theorists increasingly viewed war as the logical and inescapable result of Darwinianlike social struggle. For reasons of familiarity and simplicity, their professional thinking about war and strategy gravitated heavily to Jomini, not the more complex Clausewitz.[6] Perhaps

Clausewitz simply asked too much. His treatise, *Vom Kriege*, did not appear in complete Russian translation until K. M. Voide serialized it for the pages of the *Military Collection* in 1899 and 1900.[7] Meanwhile, those officers who read German found Clausewitz difficult to comprehend and subject to diverse interpretations. Worse, his classic definition of war as "the continuation of state policy by other means" required soldiers to confront murky nonmilitary realities.

In contrast, Jomini's thought remained pure, straightforward, subject to little variation, and readily accessible to officers. Jomini wrote in French, still the second language of choice for educated Russians; counted himself among the venerable founders of the Nicholas Academy of the General Staff; and had many years of loyal Russian service to his credit.[8] Unlike Clausewitz, who asserted that strategy was complex, Jomini held that strategy was a simple discipline, limited to the art of directing masses within theater and distinct from tactics, which did not admit to hard-and-fast rules. To retain purity of military thought, Jomini relegated political considerations to a separate discipline, military politics, thereby neatly—and dangerously—divorcing politics from strategy. Jomini also deftly sidestepped some of the more difficult questions of military art by consigning them to the imponderable realm of the great captain's genius and tact. Thus, while Clausewitz attempted to come to grips with the ambiguities of war, Jomini avoided them to retain simplicity and clarity, or as A. A. Svechin later put it, "order was attained at the expense of vitality."[9] Of more immediate importance, Clausewitz, the prophet of complexity and firm believer in the inherent strength of the defense, failed to find adherents either in St. Petersburg or within the military districts. Meanwhile, Jomini reigned supreme, advocating a strategy of the shock strike (in Russian, *sokrushenie*, or "crushing") that culminated in climactic battle, in which the commander who enjoyed the fruits of victory was the commander who concentrated the greatest force at the decisive point at the decisive moment.[10]

The sirenlike call of scientific positivism with its emphasis on method, system, and classification also figured prominently in Russian military thinking of the time. It was no accident that strategists of the era adopted the scientific trappings of civilian academia as they sought a respectable place in the intellectual sun for their own theories of strategy and military science. It was also no accident that the neatness and clarity of Jomini at least superficially lent his thought more scientific credibility than that of Clausewitz. These and similar preoccupations lent legitimacy to the quest of G. A. Leer and others to demonstrate the existence of underlying laws and principles which could be studied systematically to provide the theoretical underpinnings for what some commentators exuberantly proclaimed as military science.[11]

Strategy

The military scholar most prominently associated with this movement in Russia was Genrikh Antonovich Leer (1829–1904). Born in Nizhnii Novgorod and educated in St. Petersburg, Leer began his military career in 1848 as a field engineer, served with a reserve sapper battalion in the Caucasus, then between 1852 and 1854 attended the Nicholas Military Academy. Although he displayed a keen interest in military history, he was an indifferent military draftsman, and family difficulties interfered with his academic work. Consequently, he finished the Academy without distinction, and, after brief service as an adjutant in the Baltic provinces, he settled down to a routine assignment in the prereform doldrums of the General Staff. However, he soon demonstrated himself a talented instructor and writer, pioneering in new teaching methods as Adjunct Professor of Tactics after 1858 at the Nicholas Academy and as a frequent contributor throughout the 1860s to the *Military Collection*. By 1866, he had written a basic text on tactics for use in the military academies, but nevertheless found himself increasingly drawn to the study of strategy. In 1865, he received permanent appointment to the Faculty of Military History and Strategy at the Nicholas Academy, and by the mid-1870s he had earned a reputation both as a military historian of note and as the Academy's most prominent instructor (next to Dragomirov) of military art.[12]

Although Leer's writings identified him firmly with the newly emerging academic school of Russian military history, he is perhaps best remembered for his ground-breaking work on the development of strategy. For all practical purposes, he abandoned the tactical realm in the late 1870s to devote the majority of his theoretical work to strategy, a field which he dominated first as Ordinary Professor of Strategy, then between 1889 and 1898 as Commandant of the Nicholas Academy of the General Staff. His textbook, *Strategy*, went through six editions between 1867 and 1898, when disciples and detractors alike rose to challenge the master. Still, no other writer—save perhaps Dragomirov—exercised such a profound influence on Russian military thought between 1878 and 1904. For better or worse, Leer's teaching formed a major part of the intellectual legacy which a generation of Russian staff officers carried with them to Far Eastern battlefields in 1904–5.[13]

Leer's own intellectual baggage consisted of a devotion to Napoleon and a fixation on the seemingly diverse preoccupations of philosophical idealism and positivism. From studies of the Napoleonic campaigns he emerged with an appreciation both for military history and for individual genius as true repositories of military art. From William Lloyd and Antoine Henri Jomini he gained an appreciation for the rational element in Napoleonic strategy. From the positivists of his own era, Leer drew an understanding of classification and generalization which he would impose on his own evolving conceptions of strategy as a fledgling science. Thanks to these influences, refracted through the unique prism of Leer's own military outlook and preoccupations, Napoleon served not as a point of departure but as the touchstone against which all subsequent military developments were measured.[14]

Leer's understanding of complex military phenomena began with idea and history. For him, "always and everywhere idea came before act." When selectively and critically studied, military history enabled the discerning student to "arrive at an understanding of the idea which gave rise to the facts." In application, every military operation or sequence of operations embodied a fundamental idea from which flowed plan, lines of development, sequence of actions, establishment of priorities, and concentration of resources, all of which ultimately spelled success or failure for the commander seeking battlefield decision. From historical analysis, Leer logically arrived at an understanding of the two forces which dominated every operation:

objective (*tsel'*) as developed from idea and direction (*napravlenie*) as represented figuratively by an operational line depicting the unfolding in reality of idea and plan.[15]

For Leer, strategy in its most restricted sense treated operations within what he called a theater of military actions (*teatr voennykh deistvii*). His study of military history enabled him to classify strategic operations according to three types: main, preparatory, and supplementary. The first section of his text on strategy Leer devoted to an analysis of main, or primary, operations (*glavnye operatsii*), including selection of operational line, execution of marches and maneuvers, conduct of diversions, and the concentration of forces for combat, all of which culminated in main battle as the ultimate resolution of an operation. In the second section of *Strategy*, Leer discussed preparatory (*podgotovitel'nye*) operations, the term he used to describe the organization of armies and bases, deployment of forces in theater, and engineering preparation of the theater of military action. Finally, he outlined supplementary (*dopolnitel'nye*) operations as those involving accumulation of supplies, establishment of communication lines, and the organization of security, including planning routes of possible withdrawal and preparing fortresses and fortified lines.[16]

This intellectual framework for an understanding of operations marked one of Leer's enduring contributions to the development of Russian military thought. Subsequent students of military art at first clung to Leer's basic definition without a clear understanding that Leer's conceptual umbrella emphasized Napoleonic continuities at the expense of recalculating old verities in light of recent technological and organizational innovation. Not surprisingly, in 1891, the rising young strategist N. P. Mikhnevich penned a definition of *operations* for the Russian *Encyclopedia of Military and Naval Sciences* (*Entsiklopediia voennykh i morskikh nauk*) which stood virtually unchallenged for the next several decades. He wrote that "each war consists of one or several campaigns, each campaign of one or several operations, which represent by themselves a known, finite period, from the *strategic deployment* of the army on the departure line of the operation to the final decision of the latter by way of *victorious* battle on the field of engagement . . ."[17] The realization was that although strategy as a whole was more complex than ever before, its main task in theater was to guide the commander to a main battle which would produce decision either by encirclement of the enemy or by the energetic pursuit of his broken forces following main battle.

Leer was less successful in linking strategy within theater (which he called the "tactics of the theater of military action") with his broader conception of strategy as an all-embracing military science. He believed that strategy in its widest sense was "a synthesis of all military matters, their generalization, their philosophy."[18] Although the physical manifestations of reality might change, Leer clung to a conviction that underlying ideas remained constant and that a selective reading of military history yielded eternal and unchanging principles which existed independently of time and place. These principles he identified as four: mutual support, concentration of superior forces at the decisive moment and place, economy of force, and surprise. He was less clear about how these "eternal and unchanging principles" might apply to specific situations.[19]

Leer's approach thus left his students with two substantial intellectual impedimenta: an obsessive and exclusive preoccupation with Napoleonic precedent and the knotty problem of translating idea and immutable principle into action. Rather than ask how Moltke and his adherents varied from the French paradigm, he sought to demonstrate how they conformed with it. He taught that the campaigns of 1870–71 reaffirmed the significance of Napoleonic strategy, but he tended to ignore the campaign of 1866 because it did not neatly fit his preconceived pattern. Svechin, in addition to charging that Leer had his own "court complement of facts," later asserted that in Leer's eyes, "facts were also good children or troublemakers." If the latter required a break in consciousness, Leer's "doctrinaire thought turned away from them or ignored them."[20]

Perhaps worse, as Leer's writings of the 1890s turned increasingly metaphysical, they also became increasingly divorced from practicality. Admittedly, Leer's notion of operational direction represented an advance over earlier and more mechanistic notions of operational "line," but there remained the problem of applying Leer's idealist principles to a study of military art within an officer corps which did not greatly value philosophy. Leer himself evidently believed that strategy as a science was ill suited for teaching in an academic atmosphere. In his *Method of the Military Sciences* (1894), he argued that principles operated independently of condition and circumstance but that in approaching the resolution of concrete problems, one might act on the basis of more palpable departure points. These he labeled *rules* and *norms*, calling them conditional and warning that they were appropriate only to a given situation.[21] The difficulty was that he never seriously entered the world of concrete rules and norms, with the result that his teachings on strategy informed Russian officers *what* they must be concerned with, but not *how*. Tragically, those officers who took his views seriously left the staff academy with only a partial understanding of strategy, and they would be forced to complete the practical side of their education under fire on the battlefields of Manchuria.[22]

Nonetheless, Leer's legacy was more than ethereal. In addition to laying the foundations for a modern theory of operations, he spurred interest in the further development of strategic theory and military history. His works were translated and read by other European military establishments, and his views were sufficiently influential and controversial to earn both imitators and detractors. Perhaps his foremost disciple was E. I. Martynov, a captain of the General Staff, who in 1894 published *Strategy in the Epoch of Napoleon I and in Our Time*. Although Martynov departed from Leerist precedent to admit that each epoch might be capable of producing its own unique military institutions and practices, he warned against ascribing too much significance to technological change as an engine driving the evolution of a more advanced military art. The clumsiness of mass armies, along with the complexities and greatly enlarged scale of modern operations, rendered the contemporary military instrument cruder in comparison with that of Napoleon's era. The latter's army was a rapier, while the modern army was a bludgeon. Thus the concept of evolution gained grudging acceptance, even if it brought degradation. Martynov also saw the necessity for commanders who embodied an understanding of the theoretical and practical sides of war. Coupled with increased professionalism, he perceived the need for a common understanding of method and intent to enable commanders and their subordinates to accomplish common objectives with mass armies spread across immense theaters of operations. Still, he visualized future war primarily in Napoleonic terms, with operations unfolding by several stages within theater, leading finally to "the grandiose general engagement in which over a front of 50–60 versts [would] collide and fight for several days mass, million-man armies." These the high commands on the battlefield would direct "the same way as in a theater of war, that is, by means only of directives."[23]

Martynov was also probably the anonymous author of a series of articles, "Thoughts about the Technique of Future Wars," appearing in the *Military Collection* during 1893. These articles were significant for two reasons: the degree to which they reflected sui generis the acceptance of "future war" as a legitimate analytical category and the degree to which they continued to reflect persistent Napoleonic influences. Although earlier Russian authors treated separate aspects of technological and organizational change, this author strove to weave the parts into a coherent series of projections about the nature of future war within an entire theater. He saw that the new technologies and huge field armies would pose a series of challenges ranging from expanded reconnaissance requirements through improved training to the need for increased supply capacities. He understood Moltke's dictum about marching separately and fighting together, then interpreted it to mean that a successful commander must achieve "the greatest possible advantage in means, forces, and time." The solution to the challenge lay in seeking incremental solutions to problems which altered—but did not radically change—the accepted paradigm of strategy as the conduct of theater-level war. Thus for this anonymous writer, strategy guided military actions within a theater of war, and there was little need to look outside theater for new linkages or for original answers to pressing military requirements.[24]

New Currents

Even as the Napoleonic paradigm reached the ascendancy of its influence, Leer and the heirs of Jomini came under increasing fire, not only from particularists dressed in historian's clothing but also at the hands of the occasional amateur. As early as 1870, Leer's views on scientifically based strategy were pummeled in the Russian scholarly press. By the 1890s, his work was regularly subjected to rigorous questioning, and ironically, several of his more influential detractors traced either their support or their intellectual roots to the same flowering of military historical study of which Leer himself had always been so supportive.

Not everyone agreed with Leer's impulse to delimit either by approach, definition, or geography, and one of the gravest challenges came from thinkers who actively confronted convention by crossing disciplinary lines to ponder the relationship between politics and war. In 1892, Jan S. Bloch, a Warsaw banker and amateur student of war, embarked on a pioneering study of the relationship between a nation's social and economic infrastructure and its ability to conduct war. Unlike the adherents of *sokrushenie*, who accepted the wars of 1870–71 and, to a lesser extent, 1877–78, as models of future lightning war, Bloch envisioned future wars as costly, drawn-out contests which would eventually lead to the exhaustion of the combatants. Thus emerged in embryonic form a complex vision of linkages between fighting front and civilian rear which, with subsequent elaboration, came to support a strategy of exhaustion (in Russian, *izmor*; in German, *Ermattungsstrategie*) as an alternative to a strategy of annihilation. Abetted by A. K. Puzyrevskii, military historian and Chief of Staff of the Warsaw Military District, Bloch's opus by 1898 had blossomed into a five-volume compendium published in Russian in St. Petersburg.[25]

Bloch's ideas found only a few sympathetic listeners in the imperial capital. One of them, A. P. Agapeev, openly criticized military writers who persisted in treating military matters "as something closed [and] isolated, not having a direct connection with overall state institutions and independent of the spirit of the times and the political life of society in its entirety." Lieutenant Colonel A. A. Gulevich, another Bloch partisan and an instructor at the Academy of the General Staff, carried the argument further, asserting that "the final outcome of decisive war will depend not only on the perfection of the instrument of struggle and the art of its use [but also] on the vitality [*zhiznedeiatel'nost'*] of the state structure in general, on its ability to withstand protracted struggle and during its course to maintain a sufficiently strong and powerful armed force." Gulevich maintained that the advent of mass cadre and reserve armies meant that future war would be decided not by main blows on the field of engagement but by persistent and protracted armed struggle. Therefore the decisive element in modern war was the strength of the state's socioeconomic infrastructure, that is, the foundation of the state's ability to wage protracted conflict.[26] In a theme to which others would return, Gulevich further maintained that Russia's apparent economic backwardness was actually a strength, since the dislocations of protracted war would have far less effect on an agrarian society than on a more industrialized society. Gulevich did warn, however, that Russia's underdeveloped armaments industry and rail net would pose serious difficulties in any future European war.[27]

The Emergence of Military History

A. K. Puzyrevskii's direct and indirect participation as a historian in the intellectual ferment of the 1890s revealed the degree to which institutionally sponsored military history owed much of its origins and initial successes to the preoccupations of the period. Despite the absence of consistent official sanction, the assumption on the part of generalists and particularists alike was that history alone in all its richness offered sufficient evidence for the discerning student to discover the underlying patterns and rhythms inherent in any body of knowledge with sufficient coherence to become the foundation for a military science. Not surprisingly, conflicting opinions over the relevance of unchanging law and changing circumstance dominated the development of Russian military historiography throughout the 1890s and early years of the twentieth century. In addition to giving rise to "academic" and "Russian" schools of Russian military thought and history, debate about the nature and meaning of military history sparked wide-ranging and original research, the results of which can be read with profit even in the late twentieth century.[28]

Historical studies served as an important departure for criticism of Leer in part because of the very nature of the evolving discipline. In an age when scholars sought to unlock the recurring secrets of military art, they turned naturally to the history of warfare as the sum of military experience and wisdom, both of which might find employment in the fashioning of a true military science. Leer's successor as Commandant of the Nicholas Academy of the General Staff, General N. N. Sukhotin, summed up prevailing sentiment for military historical studies when he wrote that "the discovery of general outlines in the actions of the great captains and in their military art during various epochs constitutes the main goal of military historical science, with the result that it becomes the useful and valued foundation for other branches of military science."[29]

For military historians—then as now—the issue was how far the observer might press generalizations based on individual historical facts. The more cautious observers were extremely circumspect in judging the degree to which historical discovery might be advanced in the name of transcendental conclusions. Puzyrevskii, who attacked Leer's selective use of historical facts, wrote in 1896 that "in the history of military art there is no epoch which does not have scientific significance; the times of the slightest decline in military art are as instructive as the times for the broadest development of all elements, and only by studying both is it possible to clarify the causes favoring the advancement of military art."[30]

In the hands of Puzyrevskii and a talented group of military historians, including D. F. Maslovskii, N. F. Dubrovin, A. Z. Myshlaevskii, and A. N. Petrov, military history became an important critical instrument used to chastise Leer and his disciples for their preoccupation with an "academic" approach which yielded only dry formulae and meaningless generalizations. Although arguments were cast in scholarly terms, one of the main issues at stake was Russian national pride. Fixation on Napoleon and his interpreters had naturally led to a neglect of the Russian military tradition, and in a series of dazzling studies dating to the 1890s and early 1900s, the emerging "Russian school" strived mightily to set the historical record straight. They endeavored to demonstrate not only that Russian military art did not lag behind the West's but also that in some respects it even outstripped Western achievement.[31] Petrov even went so far as to assert that Suvorov and not Moltke must remain the Russian model.[32]

Their work had its more and less productive sides. It placed Russian military development within the context of Russian history. It also reawakened awareness of the achievements of Peter the Great, P. A. Rumiantsev, A. V. Suvorov, and other great captains of Russian history. At the hands of the Russian national school, Suvorov once again emerged as Napoleon's coequal. Dubrovin and others produced documentary studies and monographs detailing the campaigns and achievements especially of Suvorov. This trend resurrected the kind of study initiated a half-century earlier by Miliutin and culminated in 1900 with the landmark publication by the Academy of the General Staff of *Suvorov in the Studies of the Professors of the Nicholas Academy*. From that time, regardless of persuasion, military texts would highlight the maxims of Suvorov alongside those of Napoleon.[33]

In another of the ironies of the period, the studies of the Russian school led to the same kind of military dead end for which the academics were justly criticized. The findings of neither school lent themselves to systematic application. The national school emphasized particularism even as it underscored the importance of Russian military art. The issue remained one of discerning how an understanding of Russian military art was relevant to battlefield application. Russian military history provided useful context, but even a good history text was of little direct use to the officer likely to be engaged in battle. Nor did history always communicate well with strategy, a fact represented by the existence of separate chairs of strategy and military art at the Nicholas Academy. Nonetheless, by way of limited tactical studies, history would eventually make a valuable contribution to the development of tactical theory. The fledgling discipline of military history was simply asked to bear more weight than its young back could bear.

The Young N. P. Mikhnevich

Within the larger picture of seemingly chaotic intellectual diversity, one of the brighter spots was the emergence at the very end of the nineteenth century of N. P. Mikhnevich as a serious synthesizer of Russian military thought. By the late 1890s, he had already made his mark as a military analyst and historian, having completed several article-length studies during the previous decade on cavalry and partisan operations, then having acceded in 1892 to the Chair of the History of Russian Military Art at the Staff Academy. There followed in rapid succession two groundbreaking monographs, *The Significance of the German-French War of 1870–1871* (in Russian, 1892) and *The Influence of the Most Recent Technological Inventions on Troop Tactics* (in Russian, 1893). Like Mikhnevich's encyclopedia entry for "operations," these works revealed Leer's persistent influence with introductory assertions that wars were eternal and that the laws and principles of the art of war were in essence "Napoleonic." But there was also a glimmer of something new, something which was already beginning to affect many of Leer's disciples, including Martynov. This was the understanding, as Ageev later wrote, that "the phenomena to which war relates and with which it must reckon are subjected to constant change" and that "almost every epoch has its own military art, distinct from others."[34]

In the first book, after reviewing prevailing conceptions of mobilization, concentration, strategic deployments, the rear, and strategic direction, Mikhnevich examined the dependence of war upon socioeconomic factors. Contemporary wars, he felt, "ought to begin with serious historical goals in order to assure the high morale of the people and troops." From this perspective he viewed both the wars of German unification and Napoleon's invasion of Russia. In Mikhnevich's opinion, victory in war depended upon having the ascendancy in "social advantages," including superiority in numbers, armaments, economy, intellect, and morale. Mikhnevich concluded that "the application of steam, electricity, the significant improvement of fire weapons, the new system of recruitment, the new organization of armies and similar significant factors were bound to influence many questions in the spheres of strategy and tactics."[35] He continued to develop the latter theme in his second book, holding that the new technologies were changing tactics, drawing out the course of battle and enlarging its framework, while increasing the significance of fire weaponry, the cooperation of the combat arms, effective terrain use, and the application of engineering assets.[36]

Although Mikhnevich posed serious questions, he refused to throw out Leer's scientific baby with the muddied theoretical bathwater. Thus in an 1899 presentation to the officers of the garrison and fleet of St. Petersburg, Mikhnevich argued for the necessity of a well-founded military science while calling for a timely review of its focus, essence, and content, including especially the relationship of military science to other sciences. He placed military science with the social sciences and emphasized that its object was a study of "the laws of victory," the principles of military art, and the means of applying them to the concrete conditions of reality. In contrast with Leer's fixation on philosophical idealism, Mikhnevich emphasized the material foundations of military science, holding that laws and principles represented "broad empirical generalizations proceeding from a multiplicity of facts" and retaining conditional significance, but he was not quite willing to divorce them completely from Leer's emphasis on unchanging principles. Still, in contrast with Leer, who saw strategy as the essence of military science, Mikhnevich saw the latter as the philosophy of military affairs, linking it closely with the theory of military art. He emphasized the necessity for such a vision of military science as a science for application which would direct military thought to make correct decisions.[37]

Rather than negate the importance of separate strands, he attempted to draw them together in a new text on strategy that treated the main elements of war and the components of military art. He borrowed from scientific positivism to describe war as a social phenomenon, then attempted to demonstrate linkages between economic structures and resources and the conduct of modern war. Like Martynov before him, he incorporated into his work an understanding of the importance of evolution, but in a way that highlighted rather than disparaged change. He characterized strategic operations according to three types: the frontal attack, the breakthrough, and the defense. He expanded his understanding of strategy to join front and rear by attempting to demonstrate how the nature of Russia's social and economic infrastructure related to the state's capacity to wage war. For Mikhnevich, as for Gulevich, Russia's agrarian nature counted as a plus in any future war because it was less subject to severe dislocation in the event of protracted conflict. This perception opened the possibility to undertake prolonged defensive operations as a prelude to deliberate mobilization in an effort to regain the initiative and undertake decisive offensive operations.[38]

Although Mikhnevich represented an advance over Leer, his text appeared too late to make an appreciable difference on Russian conduct of the Russo-Japanese War. Mikhnevich's *Strategy* would ultimately go through three editions, with updating, correction, and additions to account for changes after 1899. His thought became more influential after 1904, when he served briefly as Commandant of the Nicholas Academy (until 1907), then went on to division and corps command (1907–11), and finally becoming Chief of the Main Staff (1911–17), all the while continuing to lecture on strategy. Mikhnevich's book remained the primary text on strategy not only for the late Imperial Russian Army but also for the Red Army until 1926, when it was supplanted by a text of the same name written by A. A. Svechin.[39]

Tactical Theory

It was against this background of intellectual ferment that tactical thought and doctrine evolved. In addition to some of the philosophical issues just discussed, two separate but related trends heavily influenced tactical thought and practice. The first was speculation over the impact of technological change on future battle. In both Russia and western Europe throughout the 1880s and 1890s new weapons employing smokeless powder were being subjected to rigid field trials, and observers rushed to draw conclusions for the future from a variety of sources, ranging from extrapolative commentary to empirical data drawn from the trials themselves. Illustrative of the former was an anonymous writer for the *Military Collection* who, in 1891, noted that "the zone of lethality will be increased at long range by a factor of more than two" and that "one of the most probable changes in the character of future engagements will have to do with the greater length and the more careful preparation for battle by long-range rifle fire." Further, he asserted that "all the more often will occur instances in which the bullet at close range will play the role of the bayonet."[40] In 1892, K. I. Druzhinin professed to see that smokeless powder weaponry would lead to future battles consisting of separate attacks by discrete parts of the armed force in a way that would presage the changing character of battle itself.[41]

Attempts to define the character of battle, together with its preparation and conduct, formed a significant theme for Russian military writers in the 1890s. In this respect theorists profited substantially from the explosion of interest in military history which had accompanied the emergence of the academic and Russian schools of military historiography. A. M. Zaionchkovskii revealed early promise for a distinguished career with the publication in 1893 of his *Nastupatel'nyi boi* (Offensive Battle), a pioneering comparative analysis of tactical ground combat. He painstakingly reconstructed the actions at Third Plevna, Lovech, and Sheinovo in an effort to determine recurring themes which might explain the success or failure of the Russian army in tactical combat. His work and that of A. K. Puzyrevskii (*Ten Years Ago*) illuminated the tactical past in the more certain light of rigorous historical analysis and provided ammunition for those who sought in the spirit of positivism to determine the underlying rationale and laws of battle.

By the turn of the century, tactical theorists were beginning to conclude that military history, or at least that part of military history narrowly conceived as combat experience, possessed utilitarian value. Writers paid increasing attention to various kinds of vicarious tactical experience, including the Boer War and Russia's own combat history. By 1901, Major General N. A. Orlov could write that "familiarity with the true character of battle can be attained in peacetime only by way of an attentive study of *military-historical examples*."[42] Although battles did not easily give up their underlying rationale, an emphasis on historically based battle analysis had long-term effects because it taught tacticians to divide battle into separate stages (two or three, depending upon distance and circumstances), including the approach (*sblizhenie*), the attack (*nastuplenie*), and (as appropriate) the assault (*ataka*).[43] These and other conclusions gradually found their way into the tactical literature and regulations of the period.

A second and formidable piece of the mosaic making up tactical development was the heavy tile of precedent. The two most popular tactical handbooks of the period were those of M. I. Dragomirov and Colonel (later General) K. N. Durop. Durop was an instructor of tactics at the Nicholas Academy, and his *Uchebnik taktiki* (Tactical Textbook) was published in ten editions between 1881 and 1901, during which time it became the official text of the military and cadet academies.[44] Both it and another widely read text, P. K. Gudim-Levkovich's *Kurs elementarnoi taktiki* (Course in Elementary Tactics), were openly based on Dragomirov's classical work. Dragomirov's own text remained a fixture of the interwar period, and although he revised it in 1881 to take into account many of the lessons of the Turkish War, his major tactical premises remained substantially unchanged throughout the period.[45]

Few commentators understood the dangers of ossification inherent in elevating the works of Dragomirov and his disciples to the status of military classics. Still fewer saw that improved answers to tactical quandaries lay close at hand, waiting for original thinkers to grasp them. One such thinker was General I. P. Maslov, who early raised his voice against the chorus of convention but with little initial effect. In 1888, writing in much the same vein as Baron L. L. Zeddeler before 1877–78, Maslov lamented that "for the good of the cause it is necessary finally to turn away from a view of the skirmish line as an accessory or a tentacle of the closed order—an understanding assimilated from the era of Napoleon; it is time to view it as the main combat force, and [view] closed formations—as the reserve of this force." Unlike Dragomirov, who insisted on seeking tactical decision through physical mass and impulsion, Maslov would choose to rely on troops in open formation to visit destruction on the enemy chiefly by fire. "With the present power of well-aimed rapid rifle fire, the latter [fire] has attained incomparably greater destructive force than the hitting power [*natisk*] of closed order infantry or cavalry."[46] For Maslov, the purpose of reserves was either to reinforce the open order or to exploit success along the main axis of attack.[47]

Interesting as Maslov's conclusions were, his method was perhaps more significant, at least over the long term. In 1887, thanks to the results of extensive field testing, he was able to publish a comparative analysis of the effects of black powder and smokeless powder rifle fire against targets at various distances. Tests revealed that reduced-caliber, smokeless powder weapons demonstrated themselves conclusively superior in accuracy, flatness of trajectory and rapidity of fire.[48] A year later, drawing from these data and citing the importance of combining them with conclusions based on combat experience, Maslov painstakingly argued the case against reliance on anything but open-order tactics against troops equipped with smokeless powder weapons. Rather than base his projections on either intuition, rule-of-thumb extrapolations, or limited individual experience, he consciously chose to employ analytical methods relying on probability theory, an empirical data base, and the insights of combat experience as demonstrated in solidly researched military historical studies.⁴⁹ Thus in embryonic form appeared a serious cross-disciplinary attempt to peer into an uncertain future with the more certain assistance of mathematical, empirical, and historical methods. Unfortunately for the Russian infantry, Maslov's approach would have little immediate effect.

Infantry Tactical Regulations

In some respects official doctrine, even when heavily influenced both directly and indirectly by Dragomirov, at times seemed to outrace the collective wisdom of conventional textbooks. In general, changing doctrine attempted to account for the increased density of fire by extending unit frontage and depth, by attempting to use cover more effectively, and by limiting as much as possible the exposure of attacking infantry to hostile rifle and artillery fire. In short, despite the picture of stagnation drawn by many commentators, especially those in the West, the Russians did actively attempt to put into effect the more palpable lessons they learned during the course of the Russo-Turkish War. But the limits of creativity within a large organization, the drag of tradition and inertia, and real or imagined problems of control and coordination all combined to prevent sweeping adjustments in tactical thought and application. Between 1881 and 1904, the Russians adjusted their tactical doctrine three times. In retrospect, it would not be an exaggeration to assert that the imperfect process of reflection, observation, ferment, and speculation produced even more change among the Russians than among their compatriots in other European armies. The problem was that not even these adjustments were sufficient to accommodate parallel and far-reaching changes in technology and organization. Worse, whatever adjustments the Russians did make were overshadowed by their dramatic adherence to the cult of the bayonet in close combat.

The first change came in 1881. Already at the beginning of 1879, the Committee on the Structure and Training of Troops had decided that any new tactical regulation had to resolve two major issues: by what formations, movements, and fires the most injury could be inflicted upon the enemy and what steps should be taken to reduce losses from enemy fire, especially in the case of friendly troops in the process of attacking an entrenched enemy.⁵⁰ Although the war had reaffirmed the efficacy of the infantry assault, the evidence of Plevna and Gorni Dubnik was too conclusive to ignore: attacks required much greater preparation and more careful control and coordination than theoreticians had earlier judged the case. To reduce losses, attack formations clearly required a different balance between the open and closed orders. At the same time, there was sufficient evidence to reaffirm the importance of the bayonet in close combat. But the attacker had to be more circumspect. Or, as Mikhnevich later wrote, "Before as now, the bullet and the bayonet will act in unison, although at present the bullet has shown itself to be far from a 'fool,' while the bayonet is a 'fine lad' only under certain [*izvestnye*] circumstances."⁵¹ The new regulations of 1881 would reveal, inter alia, what those circumstances were.

The regulation reflected the committee's consensus that all dense formations should be abandoned and that the basic attack formation should be endowed with greater depth, both by making intervals between lines greater and by increasing the number of lines. Therefore the regulation recommended that infantry deploy for the attack in skirmish formation "while keeping in mind the desirable power of fire and the requirement for a reserve."[52] Further, the regulation stipulated that the skirmish line and its reinforcements should be furnished by the tactical unit as a whole rather than according to the previous system, in which a single company of sharpshooters served as skirmishers.

Because of the increased lethality of rifled weapons, soldiers either in skirmish or line formation were to avoid presenting a mass target, and if the opportunity presented itself, they were to take cover, either by entrenching or by utilizing the terrain. Although emphasis on open formations resolved the perceived difficulty of firing from closed formation, soldiers now needed training to shoot accurately at longer ranges (800–900 paces maximum). The new regulation also recognized the necessity of individual, aimed fire from stationary positions but reaffirmed the primacy of volley fire, whenever practical. Lest fire discipline be lost, the regulation stressed that "every place when it is possible and appropriate fire by volley preferably ought to be employed."[53]

The regulation of 1881 divided the infantry attack into two phases, the approach and the attack. The approach encompassed troop movements and deployments made at distances from the enemy greater than 800 paces, which the Committee on the Structure and Training of Troops evidently believed to be the range of accurate aimed rifle fire. Upon reaching this distance from the enemy, battalion and company commanders reviewed the terrain and the situation, then made their dispositions for the attack. The combat formation (*boevoi poriadok*) for a company consisted of a skirmish line and a reserve, the latter remaining in close (but not dense) formation. A battalion was ordinarily drawn up in two lines of two companies each. Each company in the first line threw out skirmishers and formed its own company reserve, while the two companies of the second line formed the battalion reserve. Under ordinary circumstances, a company at full strength would occupy a frontage of 250 paces and a battalion (with a two-company front), about 500 paces.[54]

After deployment into skirmish line and reserve, troops entered the first zone of the attack, which included terrain within 200 to 800 paces of the enemy. Within this zone the infantry skirmish line moved forward in open formation, taking cover wherever and whenever possible and advancing by leaps and bounds, preferably 100 to 150 paces at a time. While advancing, the skirmishers aided their own attack by delivering aimed rifle fire on the objective. At the commencement of the attack, the reserve was to assume line formation about 500 paces to the rear of the skirmishers, with the distance between the two gradually closing as skirmishers neared the enemy position and were forced with increasing frequency to go to ground. Meanwhile, company reserves made good losses among the skirmishers and supported the advance either directly by fire or by making a flank attack.[55]

Within about 200 paces of the enemy, advancing infantry entered the second zone, or assault segment, of the attack, during which the reserve moved closer to pour additional fire on the enemy. If the enemy's fire was observed to grow weaker, the skirmishers alone might rush forward into a bayonet assault, followed by the reserve. If, however, the enemy's fire had not slackened, the skirmish line would maintain its fire until the reserve came up. The reserve then rose to advance in line at quick step, and when it reached the skirmishers, the "assault" was sounded. At this time, all troops got to their feet and raced forward to deliver a headlong bayonet attack. The preferable method for assault with cold steel was an uninterrupted and rapid rush to close with the enemy. "The slightest hesitation, indecision, or even worse, interruption, can cause huge losses and have disastrous consequences," the regulation warned.[56] A successful attack was followed at least with pursuit by fire. While the regulation dwelled at some length on the frontal assault, it also emphasized the importance of flank attacks and flanking movements.

If attackers enjoyed limited benefit from the new emphasis on fire, so also did defenders. The regulation of 1881 advised troops defending a position "to extract from fire action the greatest advantage, to shock with fire the attacking force and then to meet stroke with stroke [of the bayonet]."[57] If possible, defenders were to occupy trenches dug in a manner to afford a double line of fire. That is, earth from trenches was thrown to the rear, where it became a low parapet behind which a second rank was stationed, while the first occupied the trench in front. Interestingly, infantry, even in the open, was now considered sufficiently strong to withstand cavalry charges without assuming square formation for all-around defense. Once an assault of any kind was turned back, the defenders prepared immediately either to meet the next attack or to deliver one of their own.[58]

In retrospect, the regulation of 1881 incorporated at least two significant advances over previous tactical prescriptions. The first was the new emphasis on fire, which in effect elevated rifle fire to complementary status with shock action. This emphasis also found reflection in the regulation's more circumspect approach to the assault against fortified positions. The second advance lay in an emphasis on the unified action of infantry. No longer were foot soldiers functionally divided into sharpshooters and line infantry. Now all infantrymen were expected to employ at least some of the tactics and techniques heretofore reserved to the sharpshooters.[59]

On the negative side, the regulation of 1881 incorporated the worst elements of incrementalism that General Maslov would rail against in the late 1880s. Even after painstaking field tests in the early 1880s had conclusively demonstrated the overwhelming lethality of aimed rifle fire in combat, writers of doctrine failed to grasp the reform nettle to undertake truly meaningful change. Departure from closed formations diminished control in battle, and wholesale adoption of the open order seemed to magnify not only this difficulty but also that of resolving the simultaneously contradictory requirements for mass and dispersion.[60]

Several supplementary regulations appearing during the 1880s and 1890s dealt with older issues which began to receive timely attention, thanks to the lessons of the Russo-Turkish War. The first of these involved sapper training for infantry units to prepare them more fully to meet additional combat requirements, including the construction of field fortifications. Every soldier now carried his own entrenching instrument, and entire units had to be trained to employ them to the greatest advantage both in the attack and the defense. A second point of emphasis was the night attack. The successful action at Kars had taught the Russians the advantages of well-prepared night attacks as a way of permitting the infantry to operate in an environment which by day often spelled doom to the attacker charging across open spaces.⁶¹ How well the Russians would exploit this knowledge remained another matter.

By 1895, the adoption of smokeless powder weapons caused a special commission chaired by General N. N. Sukhotin to revise the infantry regulation of 1881. However, with the exception of minor alterations, Sukhotin's draft regulation—even as republished in 1897—repeated many of the basic articles of the 1881 regulation. Those differences which made their way into print appear to have stemmed from perceptions of the extended ranges and lethality of smokeless powder rifles and field artillery. Thus, while the infantry attack unfolded in two stages (the attack and the assault) as before, the infantry draft regulation of 1895 prescribed the adoption of basic tactical formations at greater distances from the enemy. In open terrain against an artillery-equipped enemy, units began deployment from march formation to tactical formation at ranges close to 5,000 meters. At 2,000 paces, the extreme distance of long-range rifle fire, the lead infantry companies assumed open-order formation and continued the advance. Between 1,000 and 1,400 paces from the enemy, battalion and company commanders made their final dispositions and pressed forward with the attack.⁶²

At this stage, the infantry attack proceeded according to the precedent laid down in 1881, with the exception that the regulation of 1895 accorded still greater importance to supporting rifle fire. If possible, companies and battalions were to form special fire support groups, the purpose of which was to lay down a fire base to facilitate the continued advance of skirmishers and reserves. These groups, termed "fire batteries," were to maintain a steady suppressive fire on the enemy until the friendly advance masked their weapons. At this point, the reserves closed with the moving fire line of skirmishers to deliver the final assault with fixed bayonet.⁶³

By century's end, however, not all officers agreed with the soundness of Sukhotin's admittedly superficial attempts at revision, and the accession to office of War Minister Kuropatkin apparently reinforced the views of those who saw the need for basic change in the infantry regulation. In consequence, Kuropatkin appointed a new commission under the chairmanship of the aging General Dragomirov. It was the regulation fashioned by this commission which governed the employment of Russian ground troops in the Far East during the Russo-Japanese War. Dragomirov's regulation was published in 1900, amended in 1902, and incorporated into the field service regulation published in 1904 on the eve of hostilities with Japan.⁶⁴

The field service regulation of 1904 assumed that the new quick-firing artillery would have its greatest effect on targets in the open, especially infantry advancing without cover in march formation. Therefore the regulation required that infantry coming within range of enemy artillery (2.5 to 3 kilometers distant) deploy into combat formation. Deployment was to be covered and supported by friendly artillery which opened suppressive counterfire on the enemy's batteries. When the division commander issued orders for the attack, the commanders of infantry regiments deployed into combat formation and designated objectives for their subordinate battalion commanders. Their task was to seize assigned objectives with the smallest possible losses while traversing the zone of effective artillery fire (three to two

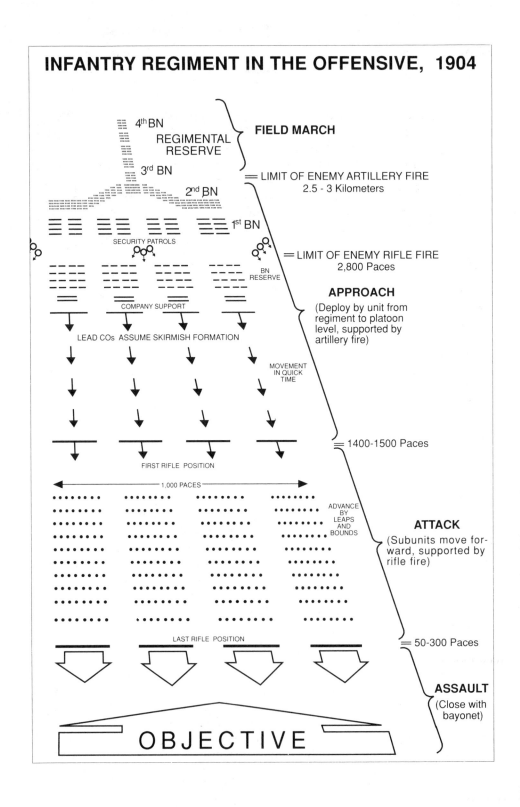

kilometers from the objective) and the zone of rifle fire (2,800 to 300 meters from the objective). [65]

The field regulation visualized offensive battle as consisting of two phases delineated by distance from the objective. As before, the first was characterized as the attack, the second as the assault. The attack had as its objective the bringing of the infantry into close proximity to the enemy in preparation for the second stage, the bayonet assault. In the attack, one battalion in the regimental battle formation was given a direction of advance and the other battalions coordinated their movement with the first. Infantry in the zone of long-range rifle fire (2,500 to 2,000 meters from the enemy) moved forward in close formation, their flanks and front covered by skirmishers whose task it was to protect the main body at distances of 300 to 500 paces from it. [66]

When attacking infantry approached the enemy closer than 2,000 meters, the forward companies deployed into combat formation and continued the advance at quick march. Only the lead companies of the battalion combat formation deployed into an open formation of battle elements (*boevye chasti*); those designated as the general reserve (*obshchii rezerv*) remained in close order. The purpose of the latter was both to replace casualties in the former and to provide mass for the bayonet assault. [67] The first battle elements occupied positions only as they approached within 1,400 paces of the enemy. Further movement forward in loose skirmish order was to occur in leaps and bounds as the battle elements occupied successive positions. The regulation of 1900 prescribed forward movement by platoons, then by squads as attacking infantry drew closer to the enemy. Depending upon the situation and the circumstances, approximately one-half of the squads or platoons remained in position to support the remainder's next rush to forward positions. Once the lead battle element was in position, it covered by fire the advance of subsequent elements. Leaps and bounds were to cover about one hundred paces at the most, depending upon circumstance and terrain. Under conditions of heavy enemy fire, successive battle element advances might be reduced in size to squads and, on occasion, even individuals. [68]

The final battle subunit position was located not farther than 300 paces from the objective. If circumstances permitted, it might be closer, but in any case not farther than the distance that troops might cover in a single sprint. Once the final position was occupied, the attackers raised their level of fire to the highest rate possible for maximum suppressive effect. Meanwhile, the general reserve closed with the battle elements to lend its weight to the attack. When in the opinion of infantry commanders enemy fire had slackened sufficiently to assure further success, they ordered the entire combat formation into the assault with cold steel. During the course of the assault both battle elements and the reserve were to form closely around their commanders in order to press forward with sufficient mass to break the enemy position. [69]

Leading Soviet commentators have condemned this regulation for being hopelessly out of step with technological developments, but it is impossible not to compare it favorably both with its predecessors and with contemporary European tactical doctrine. True, Dragomirov and his writers underestimated the effect of smokeless powder technology, but so did nearly everyone else. However, the Russian infantry tactical regulation of 1900 did represent a genuine improvement over that of 1881 and was perhaps the most advanced regulation for its time.[70] While Dragomirov has been condemned repeatedly for adherence to the outmoded cult of the bayonet, the Russian army was not alone: no other European army expected to carry a position offensively except with a final resort to cold steel. For the Russians, if progress could be measured in distance, then over the span of thirty years the bayonet assault had been reduced from 1,000 paces to 300. This, together with the reasoning implicit in terrain utilization and division of offensive battle into discrete stages, amounted to progress—albeit only incremental—in the tactical prescriptions and practices of the period.

On a more clearly positive note, Dragomirov's regulation paid more than lip service to defense. It provided for defending troops to occupy not only trenches but also, anachronistically, lunettes and redoubts. Before their positions, defenders were to construct an outpost line and lay whatever obstacles were appropriate to the situation. Great attention was devoted to obscuring the true location of the defensive position by cleverly masking it or blending it in with the terrain. As in the offense, troops were divided into the battle subunits and the reserve.[71]

On balance, the same problems dogged infantry tactical doctrine in 1900 that General Maslov had found so troubling a dozen years before. Even in the open order of 1900, the prescribed interval between attackers, one-and-a-half paces, was too little to reduce appreciable losses to massed artillery and rifle fire. Regulations and textbooks devoted little attention to the meeting engagement brought about by a collision of forces on the move. Battle itself might open with the majority of forces committed to the reserve, while battle elements probed enemy defenses to find weak spots. If enemy fire proved especially strong, by the time that the battle elements occupied their final skirmish position, as much as three-fourths of the entire combat formation's total strength might be devoted to reinforcing the battle elements, leaving only one-fourth as a general reserve to provide mass for the final bayonet assault. Meanwhile, during the opening phase of the attack, when the battle element likely remained smaller than the general reserve, the leading edge of the attack was deprived of the fire of the majority of the entire combat formation's rifles. Worse, once the whole combat formation became heavily engaged, there was no sure way in the confusion of close battle to assure that small-unit commanders would retain control of their troops.[72]

Not everyone was blind to the problems inherent in the application of new technology to modern battle. Even as field regulations came to represent the triumph of Dragomirov-style incrementalism, one anonymous Russian officer wrote in 1903 that the draft field service regulation "did not in full measure answer the contemporary requirements of battle." He found not only that the draft regulation "insufficiently emphasized the significance of fire" but also that it "did not turn sufficient attention to the extraordinary difficulty of frontal attacks." Consequently, the draft said little about shallow and deep flanking attacks, even though "one would encounter them very often."[73] However well informed this commentary was, it stood little chance of gaining sympathetic hearing under circumstances in which the Tsar himself confessed fear of Dragomirov's military preeminence.

Artillery Tactics

Another shortcoming in the field regulation of 1904 was failure to devote sufficient attention to cooperation in battle among the various combat arms. The regulation glibly noted the importance of combined arms warfare but suggested few concrete actions to bring about more effective interaction among infantry, artillery, and cavalry. Consequently, many of the same problems which had plagued the Russians in the 1870s were to reappear in the first years of the twentieth century. And the chief causes remained the same: the rapid pace of technological change simply outran the ability of artillerymen and cavalrymen to cooperate with infantry in adjusting doctrine and training to new and often surprising tactical realities.

Except for a renewed emphasis in the 1880s on officer training, tactical thought about artillery seems to have lapsed into a torpor from which it recovered only in the 1890s, when the implications of change became too blatant to ignore. One of the few exceptions to the trend was A. Baumgarten, who, already in 1887, penned a series of articles for the *Military Collection* in which he insisted that "in future wars engagements must be initiated by artillery." Further, he asserted that "by the success of the opening artillery struggle will be decided the further course of a battle." Until the attainment of decisive results in the artillery battle, Baumgarten felt that commanders should not even begin an infantry attack.[74] These views enjoyed little popularity with either infantrymen or artillerymen.

One problem was that artillerymen needed to be convinced that smokeless powder weaponry posed an even greater challenge to them than to infantrymen. Under conditions of radically changing technology, artillerymen had to determine methods of employing weapons with greater range, accuracy, and destructive power, while they themselves were becoming more vulnerable to that same power. In effect, new technology was forcing them to find new solutions to old problems of emplacement, displacement, fire control, observation, and massing of fires. Whatever solutions these problems found, one thing was certain, according to Baumgarten: "It is possible to say that with the new powder the occupation by artillery of open positions is tantamount to suicide."[75]

Central to the continued evolution of artillery tactics were problems of protection, fire control, and massing of fires. The obvious way to protect artillery was to position it so that terrain intervened between it and enemy observation and fire. New gunsights had to be devised and new methods worked out for orienting guns indirectly either by compass and map or by artificial aiming points (or a combination of the two). In addition, the emplacement of artillery in terrain-obscured positions required the development of techniques of indirect fire with remote observers, communications systems, and complicated fire direction procedures. In 1882, K. G. Guk, an instructor at the Michael Artillery Academy, wrote a book on indirect fire procedures, and in 1884 officers at the artillery school began experiments with indirect fire methods.⁷⁶ However, these early activities found little immediate application in the field.

Similarly, the prophets of a new artillery order experienced difficulty popularizing the conception of massed fires achieved through centralized control. Theoreticians recognized the value of massed fires, and as early as the field maneuvers of 1887, three batteries of the 40th Artillery Brigade massed fires on mock enemy targets. For a time, however, two conflicting requirements ran counter to the idea of mass achieved through greater centralized control. One was the conviction that new, more dispersed tactics required greater dispersion and decentralization in both infantry and artillery. The other was the practice of detaching guns from batteries to rush them forward to support the infantry in a direct fire role. Neither boded well for the future of centralized fire control.⁷⁷

These difficulties notwithstanding, thanks to the writings of Guk, Baumgarten, and others, Russian artillerists seem to have developed a firm idea of their role in battle. By the mid-1890s, they understood that it was artillery which generally initiated an engagement; likewise, they understood that artillery's role had grown since the 1870s, when it was considered only an auxiliary to the infantry attack. Thanks to increased ranges and the lethality of new weapons, artillery was now required not only to support the attack but also to cover the deployment of combat troops and their approach to the objective. To do this, artillery had to fulfill two tasks: it had to silence or suppress enemy artillery and, once the infantry had begun its advance, had to suppress the defenders' rifle fire. Observers and practitioners seem to have agreed on these basic requirements even if they experienced difficulty devising novel ways to accomplish them.⁷⁸

Attainment of fire superiority was possible only by means of massing fires, and the Russian penchant was not to look for new methods but to seek organizational solutions for this knotty problem. More meant bigger. Thus, during the 1890s, Russian artillerists created the artillery division (*divizion*), a tactical unit of two or three batteries within the artillery brigade. Its purpose was to mass fires physically under the direction of a single commander who bore responsibility for concentrating the fire of two or more batteries firing at the same target.⁷⁹

Against this background, a potential bright spot for the attack was the adoption of six-inch mortars in the mid-1880s. On the basis not only of the Plevna engagement but also of various experiments and field tests, the Russians determined the type of weapon, caliber, and explosive charge necessary for the destruction of field fortifications. However, the concept of mortars won adherents only grudgingly. As one writer observed, "the result did not come cheaply; it was necessary to sacrifice one of the most significant characteristics of field artillery—uniformity of weaponry."⁸⁰ Once the mortars appeared in service on their mobile carriages, opinion divided on where they should be located within field artillery organization, and high-angle fire weapons came to be viewed as orphans by many members of the artillery establishment.

A final innovation of promise was the development of a satisfactory nonoptical dial sight, called the *uglomer*, to facilitate indirect fire from conventional artillery. As introduced in 1902 and improved in 1905, this sight enabled artillery officers employing the new quick-firing three-inch field gun to fire readily and accurately from indirect lay. Although artillerymen were reluctant to employ the new technology in practice and field exercises, they had it at their disposal should circumstances force changes in habit.[81]

Cavalry Tactics

Theoreticians and writers of doctrine were even less united in their views on the modern uses of cavalry than they were on the employment of artillery. Although examples such as Gurko's forward detachment suggested that cavalry might utilize its mobility effectively as a strategic reconnaissance and strike force, resurgent conservative opinion stubbornly clung to cavalry's traditional role as battlefield shock weapon. Without a persistent unifying doctrine, cavalry development proceeded in two nearly opposite directions.

For a time, the Gurko precedent prevailed. In accordance with pre–1877–78 opinion, which had already begun to emphasize the importance of cavalry's strategic uses, cavalry organization was simplified and units multiplied, while troopers received training in both mounted and dismounted combat. The prophet of the new order remained N. N. Sukhotin, a serious student of cavalry in the American Civil War and a member of Gurko's staff in 1877, who at century's end would succeed Leer as Commandant of the Nicholas Academy.[82] Sukhotin and other theorists, including Mikhnevich, understood that as modern armies grew larger and more dependent upon reconnaissance, supply bases, and railroads, new opportunities for raiding and other disruptive activities presented themselves to the mounted arm. Armies and corps in the process of concentration, deployment, and march-maneuver to contact required eyes (reconnaissance) and protection (security and screening). Given the complexities of modern military organization, rear areas suddenly became even more inviting targets for cavalry marauders.[83] Mobile attacks on key rail nets could disrupt well-laid mobilization and deployment plans. As early as the 1870s, N. N. Obruchev had evidently believed that in the event of war with either Germany or Austria, Russia might buy time for mobilization by disrupting the enemy's rail movements at the outset of conflict.[84]

Throughout the 1880s, the Sukhotin-Obruchev school of cavalry as mounted infantry strike force held temporary sway, with the result that Russian cavalry tactics and deployments underwent significant change. Hussars and uhlans became dragoons, while lances were discarded in favor of rifles and bayonets. Morale plummeted as units were redeployed to the west and as officers and troopers traded shakos and colorful tunics for drab dragoon dress.[85] The cavalry regulation of 1884 for dismounted drill contained a whole chapter devoted to "individual training for action in the open order."[86] Meanwhile, a cavalry corps was formed in the Warsaw Military District to discharge strategic missions in the event of war in Europe.[87]

In view of institutional inertia, the degree to which the Russian cavalry truly assimilated these radical changes was questionable. Although dragoons received training in dismounted combat, cavalry regulations continued to stress mounted combat, especially against enemy cavalry formations. In the event that Russian cavalry confronted either infantry or artillery, every effort was to be made to attack on horseback from the flank or rear. Only slight concession was granted to new technology, with regulations of the early 1880s recommending that unsuccessful charges against infantry not be repeated more than three times "on account of the increasing confidence of the infantry and the ground being encumbered by fallen men and horses."[88] By 1890, a British reading of Russian cavalry regulations concluded that "dismounted tactics are only to be employed as an exception."[89]

Running counter to dragoon-style cavalry was a deeper, conservative strain of thought and habit which had never come to terms with proclamations for the death of classic cavalry in the mold of the Prussian Seydlitz and the French Ney. Indeed, throughout the 1880s and 1890s, apologists for the old order observed that the reason why neither the French nor the Germans had satisfactorily employed cavalry on the battlefield during 1870 was that neither side possessed adequate cavalry mounts. Reaction against the Sukhotin-Obruchev vision set in during the 1890s with the appointment of Grand Duke Nicholas Nikolaevich (the Younger) as Inspector General of Cavalry.[90] As one commentator on the Russian cavalry regulation of 1896 noted, the Grand Duke "has since been actively engaged in directing this arm in the true cavalry sense," with the result that "true cavalry action is again raised to the first rank, while dismounted action no longer has that exaggerated importance previously attached to it."[91]

On a more superficial level, the accession of the Grand Duke also meant that the annual reviews conducted at Krasnoe selo assumed the form of grand movements by large mounted masses conducted without voice command and only by signals given with sabers. One participant noted that these made a good impression, "but in a combat sense did not have any decisive significance." The Grand Duke was known as a pedant whose main concern was "appearance and order in formation" and for whom field exercises were secondary.[92]

Under the Nicholas Nikolaevich regime, cavalry regulations were rewritten to de-emphasize dismounted fire action and place stress on mastery of mounted skills and combat with the "arme blanche." Attack formation for cavalry consisted of two to four lines drawn up in either open or closed order. Attack upon cavalry was executed in close order with the utmost energy, while attack upon infantry usually called for the first line in open order followed by the second and third lines in close order. If possible, the first line began its attack between 400 and 800 paces of the enemy, and, also if possible, the attack was directed against the enemy's rear or flank. Lacking cover, cavalry deployed against infantry at a distance of 2,300 meters, crossed the intervening space at the "field gallop," then commenced the charge at 100 to 150 paces from the enemy.[93] The regulations devoted little ostensible attention to adjustments required in cavalry tactics once infantry in dispersed formation no longer presented a mass target for the classic cavalry charge. In his text on tactics as republished in the 1890s, K. N. Durop admitted that infantry in battle no longer occupied tight linear formations. Therefore he recommended that "the battle order of cavalry preparing to attack infantry should assume an irregular form."[94]

Under conditions of resurgent cavalry conservatism, two significant issues escaped the serious attention of theorists and writers of doctrine for the mounted arm: mounted scouts and the requirement for changing organizational focus to accommodate the evolving nature of larger field formations. As units occupied greater frontages and moved across wider areas, individual reconnaissance requirements exceeded the capacities of traditional cavalry formations. By 1891, the War Ministry responded to this challenge by creating four-man scout detachments for each regiment and separate battalion. Although theorists agreed that scout work required recruits with an array of special talents, few writers devoted attention to problems of linkages and employment, either with parent units or with conventional cavalry.[95]

While the appearance of mounted scouts offered partial solution to added reconnaissance requirements, cavalry-type missions multiplied on extended battlefields and within greatly enlarged theaters of operations. As early as 1893, one anonymous writer had argued that regimental scout detachments be expanded to eight officers and 200 men to cover all the gaps in employment.[96] As matters stood, although Russia was traditionally rich in cavalry, no combination of regulars, reserves, and Cossacks was sufficient to meet all tactical and strategic needs. Despite prevailing conservative views, doctrine called for both army-level (*armeiskaia*) and troop (*voiskovaia*) cavalry, with the former retained for strategic purposes and the latter for tactical purposes. However, the standard allocation of a single twenty-four-squadron cavalry division to each army corps was insufficient to serve either purpose to good effect. Consequently, despite the embarrassment of cavalry wealth, the Imperial Russian Army could not adequately meet expanding requirements for the mounted arm.[97]

Combined Arms

While cavalry continued to experience significant problems adjusting to the new tactical order decreed by changing technology, the growing importance of artillery caused many observers to turn their attention to the issue of combining arms for greater combat effectiveness. Already during the 1880s General N. I. Kokhanov proposed that even in peacetime artillery batteries should be subordinated to the commanders of infantry and cavalry divisions. His proposal aroused a storm of controversy in the pages of the *Military Collection* and the Russian army daily, *Russkii invalid* (The Russian Veteran), as irate artillery officers rejected the prospect of peacetime subordination to the officers of other combat arms. Other writers asserted that infantry and cavalry commanders might learn to employ artillery in wartime if they devoted themselves to peacetime study of artillery's characteristics, capabilities, and application.[98]

Despite intraservice animosity, the Russians turned increased attention to combined arms warfare in the 1890s. One writer, V. N. Sviatskii, asserted that "success in war is in the main achieved by the cooperative work of all three types of arms." And there was evidence to indicate that officers paid more than lip service to the concept of combined arms operations. Annual maneuvers provided the opportunity to work closely among the arms. In addition, at least a few officers attempted to work out techniques which would foster greater cooperation. A Captain Mukhin, for example, on the basis of experimental artillery data, worked out a method to enable artillery and infantry to train together under conditions of live artillery fire.[99]

The issue of combined arms cooperation and coordination raised the collateral issue of battlefield control. In the literature of the period, General Durop's text devoted the greatest attention to problem of controlling troops in battle. Among the Russians he was one of the first to describe how officers might permit the maximum development of initiative while in pursuit of the common objective. To this end he wrote that commanders must develop in their subordinates the capacity to understand correctly both the general context and the specific situation, to accurately convey to their subordinates in a timely manner a complete orientation on the general situation, and to assure uninterrupted communication among commanders, subordinates, and adjacent commanders.[100]

Exercises and Maneuvers

During the period 1881–1904, exercises and maneuvers reflected a diversity of approaches and preoccupations and provided important clues on how the Imperial Russian Army might perform in future combat. At their worst, exercises and maneuvers mirrored the throne's preoccupation with the external manifestations of militaristic fetishism. At their best, they afforded commanders and troops rare opportunities to develop in the field those skills and habits which could make important differences on future battlefields. The experience under Alexander III and Nicholas II demonstrated elements of the worst and the best.

One of D. A. Miliutin's last activities as War Minister was to devise a systematic annual program for troop training conducted under the supervision of senior commanders. As promulgated in 1881, Miliutin's scheme divided the training year into two periods, winter and summer, with each of these broken down into smaller increments. The summer cycle was itself divided into two large segments, with the first devoted to training at various levels from company to regiment and the second devoted to training the three combat arms in unison.[101] Thanks to emphasis on the latter, by 1894 the War Ministry could report that the number of troops participating in joint exercises had risen to more than 90 percent of the total complement of the three combat arms.[102]

The annual culmination of training involved either small or large maneuvers, depending upon the scale of forces committed to the effort. Maneuvers (*manevry*) remained strictly differentiated from exercises (*ucheniia*), with the latter limited in scale, size of formations involved, and tasks assigned. Commanders strictly controlled exercises, setting training objectives and prescribing methods of implementation. Exercises were usually small in scale and were conducted against simulated opposition, although on occasion they might involve genuine opposing forces. In contrast, maneuvers could involve large forces, were always two-sided in nature, and were conducted under circumstances permitting a maximum degree of free play. As combat experience receded into the past, maneuvers assumed greater importance, not only as devices for testing readiness and training but also as means of assessing the impact in the field of changing technology, training, and organization.[103]

With the emphasis on training that came during Alexander III's reign, exercises and maneuvers assumed increasing frequency, at least until the mid-1890s. In 1885, the War Ministry instituted rotating encampments (*podvizhnye sbory*) to familiarize troops with conditions in various localities. By 1894, two-thirds of Russian troops had participated in them. In 1886, the army instituted live-fire exercises in conjunction with maneuvers, and by the mid-1890s, foreign military attaches reported that the Russians regularly practiced advancing while artillery fired over their heads. Finally, as the army grew in size, the War Ministry recognized the requirement to conduct maneuvers with larger formations, incorporating into them reserve troops. Beginning in 1886, the army held large annual maneuvers to combine the forces of several military districts. In addition to the annual maneuvers near Krasnoe selo, other large maneuvers rotated annually among the districts, since their prohibitive cost (half a million rubles) precluded yearly large maneuvers within each military district.[104]

The composite image which emerged from the larger exercise and maneuver mosaic was variegated. In part, this was because training emphases varied among the districts, the existence of uniform regulations notwithstanding. Troops from I. V. Gurko's Warsaw Military District regularly trained in cold weather, an emphasis stemming from Gurko's experience with winter warfare in the campaigns of 1877–78. Some district commanders pressed their troops to the limits of physical exertion, while others preferred to husband troop strength for "real combat." In the Kiev Military District, Dragomirov's troops paid less attention to aimed rifle fire than to the development of élan in the bayonet assault. At Krasnoe selo, the Emperor's frequent presence created an atmosphere in which external appearances often counted for more than realistic training.[105]

In consequence, both contemporaries and historians have commented variously on the value and impact of maneuvers. British military attaches were consistent critics of the exercises and maneuvers held near St. Petersburg, to which they had relatively free access, limited only by lack of access to exercise maps and communications traffic. They were, however, sharp-eyed observers of equipment and armament and their application. They noted the rapid rearmament of Russian troops with the Mosin rifle.[106] They also noted the strength and physical fitness of the troops, something they monitored closely by trailing after road march columns. They were critical of commanders who failed to maneuver their troops skillfully and deploy their artillery to best advantage. They were also critical of the Russian penchant to underestimate the effects of rifle and artillery fire.[107] They noticed that artillery officers preferred to mass their guns physically rather than to mass their fires through better control procedures.[108] They also noticed a tendency "on the part of the supports and reserves to close too rapidly upon the skirmishers, thus giving rise to heavy losses."[109] With one voice, the British found the Russian soldier stolid in the field but lacking the individual initiative so prized in western Europe.[110]

Not surprisingly, many of these observations have been noted by historians and other commentators, but not necessarily with the same uniform emphasis. P. A. Zaionchkovskii, for example, has castigated the maneuvers at Krasnoe selo for their stereotypical rigidity and for their stress on appearance. At the same time, he has recounted Kuropatkin's lengthy critique of his own many maneuver experiences.[111] M. A. Gareev, Chief of the Military Science Directorate of the Soviet General Staff during the 1980s, has taken Zaionchkovskii to account, not only for selective citation from Kuropatkin but also for failing to note the positive side of the Imperial Russian maneuver experience. Thus Gareev has acknowledged that while many serious flaws existed, "the method of conducting exercises and maneuvers during this period attained significant development in comparison with the preceding period."[112]

Diverse perceptions aside, the exercise experience sounded at least one very ominous note. In 1902, the army held large maneuvers near Kursk, and their course and outcome provided important implications for the coming war with Japan. Two "armies" were put together for these maneuvers, the "Moscow Army" consisting of troops from the Moscow and Vilnius districts under Grand Duke Sergei Aleksandrovich, and the "Southern Army" consisting of troops from the Kiev and Odessa districts under War Minister Kuropatkin (a total of 165 battalions, 85 squadrons, and 414 guns).[113] Kuropatkin possessed overall numerical superiority, with the ability after reinforcement to pit approximately 54,000 troops against the Grand Duke's 44,000. The mission of the Southern Army was to advance northward to seize Kursk, then prepare for offensive operations against Moscow. The Moscow Army's mission was to delay, then repulse the Southern Army.

Although Kuropatkin was the attacker, from 27 to 29 August he chose to occupy defensive positions along the River Reut, awaiting the necessary reinforcements to press forward into the advance. Meanwhile, the Moscow Army pushed cavalry and supporting infantry forward both to check the Southern Army's advance and to prepare the way for an envelopment of Kuropatkin's left and rear. On 30 August, the Grand Duke conducted a pinning attack and demonstration against Kuropatkin's front on the Reut, while simultaneously sending cavalry and one army corps on a deep envelopment of the Southern Army's left. Although Kuropatkin's forces slowly gave way before the Moscow Army's frontal blows, the Grand Duke's envelopment was foiled by deep reserves and advancing reinforcements. On 31 August, the front restabilized along the Reut. After resting on 1 September, both armies repositioned themselves to the north along opposite banks of the River Seim. September 2 witnessed the Grand Duke's slow withdrawal before Kuropatkin's superior numbers, and finally, on 3 September, the last day of maneuvers, the Southern Army broke through to the Orlov-Kursk rail line.[114]

The Kursk maneuvers were instructive both for what they revealed and for what they failed to incorporate. Except for general guidance, neither side received more complete or modified missions as the situation changed. Neither side was evaluated on its overall success or failure at the completion of various maneuver stages. Because neither army ever achieved an absolute preponderance of force, success remained a function of limited actions and incremental changes in force ratios for a given time.[115] More ominous for the future were Kuropatkin's actions. Although charged with conducting an offensive, he surrendered the initiative while waiting for reinforcements. Then, rather than resolutely going over to the attack, he advanced slowly while failing to mass his cavalry and while retaining large forces in reserve to meet the ever-present danger of envelopment. These were precisely the tendencies that Kuropatkin would demonstrate two years later as commander of the Tsar's armies in Manchuria. The problem in the Far East was that the Japanese were to prove themselves less forgiving foes than the Russian Grand Duke at Kursk. When M. I. Dragomirov, Commander of the Kiev Military District, learned in early 1904 that Kuropatkin had been named commander in the Far East, the aging tactician's comment was that "for him this will not be the Kursk maneuvers, where any kind of sleight of hand can be gotten away with."[116]

5.

Russo-Japanese War, 1904–1905

> The capitulation of Port Arthur is the prologue to the capitulation of tsarism.
>
> V. I. Lenin[1]

WAR CAUGHT THE Russians unprepared in the Far East. For a combination of reasons, including contempt for orientals, novelty of Far Eastern military commitment, corruption in high places, and difficulty of communications, the Imperial Russian Army and Navy were far from ready to wage war against the Japanese. Although the Russians had originally ventured into Siberia during the sixteenth century, the Tsar's window on the Pacific dated to more recent times. Only in 1860 had Russia founded Vladivostok, while the twenty-five-year lease on Port Arthur and nearby Darien (Russian Dal'nii) had been concluded in 1898. The Russians needed additional time to secure their toehold in the Far East, but their cause was ill served by inept diplomacy, great power rivalry, and corrupt political and commercial interests which reached into the inner circles of the imperial court. During the decade preceding hostilities, the Russians at first had denied and then only grudgingly had begun to contend with the rising tide of Japanese power. Thus, even after the Sino-Japanese War of 1894–95 had demonstrated a remarkable Japanese military prowess, the Russians were slow to change their basically racist conceptions of oriental ineptitude. Privately the Tsar referred to Japanese as "short-tailed monkeys." And official St. Petersburg apparently felt comfortable with the judgment of the Russian military attaché in Japan, Lieutenant Colonel V. P. Vannovskii, who wrote in 1900 that the Japanese were decades, perhaps a century, away from possessing an army with the "moral foundations" which served as the basis for modern European armed forces. Meanwhile the Japanese might acquit themselves well only against the weakest of the European powers. To this analysis, War Minister Kuropatkin affixed his own comment, "a sober view."[2] In December 1903, when Colonel V. K. Samoilov with the Russian legation in Tokyo reported that the Japanese army was in complete readiness to undertake military operations, Kuropatkin responded by commenting, "Colonel Samoilov is mistaken."[3] Two months later the Japanese conducted their "first Pearl Harbor" by executing a surprise attack on the Russian Pacific Squadron anchored at Port Arthur. Little more than a year after that, the Russians would acknowledge Japanese victories by acceding to the Peace of Portsmouth, and, as one writer has observed, "the coolies had become conquerors."

Contending Plans and Forces

What went wrong for the Russians? To begin with, their plans for war in the Far East were incomplete and subject to frequent change. Officers of the Main Staff had begun planning only in 1895, when the Japanese suddenly showed military promise. Subsequently, the original sketchy war plans were altered in 1901 and 1903 to reflect the realities of the changing military and political situation in the Far East. As approved by Kuropatkin and Admiral E. I. Alekseev, Viceroy of the Far East, in the event of a Japanese attack the 1903 war plan variant called for the Russians simultaneously to defend Port Arthur and Vladivostok and to establish a screen behind the Yalu River while concentrating forces in the vicinity of Mukden. There the Russians would gradually increase their troops to achieve the superiority in numbers necessary to launch a successful offensive. Once the Russians began to move, they envisioned concluding the war with an invasion of the Japanese home islands. However, no details had been worked out beyond the first deployment of ground forces to meet a Japanese ground attack.[4]

In addition to lack of completeness, Russian war plans had at least three shortcomings. First, they relied on Russian control of the sea at a time when the Japanese actually possessed a clear quantitative and qualitative naval superiority in the Far East. Second, despite Admiral Alekseev's position as overall commander of land and sea forces in the Far East, there was little coordination between the army and navy in defensive planning. Finally—and fatally—Russian plans were predicated on the availability of time to mobilize sufficient forces to take the offensive in the Far East.

It was the necessity to concentrate forces that revealed more than anything else the precariousness of the Russian military position in the Far East. In the event of war with Japan, the Russians would have to make do with resources on the spot until reinforcements could be laboriously brought from European Russia either by the single-track Trans-Siberian Railway or by sea around the Cape of Good Hope. Local resources amounted to forces within the Priamur and Siberian Military Districts and the Kwantung Fortified Region. By 1904, these numbered 98,000 troops in sixty-eight infantry battalions, thirty-five squadrons of cavalry (chiefly Cossack), thirteen engineer companies, five fortress engineer companies, and four and one-half battalions of fortress artillery. These troops were equipped with 120 field guns, a small number of mountain guns, and eight Maxim machine guns.[5] Local security detachments added another 24,000 troops, the chief purpose of which was to guard the fragile transportation net. Any additional forces and armaments would have to come from the Kiev, Moscow, and Kazan Military Districts. Because of the ever-present danger of European conflict, war plans called for the mobilization first of noncadre reserves, and second, of select cadre reserve units. They could make their way to the Far East at the rate of 20,000 men per month. Supplies would take fifty days to reach Mukden from St. Petersburg.[6]

Implicit in the plans were a number of alarming assumptions. One was blind acceptance of estimates which placed the size of the mobilized Japanese army at 200,000, when in reality the total figure was more than three times that. Another was surrender of initiative, especially in the event the navy failed to control the sea approaches to Manchuria and Korea, thus permitting the Japanese to come ashore in strength to maneuver against objectives far inland. Another was that the army in the Far East could hang successfully by the uncertain thread of an 8,000- kilometer-long supply line. Still another was that with the exception of garrisons and regulars stationed in the Far East, the Russian high command would deliberately choose not to employ its best troops in what was considered a theater of secondary importance. The implications of this decision were large: reservists lacked sufficient training with their new rifles and quick-firing artillery to take the field with confidence against a highly motivated enemy. Finally, war plans assumed the completion of defensive positions, including Port Arthur, when fortress armament and defenses stood incomplete and understrength.[7]

The Russians might have overcome all these difficulties with talented leaders, but they were a scarce commodity in the Far East. Except for Admiral S. O. Makarov, the naval hero of 1877–78, who arrived at Port Arthur in time to die, Russian naval leaders from Admiral Alekseev on down presented a picture of incompetence and sloth. The army seemed to possess more talented men in second-level positions, but too often the best officers were wasted by inept seniors. Kuropatkin himself left the War Ministry to assume command of the field army, but he was far from an ideal troop commander. Although he was a man of unquestionable courage, charm, and intellect, he lacked the resolution and decisiveness to command large forces in the field. Skobelev, whom he had once served as chief of staff, was reputed to have said to Kuropatkin, "Remember, you are good for secondary roles. God keep you anytime from accepting the role of the principal commander because you lack decisiveness and firmness of will."[8]

The problem of personalities was worsened by a failure to designate a single commander for the Far East. Although Kuropatkin commanded the field army, Alekseev remained nominally commander-in-chief of all forces in the Far East. The two did not get on well, and the problem was exacerbated by interference from St. Petersburg. Although the Tsar himself did not come to the Far East to assume personal command, the prospect of court interference was a constant consideration in the command calculus.[9]

In contrast with the Russians, the Japanese presented a picture of resolution, preparedness, and decisiveness. Since Russian intervention in 1895 to deprive them of the fruits of victory in the Chinese war, the Japanese had thirsted for revenge. When for various reasons by 1903 the Russians seemed less subject than ever to compromise on issues of mutual contention and concern, the Japanese determined to further their interests by going to war. Since 1895 they had embarked on a naval modernization program which gradually tipped the scales at sea in their favor. They flooded the Russian Far East with spies, with the result that they were relatively well informed of Russian dispositions in the event of war. In Europe and European Russia their overt and covert spies reached into Russian embassies and possibly even into the planning section of the Imperial Russian Main Staff.[10]

While gathering information and strengthening the fleet, the Japanese continued to expand and modernize their ground forces. In 1897 they adopted the Arisaka rifle, in 1899 they purchased Krupp 4.6-inch howitzers to complement their field artillery, and from 1898 they embarked on an ambitious training program and force expansion. Consequently, by 1904 the Japanese possessed a peacetime ground force numbering nearly 400,000 men led by seasoned officers who had gained combat experience in the Sino-Japanese War of 1894–95.[11]

Japanese war plans were calculated to exploit Russian weakness while maximizing the advantages of surprise, initiative, careful preparation, and flexibility. While in 1904 V. K. Pleve, the Russian Minister of the Interior, had spoken of the benefits to be gained for Russia from "a short glorious war," it was the Japanese who were more prepared than the Russians to wage one. Their plans called for neutralizing the Russian Pacific Squadron, then putting ashore two field armies, one to seize Port Arthur and the other to seize Liaoyang, the Russian communications hub for Manchuria. In all instances the watchword was speed, for the Japanese knew they had to seek decision before the Russians could concentrate superior ground forces in Manchuria. In addition, reasons of finance and diplomacy dictated that the war be short and decisive. Possibly at no time in the twentieth century would a regional power go to war in such a rational and cold-blooded manner.[12]

Attack on Port Arthur and Advance to the Yalu

On 24 January 1904, the Japanese broke diplomatic relations with St. Petersburg, and on the night of the following day Admiral Togo conducted a surprise torpedo attack on the Russian fleet anchored outside Port Arthur. The Japanese had correctly surmised that the success of their entire effort depended upon mastery of the sea. Despite initial Russian confusion and unpreparedness, however, the fleet and the defenses at Port Arthur put up stouter resistance than the Japanese had originally expected. Admiral Makarov soon arrived to instill spirit and discipline in the Russian fleet, and although the Port Arthur defenses were far from complete, they were sufficiently stout to prevent the Japanese from advancing as speedily as they had against the Chinese in 1894. After initial dueling between the fleets and the tragic sinking of the battleship *Petropavlets* with Makarov on board, the Japanese settled down to blockading the Russian fleet and advancing overland to neutralize Port Arthur and to menace Russian concentrations in Manchuria.

With the Russian fleet bottled up in Port Arthur, the Japanese moved quickly

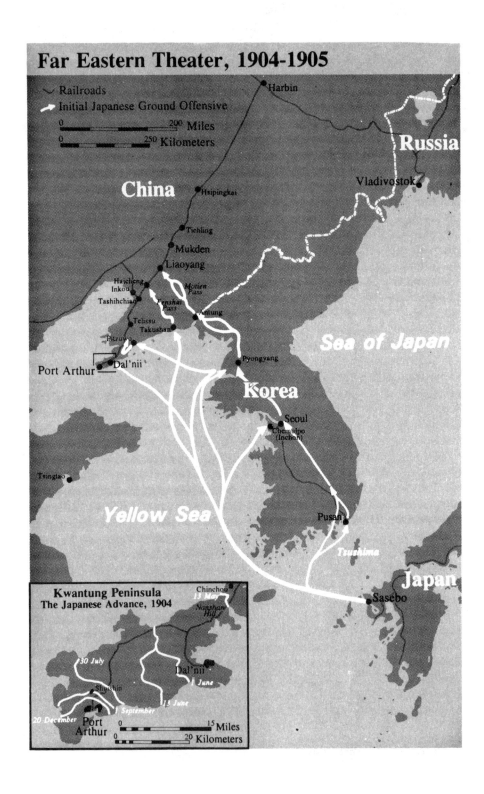

to put ground forces ashore in Korea at Chemulpo (Inchon), the port city of Seoul, to begin their advance to Manchuria. After clearing the harbor of a Russian cruiser and an auxiliary, lead elements of General Kuroki's 1st Army landed and marched to Seoul. To speed the advance northward, on 29 March the Japanese Guards Division landed at Pyongyang. Kuroki's plan was to march behind an advance guard to the Yalu, which he hoped to cross somewhere in the vicinity of Antung as the Japanese had done in 1894 against the Chinese.[13]

Meanwhile, on 15 March, Kuropatkin had arrived in Manchuria to begin a buildup which he anticipated would conclude only in September with superiority needed to defeat the Japanese on land and reassure the security of Port Arthur. He moved his area of concentration from Harbin to Liaoyang. His forces were divided into three detachments, Eastern, Western, and Southern.

The Eastern detachment under General M. I. Zasulich had responsibility for delaying the Japanese as long as possible along the line of the Yalu River. Kuropatkin's orders were "to delay and not to become decisively engaged."[14] Meanwhile, Alekseev had wired Zasulich to hold at all costs. Zasulich, the brother of a one-time revolutionary, Vera Zasulich, seemed psychologically incapable of conducting anything except a rigid defense. He remained largely ignorant of Kurokoi's movements and dispositions, and in the end reacted woodenly to Japanese initiatives. Consequently Zasulich presided over the first Russian defeat in the land campaign.

Ignorance of the enemy resulted in part from improper use of Russian cavalry for reconnaissance. Russian cavalry in the Far East consisted primarily of Cossacks, but they were drawn from the Far Eastern hosts and lacked the discipline and training of their counterparts from European Russia. General P. I. Mishchenko commanded a detachment in Korea, the mission of which included vague reference to disrupting the Japanese advance northward. Kuroki's advance guard element was not especially strong, and Russian cavalry used with skill and tenacity could have delayed him for weeks. Instead, after several minor skirmishes, Mishchenko retreated northward, and Kuroki advanced unhampered to the Yalu.[15]

Because the upper reaches of the Yalu were too rugged and the lower reaches too wide to facilitate easy crossing, Kuroki settled on forcing his way across somewhere in the vicinity of Antung. Zasulich had guessed the location of the crossing and had accordingly arrayed his forces opposite Yanzin on the Yalu. He deployed a line of outposts on the south side of the river, a thin line of dispositions on three islands in the river, and his main forces on a range of hills on either side of a small stream entering the Yalu. An outpost on a local prominence, Tiger Hill, offered potentially excellent observation of the entire river valley for many kilometers upstream and downstream.

To accomplish his crossing Kuroki relied on a combination of careful planning, deception, surprise, fire, and maneuver. Zasulich assisted him by failing both to conceal his own dispositions and to learn more about Kuroki's. First, Kuroki drove in Zasulich's pickets and gained direct access to the south shore of the Yalu. Next, he employed elaborate camouflage measures, including the construction of screens and the transplanting of vegetation to hide his crossing preparations from Russian observers stationed on Tiger Hill. Other observers, including the British officer Ian

Hamilton, were amazed to see entire groves of trees appear and disappear overnight. While the Russians were pondering the reasons behind Japanese secretiveness, Kuroki began construction of movable bridges and moved up his secret weapon, several batteries of 4.6-inch howitzers, to positions from which they could engage Russian light artillery batteries. Finally, he made plans to put two divisions, the Guards and the 12th, across the Yalu upstream in preparation for an advance over rugged terrain to turn the Russian left, which was commanded by Major General N. A. Kashtalinskii. Kashtalinskii had anticipated just such a move, but when he requested permission from Zasulich to adjust his dispositions, Zasulich refused, citing Alekseev's orders not to give ground. Except for Kashtalinskii's premonitions, complacency reigned among the Russians.[16]

On 17 April, Kuroki began his attack by sending the Guards and 12th Divisions upstream and driving the Russians off the three islands in the Yalu. When morning mists lifted on 18 April, the Russians were dismayed to find three Japanese divisions across the river and arrayed against them in assault formation. When the Russians opened direct fire on them with their light guns, Kuroki's heavier howitzers immediately silenced the Russian batteries with a hail of counterbattery fire. As Japanese divisions went into a frontal pinning assault, the Guards and 12th Divisions suddenly appeared from nowhere to roll up the Russian left. For a time Russian rifle fire by volleys sliced ragged holes in advancing Japanese lines, but gradually the Russians on the left gave way, uncovering the center. When Russian officers tried to rally their troops in a valley to the rear of their main positions, the Japanese broke through to occupy the surrounding heights and slaughter the Russians below.

Altogether the Russians lost 5,000 men, while the Japanese lost 2,000. For the first time in open modern battle, Orientals had bested Europeans. The Japanese simply outgunned, outmaneuvered, outworked and outthought their one-time tactical masters. More quickly than the Russians, they had grasped the limitations of artillery in a direct fire role, and they exploited those limitations to gain the upper hand by utilizing indirect fire. More quickly than the Russians, they had also appreciated the flexibility of independent fire over volley fire and the importance of speed, maneuver, and deception in bringing about tactical decision. Zasulich lost his command, the remnants of which halted their retreat only after having cleared the mountain passes into Manchuria.[17]

The Battle of Nanshan

With a success in Korea on the right, the Japanese high command now turned to the left to reinforce that success and to deal with Port Arthur. No sooner had Kuroki forced the Yalu than the Japanese 2d Army, waiting patiently at sea for news of the victory, put itself ashore not sixty kilometers from Dal'nii. The Japanese also made preparations at Takushan to debark an additional army, the 4th under General Nodzu, to cover Kuroki's left in a general advance through mountainous terrain against Russian communications at Liaoyang. Meanwhile, after consolidating his position ashore, General Oku, commander of the 2d Army, directed his forces west between 2 and 7 May to cut the rail line between Port Arthur and Liaoyang, then southwest to penetrate the first line of Russian defenses for the Kwantung Peninsula.[18]

Although the Russians had been unable to halt the 2d Army's landing, the situation on the Kwantung Peninsula was far from hopeless. True, Admiral Togo held the fleet at Port Arthur under close blockade, but Russian mines had unexpectedly destroyed two of his battleships, reducing the number of Japanese capital ships from seven to five. Although the Imperial Russian Army and Navy had their differences over how best to defend Port Arthur, there was a reasonable supply of provisions and ammunition.

In particular, the Russians held what was a potentially strong position at Nanshan on the narrow waist of the Kwantung Peninsula. There, some twenty kilometers north of Dal'nii, a set of heights broken by ravines dominated a neck of land five kilometers wide. A solid line extending from bay to bay constituted a strong natural position, its chief disadvantage on either flank being shallow water, which even at high tide might permit a man to wade waist deep around the main position as far as a kilometer and a half from land.

A second weakness, a shortage of troops and artillery ammunition, was self-imposed. Despite the presence of the nearby 4th East Siberian Rife Division (18,000 troops, 131 guns, and 10 machine guns), General A. M. Stessel from Port Arthur had authorized only a single reinforced regiment to defend the three-mile front. The task was assigned to an unusually competent officer, Colonel N. A. Tretiakov, and his 5th East Siberian Rifle Regiment. The day after the initial attack on Port Arthur he had begun improving his position, even while the ground was still frozen. Reinforced by 5,000 Chinese coolies, his men dug deep defensive positions, laid wire and several mine fields, and constructed overhead cover for their strongpoints. Because his assets consisted of only 3,500 men, 65 guns, and 10 machine guns, Tretiakov stretched thin his line, holding only one company in reserve, and requested that General A. V. Fock, his immediate superior in Dal'nii, send reinforcements from the nearby 4th Division. Luckily for Tretiakov and his Siberians, bad weather and the necessity for a cumbersome overland march to Nanshun held up the Japanese attack for a time.[19]

On 12 May at sunrise General Oku jumped off with his attack to seize Tretiakov's outposts in the settlement of Chinchou. Thanks to the stubborn resistance of several Russian scout companies, the outposts held until early the next morning, at which time their fall enabled Oku to deploy his three divisions (the 1st, 3rd, and 4th) directly opposite Tretiakov's defenses. The 5th East Siberian Rifles, reinforced with 50 light guns, now faced 35,000 Japanese troops, 216 guns, 48 machine guns, and supporting fire from three gunboats in Chinchou Bay. Oku arrayed his divisions in a semi-circle and launched them at daybreak on 13 May in a series of probing attacks. These the Russians answered with accurate rifle and artillery fire to leave many khaki-clad corpses strewn on the approaches to Nanshan Hill. However, because Colonel Tretiakov had been allocated only 150 rounds per gun, by 0800 hours his artillery began to run low on ammunition. By 1100 only two guns remained in action, the rest victims of ammunition shortage or accurate Japanese counterbattery fire. Still, the situation was not desperate for the Russians. Everywhere along the line Tretiakov's men held, and even without supporting artillery fire the Mosins and Maxims exacted a fearful toll on the attackers. By 1530 Oku was no closer to penetrating the Nanshan position than he had been at 0800. His men were pinned down in hastily dug holes before the main Russian defensive position, and every time they tried an advance against the Russian barbed wire they were swept away in a hail of rifle and machine-gun fire. Even worse, Oku's troops were beginning to run short of ammunition.[20]

It was at this time that Oku decided to turn his 4th Division into "human bullets" by launching them through the surf against Tretiakov's left in a daring enveloping movement. The Japanese commander had an uncanny sense of timing. His adversary had just committed his last scarce reserves, and General Fock not only refused to send additional battalions forward but, unknown to Tretiakov, had issued orders for a general withdrawal. With no prospect of help but with the situation ostensibly still under control, the troops of the 5th East Siberian Regiment fought well until 1800, at which time Oku's 4th Division unexpectedly came inland to take the Russian center and left under flanking fire. Even then Tretiakov refused to admit defeat, ordering his men to withdraw to a second defensive position. Only as the officers and troops of the regiment were preparing their new positions during the early hours of darkness did they discover that General Fock had abandoned them. In the confusion and rout that followed, some companies fought to the last man, while others ran. During the daylong fight on the 13th, Tretiakov had lost only 450 men. In retreat another 650 fell. When the remnants of the regiment reached Port Arthur several days later, they were treated as cowards. Only the Tsar's personal intercession earned decorations for sixty of the most severely wounded. Meanwhile, Fock was also decorated for heroism.[21]

In contrast with the Russians, the Japanese had taken losses that were alternately heavy and light. During the hours of daylight attack, Oku lost 739 killed and 5,459 wounded. In this one action his men had expended more ammunition than all of the Japanese armies during the Sino-Japanese War ten years before.[22]

Despite losses, Oku pressed the initiative, occupying Dal'nii without opposition on 14 May as Stessel and Fock withdrew their forces in confusion to cover the outer approaches to Port Arthur. During the ensuing lull, the Japanese high command dispatched Oku's 2d Army (minus its 1st Division) northward to meet a belated Russian advance along the railroad from Liaoyang. Meanwhile, Oku's 1st Division, heavily reinforced through an undamaged Dal'nii, became the nucleus for General Nogi's 3rd Army, the mission of which was to seize Port Arthur.

The Siege War at Port Arthur

Three considerations, two operational and one psychological, governed the development of Japanese operations against Port Arthur. In an operational sense, the need to concentrate forces in Manchuria for a speedy offensive against Kuropatkin dictated a requirement for the rapid reduction of the fortress before the Russians could bring additional reserves and resources via the Trans-Siberian Rail Line. Second, if the Russians chose to reinforce their Pacific Squadron by dispatching the

Baltic Fleet to the Far East, then the best way to foil Russian naval designs would be to deprive the Baltic Fleet in advance of an operating base.

Once the Japanese had bottled up the Russian Pacific Squadron in Port Arthur, the psychological factor also came into play. Japanese field armies had marched to a succession of victories, and the public and government now impatiently awaited news of Port Arthur's fall. More directly in the field, the remembrance of past victories, including General Nogi's easy assault on Port Arthur in 1894, played on the consciousness of the Japanese, especially their commander. Until it was too late he did not bother to consider improvements which the Russians had made in Port Arthur's defenses. At the outset he lacked the siege artillery necessary to reduce a modern fortress defended by a determined garrison equipped with magazine rifles and machine guns. In consequence the Japanese managed to repeat a quarter century later the Russian experience at Plevna.[23]

Russian lassitude on the far approaches to Port Arthur offered little to challenge Nogi's initial complacency. After the fall of Nanshan in mid May, Oku had turned north with the majority of his 2d Army, leaving a rump at Dal'nii around which the veteran Nogi assembled his own 3rd Army with reinforcements from freshly landed troops. Meanwhile, the Russians determined that they must concentrate on improving the inner defenses of the fortress rather than actively defending its remote approaches. During the six weeks of calm which ensued, Nogi put his operational base in order, then advanced to penetrate the fortress's outer defenses. On 13–15 July at the Battle of the Passes, he easily turned the Russians out of their first defensive line in the Green Hills. With scarcely a pause Nogi hammered straight ahead to attack a second defensive line anchored on the Wolf Hills, which the poorly fortified Russians gave up on 17 July after a three hour battle. By the last week of July, Nogi's army stood at General Stessel's doorstep, and only several prominent elevations adjoining the immediate defenses of the fortress lay outside Japanese control.

General Stessel, the Russian commander at Port Arthur, had at his disposal nearly 42,000 men in nine infantry regiments (the 4th and 7th Rifle Divisions), three reserve battalions, and a number of smaller separate companies and detachments. As of 30 July, the fortress defenders had 646 guns and sixty-two machine guns to cover a twenty-five-kilometer perimeter. Many of the guns were of outmoded design, but sixty-seven were of the quick-firing variety, and naval weapons were available to augment fortress defenses. Of the total, 514 guns and forty-seven machine guns were allocated to the defense of the landward side of Port Arthur. Another nine guns and ten machine guns were held in reserve, while the remainder were allocated to defense against attack from the sea. Stessel's supplies were far from satisfactory, although blockade runners managed on occasion to slip through the Japanese naval cordon with provisions.[24]

Complacency and the recency of the Russian occupation of Port Arthur meant that fortifications were incomplete. However, defenses were still formidable, and, as the reality of war had loomed closer and clearer, the Russians worked feverishly to strengthen their positions. One of the defenders later wrote, "That which for seven years was not done Kondratenko [the ground commander] in so far as possible did

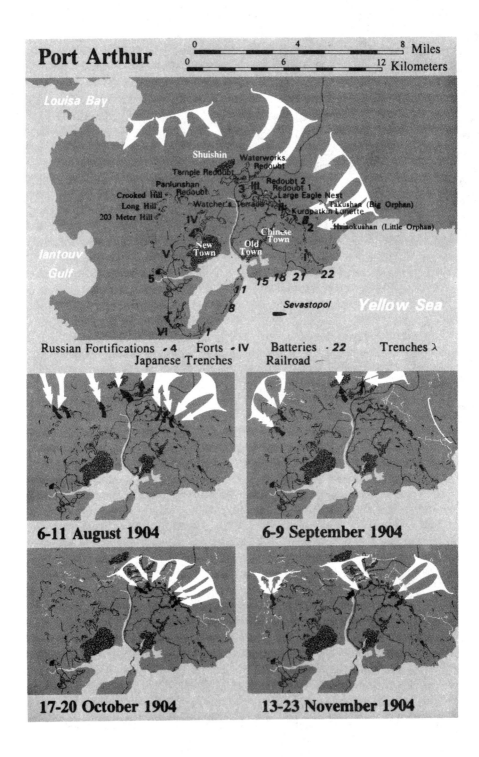

in several months."²⁵ Fortress defenses were arrayed in three main lines. The outer line consisted of a series of fortified hills and positions, including Takushan (Big Orphan Hill) and Hsiaokushan (Little Orphan Hill) to the east of the fortress and 174 Meter (Crooked) Hill and 203 Meter Hill to the west. The middle and what would become the most important line consisted of a series of concrete forts joined by an extensive system of entrenchments and supplementary positions. These extended in an arc along the Chinese Wall about 4,000 meters from the center of the old town of Port Arthur. The inner defense, consisting of an aging line of strong points encircling the old town, was of little tactical value. Many of the positions within the various lines were incomplete, and there was a shortage of barbed wire, for which telegraph wire served as a poor substitute.²⁶

Against the Russians General Nogi deployed his 3rd Army in a semicircle from Louisa Bay in the west to a series of heights on the Russian right. Nogi had more than 80,000 men in 63 battalions, 3 squadrons, 17 companies of engineers, and 474 guns. He and his staff estimated that approaches on the Russian right offered the best prospect of success, and in anticipation of a general assault on the fortress Nogi ordered the capture of Big Orphan and Little Orphan Hills. The attack on the two hills began on Sunday, 25 July, with a three-hour artillery bombardment commencing at 0430. A combination of heavy rain, the necessity to cross the swollen Ta-ho, and the rugged nature of the terrain slowed the advance of infantry from the Japanese 11th Division, which had jumped off at 0730. Still, a regiment managed to wade the stream, then swarm up the steep northwestern and northeastern slopes. One attacker remembered that "above us the steep mountain stood high, kissing the heavens—even monkeys could hardly climb it."²⁷ Russian artillery from neighboring hills and fortified points flailed the Japanese, and the defenders even rolled large rocks down the hillside against the attackers. Rain and darkness temporarily halted the assault, with the attackers still on the lower slopes. When the rain lifted the following day at 1530, the Japanese renewed their attack with a heavy artillery preparation which drove many of the defenders from their hastily dug trenches. By this time the Japanese had gained artillery superiority, and Nogi had twelve infantry battalions committed against the Russians' three. Yet before the attackers reached the summit of Big Orphan at 2000 on 26 July, they had to use their sheer weight in numbers to push the last Russians from their positions in hand-to-hand combat. The next morning at 0430, Little Orphan fell. At the cost of 1,280 men against Russian losses of 450, the Japanese now possessed an unobstructed view westward to Louisa Bay. Although an inner ring of hills and mountains still obscured Port Arthur and its anchorage from direct Japanese observation, Nogi could now approach the fortress's eastern defenses directly.²⁸

When the Tsar heard of the loss of the two hills, he ordered the fleet to sortie in an effort to rejoin the remainder of the Pacific Squadron at Vladivostok. Admiral V. K. Vitgeft, who had replaced Makarov, took the fleet out on 28 July and was killed by a twelve-inch shell before the fleet was cut off from its escape route and forced to return to Port Arthur. The Japanese now held absolute command of the sea until the arrival of the Russian Baltic Fleet.

To forestall the arrival of that fleet, Nogi planned to overrun the defenses at Port

Arthur. After the Russians had refused a formal surrender ultimatum on 3 August, Nogi commenced his first general assault on 6 August. His intention was to drive through the outer ring of Stessel's defenses for an attack on Watcher's Terrace in the center of the northeastern semicircle of forts, then use the captured position as a springboard for an assault against the fortress's inner defenses. Consequently, the Japanese would concentrate their primary attacks on a narrow sector of the front running between Fort No. II and Fort No. III on the northeastern semicircle of Russian fortifications. However, to obscure his true objective Nogi made plans for a series of supporting attacks along much of the fortress's central frontage, from Crooked Hill in the west to Battery B in the east. All these attacks unfolded between 6 and 11 August, and together they comprised the first general assault against Port Arthur.[29]

On the Russian side, Major General V. N. Gorbatovskii commanded the sectors earmarked for attack. On the eastern front running from Battery 22 to Fort No. II and Fortification No. 3, he defended with sixteen companies of the 15th, 16th, and 25th regiments, a detachment of volunteers and three companies of sailors. On the northern front between the Waterworks Redoubt and Fort No. V, he defended with twenty-five companies of the 15th, 16th, and 26th regiments, three detachments of volunteers, a railroad battalion company, and the 4th Artillery Brigade. Gorbatovskii held eight companies in reserve. The western front was held by fourteen companies of the 13th and 27th Infantry Regiments. The high ground covering parts of the western front and the northern front was held by a total of twenty-six companies of infantry, five detachments of volunteers, five batteries, and four companies of sailors. All were under the command of Colonel Tretiakov of Nanshan fame. His bête noir, General Fock, commanded the fortress reserves (sixteen companies and two batteries), while General P. I. Kondratenko commanded all forces committed to the ground defense of the fortress.[30]

The attack opened before dawn on 6 August with a tremendous cannonade against the western front which served as a prelude to an infantry attack against Crooked Hill. The Russians suffered considerably, but General Kondratenko rushed personally to the endangered sector to bolster the defenses, and the Japanese were soon pinned down on the approaches to the position. Heavy artillery attacks followed against the Waterworks Redoubt and Fortification No. 3. Artillery fire continued until 1500, by which time the attackers evidently believed the defenses had been destroyed. Wave after wave of Japanese infantry went into the attack, only to dash themselves against a metal wall of rifle and machine-gun bullets. When nightfall came, a lone battalion found itself pinned in a dead space before the Waterworks Redoubt, unable to retreat or advance. On the right, Russian positions suffered considerably from artillery fire, but no assaults were attempted on what General Nogi evidently felt was the strong part of the Russian defenses.

The morning of the 7th brought renewed artillery fire and attacks on the Crooked Hill complex. By noon the defenders began to break, with one company straggling to the rear. When Kondratenko arrived to shore up the defense, he found there was nothing to do: "Our departure from [Crooked Hill] was elicited by the complete destruction of our bunkers and horrible losses from shrapnel, since the troops were forced to defend standing completely in the open."[31] The defenders had lost nearly all their officers and 500 men.

The situation was only slightly better at the Waterworks Redoubt and worse at the Panlunshan Redoubt. At the former, the local commander had organized a counterattack which drove the Japanese away from the dead space before the redoubt. At the Panlunshan Redoubt, a Japanese battalion finally overwhelmed the defenders, who conducted an immediate counterattack under pain of field court-martial. One company momentarily regained its position, but in the process the defenders lost all their officers, twenty-six noncommissioned officers, and half their complement of soldiers.[32]

The eastern front suffered only from artillery fire. At this time, the fortress commandant of artillery, General V. F. Belyi, in the interest of conserving ammunition and masking the location of his guns, ordered his troops to limit their fire, using it primarily against enemy infantry attacks. By the second day of the storm the Japanese had conquered Crooked Hill and Panlunshan Redoubt, neither of which was crucial to the Russian defense of the fortress. Through the morning of the third day, 8 August, Nogi focused his efforts on the Waterworks and Temple redoubts. They withstood heavy artillery fire, and, although they were now half ruined, they also withstood a renewed onslaught by 1,000 Japanese infantry.[33]

The early morning hours of 9 August brought the Russians their most severe challenge of the first storm. The Japanese managed to creep forward unseen to fall on the Kuropatkin Lunette in a surprise attack which the Russians repulsed only at daylight. More serious were developments between Redoubts Nos. 1 and 2, where the attackers managed to throw three battalions in succession at the Russians in an effort to pry open a gap in the defenses. Gorbatovskii had to use all but his last company of reserves to reinforce the position. Meanwhile, the fleet released several thousand additional sailors to join the defenders, especially on the western front. This enabled ground commanders to shift infantry to the eastern front, which by the 9th the Russians had surmised was Nogi's main objective.

All through the daylight hours of 9 August Gorbatovskii fed reinforcements into the defense of various strongpoints along the eastern front. Finally, he even ran out of sailors, informing the fortress commandant that his reserve was used up, that "Redoubt No. 1 four times had changed hands, that at the moment one-half of Redoubt No. 1 is ours, the other half is occupied by Japanese." Somewhat later, Gorbatovskii informed the commandant that the enemy was perilously close to breaching the defensive line along the Chinese Wall, since the Russians had no reserves left. Unknown to him, at about the same moment Nogi had concluded that further attacks would be fruitless and issued orders to call off the attack—at least temporarily.[34]

For the Russians the problem was that not all the Japanese commanders received Nogi's word. General Fock refused to send reinforcements to the beleaguered posts, and Gorbatovskii and Kondratenko were forced to stand by helplessly as remnants of Japanese assault formations inched ever closer to key defenses on the Chinese

Wall. Meanwhile, the sailors, over 1,000 of whom had been thrown into the defenses, returned to their ships with three officers and 146 ratings dead and 225 wounded.

The 10th was a relatively calm day, with the battleship *Sevastopol* firing sixty-seven rounds against Japanese attackers in the Takushan Valley. The calm was deceptive, however, for it only masked Nogi's plan for a night attack on the Large Eagle Nest slightly behind Russian lines between Fort No. II and Redoubt No. 1. Between 2300 on the 10th and 0300 on the 11th, Nogi threw seven battalions against the line in a vain attempt to drive a wedge into the Russian defenses. Russian searchlights played across the sector, and the rattle of machine-gun and rifle fire mixed with explosions of artillery shells to turn the landscape into what one observer called a mad artist's version of hell. By dawn of the 11th, the line still held, and Nogi was forced to call off the attack.[35]

Thousands of corpses lay for months where they had fallen, permeating the air with the stench of decaying flesh. Nogi had lost about 16,000 men (against Russian losses of 3,000) to discover that the eastern front could be taken only by traditional siege methods. Meanwhile, some several hundred kilometers to the northeast, Marshal Oyama concluded that Nogi's troops would be of no help in pinning down Kuropatkin's field army. Nonetheless, Oyama launched his men into an enveloping attack against Kuropatkin at Liaoyang, but the Russians just managed to elude the trap. Had Nogi's 80,000 men been present, the story might have been quite different.[36]

The Second Storm

Following the unsuccessful first Japanese storm, both sides set to work to improve their respective situations. Japanese sappers began running parallels in the direction of the Russian lines in the vicinity of the Waterworks Redoubt. Huge eleven-inch siege guns began arriving at Dal'nii. Closer to the Russian lines, engineering troops began clearing areas for their emplacement. For their part, the Russians improved their fortifications by stringing more wire and attempting to repair the worst damage of the previous attack. Thanks to ordnance taken from the fleet, the number of fortress guns actually increased to 649 tubes, even though forty-nine had been lost to enemy action during August. The loss of thirteen machine guns left the garrison with fifty-three. Artillery officers completed elaborate fire support plans which divided the fortress into defensive sectors, to each of which were allocated the fires of specific weapons.[37]

In contrast with the first Japanese plan of attack, the second called for penetration of the Russian defenses between the Waterworks Redoubt and the Temple Redoubt in the north and between 203 Meter Hill and Long Hill in the west. The 9th Division and the 1st Division with its attached reserve brigade were assigned to conduct the main attack. They possessed a three-to-one superiority over the Russians, and in select places the Japanese could concentrate sufficient troops to raise the odds against the Russians to ten-to-one.

The attack opened at daylight on 6 September with a six-hour artillery preparation, first on the eastern defenses to deceive the Russians, then on the northern defenses, especially against the two redoubts designated as primary objectives. About forty siege guns and forty-eight mountain howitzers concentrated their fire against the two redoubts. A thousand rounds fell on the Waterworks Redoubt alone, reducing it to rubble. During the artillery preparation, the Japanese moved attacking infantry forward with field guns and machine guns to jumping-off positions.

When the artillery preparation lifted, the Russians greeted their attackers with interlocking crossfires which drove the Japanese infantry back into their trenches. One hour later the Japanese attacked again, this time moving their field guns closer to knock out Russian machine guns. Although the defenders within the redoubt were reduced to thirty men, they received two companies of reinforcements, and together they drove the Japanese back to the dead space before the position. However, the respite was only momentary, and the defenders were soon pressed back by the sheer weight of numbers. With the redoubt nearly overrun, the commander ordered a withdrawal, taking with him the one surviving artillery piece. The withdrawal uncovered the neighboring Temple Redoubt, which was overrun on the morning of the 7th. The Japanese had lost 1,500 more men, but they now held possession of the two redoubts.[38]

Simultaneously with the attack on the redoubts, the Japanese 1st Division and its reserve brigade began their assault on Long Hill and 203 Meter Hill. Strangely, until recently neither side had paid much attention to these prominent terrain features, the tallest of which, 203 Meter Hill, commanded a view of Port Arthur and its harbor. Following the failure of Nogi's first assault in August, the Russians had belatedly realized the importance of the two prominences and began feverish work to fortify them. By 6 September, both hills were protected by two chains of trenches and numerous covered and hardened positions.

The defensive fate of the two hills differed markedly. After an all-day artillery preparation, much of which missed the mark, four Japanese battalions launched an evening attack on Long Hill. All night the defenders (five infantry companies and a volunteer detachment) and attackers fought one another in the flickering glare of searchlights and magnesium rockets. Separate groups of Japanese drew to within seventy or eighty meters of the Russians, with the result that artillerymen ceased fire for fear of hitting their own troops. By evening the Russians, now reduced in strength to ten or twenty men per company, began to retire from the position. Artillerymen who remained to cover the withdrawal were killed at their guns.

The loss of Long Hill exposed 203 Meter Hill to fire from its northern flank, but the defenders of this height displayed even more tenacity than their neighbors. Although the Russians numbered only three rifle companies and a company of sailors, they possessed better positions and had at their disposal seven field guns and four machine guns. During the night of 6–7 September, 203 Meter Hill withstood two serious attacks, and fighting continued to flare throughout the daylight hours of 7 September. Toward evening the commandant of the hill informed Kondratenko that the defenders were barely hanging on and requested the General to inspect the position in person. After personally familiarizing himself with the situation, Kondratenko reinforced 203 Meter Hill with three companies drawn from lesser engaged sectors.

The reinforcements came just in time, for during the night of 7–8 September Nogi threw an additional 2,000 infantry into the struggle for the hill. They took the first line of trenches, and by daybreak of the 8th, with the aid of additional reinforcements, they broke through the second. When the whole position seemed in danger of falling, the Russians counterattacked with cold steel and threw the Japanese back. By now, most of the Japanese officers were dead and their soldiers had lost contact with their own units. As they fell back in confusion, however, a new rain of shells fell on the Russian defenders. As the firing continued, no one slept for three days, and some companies of the 5th Regiment were reduced to twenty and even fewer men.[39]

The Japanese continued to infiltrate around the position, and Russians and Japanese frequently fought at close quarters with little more than a few meters separating the defenders' trenches from advanced Japanese positions. The Russians fought like demons, utilizing every advantage of the terrain and loosing a rain of hand grenades downhill against the attackers. To fashion these instruments of destruction, the Russians used empty 37- and 47-millimeter shell cases, filled them with peroxide, and fitted them with Bickford fuzes. These grenades the Russians put together in several idle Port Arthur factories, and at the height of the attack the defenders were using 1,500 per day.[40]

The last act in the horrifying spectacle of the second storm occurred on the evening of 9 September, when Russian observers noticed a large enemy infantry force gathering on the approaches to 203 Meter Hill for another attack. Kondratenko quickly ordered a battery of field guns to a position northwest of Fortification No. 5, where, masked in a stand of millet, it unlimbered to fire shrapnel against the troop concentration, later identified as three Japanese battalions. In the space of five and one-half minutes the battery fired fifty-one rounds with deadly effect. An English observer on the scene testified that not one Japanese soldier escaped the deadly hail of fire. With this rare instance of the textbook employment of field artillery the second storm ended.[41]

The Japanese failure to take 203 Meter Hill concluded four days' fighting, during which Nogi at the cost of 7,500 troops had taken the Waterworks and Temple redoubts and Long Hill. Although the principal object of the assault had eluded Nogi's grasp, the Russians had lost another 1,500 men, whom they could not replace. In addition, Long Hill afforded the Japanese partial observation of the harbor and a more convenient jumping-off place for subsequent attacks on 203 Meter Hill.[42]

The Third and Fourth Storms

After the second storm, the operations at Port Arthur began to assume the nature of a classic siege. Japanese troops above ground drove their trenches closer to the Russians, while sappers underground began extensive mining operations. The Russians, in turn, began countermines, and occasionally the two sides attempted to blow up each other's excavations. At times the adversaries broke into each other's galleries, and there ensued hand-to-hand fighting with dynamite grenades, entrenching tools, knives, and pistols. "For an entire month," wrote the correspondent Ellis Ashmead-Bartlett, "in the fetid atmosphere of narrow concrete cellars, with the ever-present danger of mines, amid the bursting of dynamite hand-grenades, and exposed to death from bullet and bayonet, the Japanese sapper struggled, unobserved by the world to drive his equally stubborn opponent out of these underground works."[43] Japanese mining operations were directed against Fort No. II, Fort No. III, and Fortification No. 3, all likely objectives for Nogi's next storm.

Siege artillery also lent its fury to the attack above ground. The Japanese now had 474 guns, of which eighteen were eleven-inch Krupp howitzers brought through Dal'nii from Japan especially for the reduction of Port Arthur's defenses. These howitzers had a range of 10,000 meters and threw shells weighing 550 pounds. Altogether the Japanese would drop one and a half million shells on Port Arthur, of which 35,000 were of the eleven-inch variety, called by the Russians "train shells" for the sound they made in flight. With heavy siege artillery the Japanese now possessed the capacity literally to pound to pieces Russian defenses both above ground and underground. The Japanese controlled their artillery fire by means of a single telephone net, but they now needed better observation posts and improved coordination with the attacking infantry.[44]

For his third storm Nogi selected a narrow sector of the northeastern front between Battery B on the Russian right and Fortification No. 3 on the left. This sector included Forts II and III. The attack would begin on 13 October with an uninterrupted four-day artillery preparation. Under cover of the preparation, sappers would move forward to fill the ditches before Fort No. III. The 9th and 11th divisions, together with elements of the 1st, would launch the primary attack with five columns of regimental size backed by a two-column secondary attack. Japanese guns severely damaged the Russian defenses, but defenders were able to empty their ditches of sandbags as quickly as the Japanese were able to bring them forward.

At noon on 17 October, the Russians crawled from the rubble of their defenses to meet the Japanese attack with twenty-six companies, many of which were understrength. Again the Maxims and Mosins did their deadly work, and again the ditches were filled with bodies and the breastworks and glacis areas were littered with dead and wounded. For a few minutes the attackers managed to plant a Japanese flag on the breastworks of Fort No. II, but a furious bayonet attack drove them back. Similarly, several groups of Japanese broke into Battery B and were hurled back. On the Russian right an assault column reached the ditch before Fort No. III, but the attack died when its participants discovered that their ladders were too short to reach the parapet. At 1530 the Japanese artillery ceased firing, the attack came to a halt, and Nogi refused to commit his reserves. Remnants of the five attacking columns made their way back to their jumping-off positions. The Japanese Emperor would not receive the surrender of Port Arthur for his birthday on 3 November (N.S.).[45]

For the next month the Japanese continued their sapping activity, while their heavy siege artillery drove the Russians into underground shelters. Each of the adversaries was driven by a special kind of desperation. The Russians were beginning to run short of food, supplies, and large-caliber ammunition. Squabbling continued within the fortress high command, but there was no open talk of surrender. The Russians stolidly worked to improve their defenses and make the best of their situation even after mid-October, when they received word of Kuropatkin's reverse on the Sha-ho. On the Japanese side, the high command was growing restless and impatient because of the way that the siege was tying up men and materials which were urgently needed elsewhere. In addition, news reached Tokyo in late November that the Russian Baltic Fleet, which had sailed for the Far East on 2 October, had reached the Indian Ocean.[46]

The fourth storm, the one that would pave the way to capitulation, was set for 13 November. Except for a shorter artillery preparation and the hope that exploding mines would open gaps in the defense, Nogi's plans called for a duplication of the third storm. That is, the objectives would run along the eastern front from the Russian right at Battery B to the left and Fortification No. 3. In the event, it was the failure of these plans and the search for alternatives which eventually brought the attackers closer to final victory.

At 0800 on 13 November the Japanese began a three-hour artillery preparation against the eastern front of Port Arthur. At 1100 infantry from the 11th, 9th, and 1st Divisions began concentrating in trenches forward of Redoubt No. 1, Fort No. III, and Fortification No. 3. They jumped off into the attack at noon, with their assault timed to follow the explosion of a huge mine planted under the breastworks of Fort No. II. Subsequently, several other charges exploded under the defenses along the Chinese Wall. The attacks were stereotypical. Japanese infantry would creep forward in their trenches, many of which reached to within fifty meters of the Russian defenses. They would launch themselves from the trenches in a frantic dash for the Russian breastworks or casemates, in effect staking the success of the attack on numbers and the destructiveness of various kinds of artillery and engineering preparations. Although the Russians were groggy and understrength, they met the Japanese with machine-gun and rifle fire, grenades, and direct fire from guns assigned to sector defense. By 1500 the Japanese attack had failed everywhere along the line.[47]

Now was the time for desperation. That evening General Nakamura formed a special detachment of 2,600 volunteers whose task was to infiltrate the Russian position along the valley of a small stream, then turn to strike the positions at Fortification No. 3 from the flank. Unfortunately for the Japanese, Russian observers spied them as they neared the vicinity of Fort No. III, and the attackers were taken immediately under fire with shrapnel and canister. Next, General Kondratenko counterattacked with three companies of sailors who converged on remnants of the enemy column and annihilated it with cold steel. This action closed the first stage of the fourth storm of Port Arthur. The Japanese set their dead and wounded at 4,500, while the Russians had lost an additional 1,500.[48]

Rather than cancel the attack, Nogi now shifted his attention to 203 Meter Hill and the western front. There the redoubtable Colonel Tretiakov held strong defenses with about 2,200 men from three of the East Siberian rifle regiments. At dawn on 14 November, Nogi opened fire on 203 Meter Hill with his eleven-inch guns. By 1700 they had destroyed twenty-two bunkers and a major portion of the defenses on the hill's southwestern slopes. When a late afternoon and early evening attack of the 1st Division failed, the Japanese renewed their heavy artillery fire the following morning. At noon on the 15th the second attack of the 1st Division failed. The next morning Nogi had to introduce the 7th Division into the attack because the ranks of the 1st were utterly depleted.

The battle raged between 17 and 21 November. A succession of attacks blurred one into the next, and a storm of artillery fire engulfed the defensive positions on 203 Meter Hill. The Russian correspondent Nozhin eloquently summed up the action when he wrote that "it was hardly a fight between men that was taking place on this accursed spot; it was the struggle of human flesh against iron and steel, against blazing petroleum, lyddite, proxyline, and melinite, and the stench of rotting corpses."[49] What became the final Japanese assault began at 0815 on 22 November, when Nogi sent a brigade against the hill, which the Russians could no longer reinforce without dangerously weakening their defenses elsewhere. The last redoubt was captured shortly after noon, and by 1700 the attackers were at last able to plant the rising sun on the crest of 203 Meter Hill.[50]

Not until Verdun in 1916 would a single piece of land claim so many lives. The Russians counted more than 400 dead and another 5,000 wounded. The Japanese lost more than 10,000 killed and wounded. Ashmead-Bartlett wrote that "there have probably never been so many dead crowded into so small a space since the French stormed the great redoubt at Borodino. . . . There were practically no bodies intact; the hillside was carpeted with odd limbs, skulls, pieces of flesh, and the shapeless trunks of what had once been human beings intermingled with pieces of shells, broken rifles, twisted bayonets, grenades, and masses of rock loosed from the surface by the explosions."[51]

The fall of 203 Meter Hill sealed the fate of Port Arthur. With the harbor in full view, the Japanese siege artillery commenced to pound the Russian Pacific Squadron to pieces, thus releasing Admiral Togo's fleet for a badly needed refitting prior to its engagement with the Baltic Fleet at Tsushima. Sapping and local attacks continued for another month, while the garrison began to run low on provisions. On 20 December 1904, General Stessel hoisted the white flag. Two days later he surrendered 546 guns, 82,000 shells, 2.25-million small-arms rounds, 878 officers, and 23,481 men. The 3rd Japanese Army was now free to join Oyama in the field against Kuropatkin.[52]

The Maneuver War in Manchuria

For General Kuropatkin and the Russian Manchurian Army, Port Arthur was at once an asset and a handicap. On one hand, already in the spring of 1904 Kuropatkin had correctly perceived that the fortress was not in imminent danger of collapse. Therefore it might serve as an important diversion of Japanese troops and resources while the Russians themselves fought for time to mobilize a credible field army in Manchuria. On the other hand, Port Arthur exerted an inexorable psychological pull on St. Petersburg and the Russian high command. Neither the Tsar nor his closest military advisers could easily resist the sirenlike call of a beleaguered garrison, and

Kuropatkin came under considerable pressure to speed up his preparations for an offensive that would rescue the fortress. Foremost among those advocating an early relief was Admiral E. I. Alekseev, who had fled Port Arthur at the last moment to take up residence in Mukden, where he commenced to issue a flood of orders contradicting and confusing those of Kuropatkin. Given the difficulties of mobilization and transportation across extended logistical lines, Kuropatkin had concluded that he could not take the field for offensive action until at least September.[53]

In large measure, the Russian fortunes of war in the Far East during the summer of 1904 came to depend on Kuropatkin's resolve and the way that he would choose to use time. Under pressure from St. Petersburg and Alekseev, he eventually compromised. Under pressure from Port Arthur, he chose not to trade space for time. Under pressure from the Japanese, his forward-deployed units failed to win time by conducting adequate delaying operations.

No one was more aware of Kuropatkin's need for time than the Japanese high command. Although Port Arthur had become a major nuisance, the Japanese with single-minded determination pursued their quest against Kuropatkin for a gigantic battle of envelopment on the pattern of Sedan in the Franco-Prussian War of 1870–71. If Kuropatkin came south to relieve Port Arthur, the Japanese would not have to march so far north to accomplish their objective. The Japanese high command firmly believed that this battle would take place in the vicinity of Liaoyang.

Immediately after the debacle at Nanshan the Japanese put in motion the movements necessary to fulfill their quest. During May and June, sufficient troops arrived by sea to reinforce one field army and create two new ones. While General Nogi assumed command of the newly created 3rd Army at Port Arthur, General Oku turned northward with the 2d Army to advance along the Port Arthur–Mukden Railroad. The 10th Division, which had come ashore at Takushan on 19 May, was reinforced with the 5th Division, and the resulting 4th Army was placed under the command of General Nodzu. Northwest of the Yalu, General Kuroki's 1st Army occupied Fenghuangcheng. The idea was that the 1st, 2d, and 4th Armies would conduct a simultaneous advance to Liaoyang, where part of the total force would launch a pinning attack while the other part overran the Russian flanks and rear.[54]

Against this threat the dispositions of the Russian Manchurian Army gave Kuropatkin some cause for worry. Although by the end of April reinforcements had begun to stream in from the Priamur and Transbaikal Military Districts, Kuropatkin awaited arrival of IV Siberian Corps and X and XVII Army Corps. His troops were spread over an enormous distance along the rail line extending south from Harbin. In early May the Russian Manchurian Army was divided into three segments, the Southern and the Eastern Detachments and a general reserve. The Southern Detachment, consisting primarily of I Siberian Army Corps, was deployed along the lower reaches of the Liao-ho delta. The Eastern Detachment, consisting primarily of III Siberian Army Corps (18.5 infantry battalions, 14.5 Cossack cavalry squadrons, and 32 guns), was deployed to cover the mountain passes against Kuroki's 1st Japanese Army. The reserve consisted of 28.5 infantry battalions, 2 Cossack squadrons, and 84 guns.

With these dispersed and inadequate forces at his disposal, Kuropatkin came under pressure to parry the advancing Japanese tentacles, and the result was defeat at Telissu and the Motien Pass. On 27 May, under pressure from Alekseev and St. Petersburg, Kuropatkin agreed to push troops southward both to offer some promise of relieving Port Arthur and to stop Oku's march along the rail line to Liaoyang. The Russian commander was aware not only of Japanese reinforcements but also of the possibility of an unfolding envelopment. Still, he dispatched General G. K. Shtakel'berg's I Siberian Corps for a countermove against Oku, and Shtakel'berg chose to move south through a series of defensive positions—a strange tactic for a cavalryman—to halt the Japanese at Telissu. Because of poor information and inadequate reconnaissance, Shtakel'berg was unaware that his 32 battalions and 48 guns faced General Oku's whole Japanese 2d Army with 48 battalions and 216 guns.[55]

Telissu was located five kilometers south of a rail junction at Wangfangkou, about 135 kilometers north of Port Arthur and 225 kilometers from Liaoyang. Shtakel'berg occupied a defensive position approximately twelve kilometers in width along a string of hills astride the railroad. His strongest dispositions were east of the railroad, where General A. A. Gerngross defended with twelve battalions of infantry and their supporting artillery. General F. F. Glasko commanded a strong reserve of eight battalions located to the north in the vicinity of Sisan. A cavalry screen covered Shtakel'berg's open right flank. His left rested on terrain that would not easily admit a flanking movement. His artillery was deployed in open positions on the orders of officers who failed to understand the importance of firing from concealed positions. Shtakel'berg himself sited batteries on high, open prominences.

The battle began at 0830 on 1 June, when advance elements of Oku's army came under Russian artillery fire. As the Japanese moved forward in three columns along the railroad embankment, the Russians failed to notice a fourth column several kilometers west of the others. This was the Japanese 4th Division, and with it Oku planned to conduct a deep enveloping movement to fall on Shtakel'berg's rear approximately ten kilometers from the line of contact. Meanwhile, Oku developed a frontal attack with his remaining two divisions, the 3rd and 5th.[56]

From the beginning, the battle went badly for the Russians, although they managed to recover their fortunes briefly before the 4th Division appeared in their rear. The Japanese opened their own attack after noon with a storm of artillery fire which caught the Russians either in the open or in inadequate cover and caused many casualties. As had been the case on the Yalu, Russian batteries, which had been so casually located in the open, suffered heavily from accurate, indirect Japanese fire. On Shtakel'berg's left the Japanese 3rd Division delivered a spirited attack which nearly broke through the thin Russian line before General Gerngross personally counterattacked with the 2d Infantry Regiment to restore the situation. After four hours' heavy fighting east of the rail line, darkness halted the battle. Heartened by the outcome of the day's action, Shtakel'berg intended to attack early the next morning on his left, drive in Oku's right flank, then roll up the middle of the Japanese line. The plan required close cooperation between Gerngross and Glasko, neither of whom was able to find the other during the night of 1–2 June. Consequently the sun rose on 2 June with each general knowing vaguely that he was supposed to attack but remaining undecided about timing, cooperation, and objectives. After coming once again under damaging artillery fire, at 0800 Gerngross decided to attack with two of his own regiments. Thirty minutes later he requested support from one of his brigade commanders, and at 1000 Gerngross attacked with a third regiment. By 1130 he had suffered heavy losses, but his pressure also threatened to rupture the defensive line of Oku's 3rd Division. Glasko's appearance with the reserves at this critical point might have won the battle for Shtakel'berg.

However, Glasko was nowhere in sight, and at 1130 news came to corps headquarters that large formations of Japanese had been sighted north of Wangfangkou. For nearly an hour Shtakel'berg refused to believe that an additional Japanese division was now located in his rear area. Finally, when reports filtered through that his reserves were being attacked and rear firing positions were being overrun, Shtakel'berg ordered a withdrawal. Glasko now redeemed himself by reinforcing Gerngross's right, thus permitting a large number of Russian units to escape annihilation. Fortunately for Shtakel'berg, the Japanese 4th Division had struck too deeply, and Oku was momentarily unable to coordinate the blow of hammer against anvil. In addition, a blinding rainstorm at about 1500 afforded the Russians valuable cover for their withdrawal. Had conditions been different, the Russians might have lost more than 3,500 men against Japanese losses of 1,163.[57]

While Oku's 2d Army was advancing against Shtakel'berg, Kuroki's 1st Army stood motionless at Fenghuangcheng, the victim of supply difficulties. Until Kuroki's engineers could construct a tramway to help shuttle supplies forward from Antung, the 1st Army was unable to press into the mountains to menace the Russian flank at Liaoyang. Once the tram opened, in mid-June Kuroki swung forces into the mountains astride the old Mandarin Road with the Guards on the left, the 2d Division in the middle, and the 12th Division on the right. The route wound its way through green and terraced hills to the Motien Pass, which one modern writer has called "the Thermopylae of Manchuria." Having committed his troops to stopping Oku to the southeast, Kuropatkin now had no Leonidas to halt Kuroki in the mountains. Consequently, Kuroki rolled through the pass with unexpected ease. Where only small detachments could have arrested his march for days and weeks, Kuroki encountered little or no opposition.[58]

Too late Kuropatkin realized the significance of the lost passes, then compounded his mistake by giving in to pressure from St. Petersburg to retake them. The mission was assigned to Count Fedor Keller, a close friend of Kuropatkin's, who had replaced Zasulich in command of the Eastern Detachment. On 4 July, with six regiments of infantry, including the 9th Army Division from European Russia, a battery of artillery, and a small cavalry force, Keller attempted a surprise attack on the 2d Japanese Division holding Motien Pass. Under cover of a morning ground fog, the Russians forced their way through the Japanese outposts, then ascended to the ridge line in an attempt to dig in. Before they had established their positions, the fog lifted to reveal the Russians to the Japanese artillery observers. With no pack artillery of their own to silence the enemy's deadly guns, the Russians suffered heavy casualties. Keller lost about 1,000 men before he withdrew his forces west of the Lan-ho.[59]

To impose a greater degree of unity on its operations in Manchuria the Japanese high command appointed General Oyama commander in chief of all Japanese forces in Manchuria. Soon after his arrival at Oku's headquarters on 9 July, Oyama helped orchestrate the 2d Army's advance to attack I and IV Siberian Army Corps at Tashihchiao, about eighty kilometers north of Telissu. The idea was to conduct a preemptive attack that would prevent the Russians from resisting the advance of Nodzu's 4th Army through Fenshui Pass. A battle raged all day on 11 July, but Kuropatkin broke off the action at nightfall when he became apprehensive over the security of his extended left flank. Subsequently the Russians fell back to Haicheng, about thirty-two kilometers north of Tashihchiao. Nodzu crossed the Fenshui Pass without incident.[60]

Liaoyang

By late July, Kuropatkin had suffered a number of setbacks, none fatal. Against his judgment he had attempted to keep open a door to Port Arthur, and that door had slammed shut at Telissu. More ominously, his forces remained scattered in the face of an enemy who was persistently concentrating his own. At first, Kuropatkin had hoped to conduct his own concentration at Haicheng. Now it would be farther north at Liaoyang. All of his delaying actions had failed, thanks either to inadequate and incomplete instructions or command incompetence. He lacked satisfactory intelligence on the enemy, and there was little hope that the assets available in theater—a combination of cavalry, scouts, and spies—would compensate for this important shortcoming.

Above all, Kuropatkin needed additional time to build the strength of the Manchurian Army, but neither the Japanese nor St. Petersburg would permit delay. Under pressure Kuropatkin decided to give battle at Liaoyang. Although the Trans-Siberian Railroad permitted only a modest stream of supplies and troops, he had been able to assemble five Siberian army corps and two army corps from European Russia, in all fourteen divisions. Altogether Kuropatkin now had 158,000 men in 155 infantry battalions and 141 cavalry squadrons with 483 guns. Against him Oyama counted the eight divisions of his three field armies and a number of smaller separate units of the reserve. His 125,000 men were arrayed in 115 infantry battalions and 33 cavalry squadrons with 170 guns. The combined forces that would soon face each other totaled slightly fewer than 300,000, a figure exceeded up to that time in history only at Sedan on 1 September 1870.[61]

Two considerations governed Kuropatkin's approach to the battle. First, his intentions were primarily defensive, since his objective was to win time for the accumulation of reserves and resources necessary for a full-scale offensive that would drive the Japanese from Manchuria. Second, he was obsessed by a preoccupation with Japanese numerical superiority. His staff assistants consistently overestimated the forces at Oyama's disposal, adding more than 25,000 imaginary troops to their adversary's already impressive field forces. Consequently, Kuropatkin lived in fear of a ghostly enveloping force that rendered him alternately negative and passive at critical points during the operation.[62]

Russian deployments and defensive positions were basically sound. While Oyama awaited the outcome of Nogi's storm against Port Arthur on 7 August, Kuropatkin had withdrawn along the rail line toward Liaoyang to establish three successive defensive positions. The first, the "Rear Guard" position, was a line stretching seventy kilometers in length between defensive strong points thirty kilometers to the south and southeast of Liaoyang. Approximately twenty kilometers to its rear stretched the second defensive position, called the "Vanguard." It was twenty-two kilometers in length. Terrain along the first line was hilly, a factor which limited visibility and the effect of defensive fires. Terrain along the second was rolling, permitting fair visibility and defensive fires to ranges of 500 meters. Kuropatkin's third position was four to six kilometers inside the second line. It extended fifteen kilometers and was organized to a depth of four kilometers with three successive lines of its own. It also included eight forts located at distances of two to three kilometers from each other and covered by elaborate trench systems.[63]

The nature of these dispositions testified both to Kuropatkin's willingness to surrender the initiative and to his army's growing appreciation for field fortification. Two other changes were coming about more slowly. One was a growing preference among artillerymen to fire from concealed positions. After failures on the Yalu and at Telissu, battery commanders were beginning to insist that their guns be employed increasingly in indirect fire roles. In skirmishing around Tashihchiao during mid-July, guns had occupied dug-in or reverse slope positions. Officers observed fire from nearby elevations and passed corrections to the guns either by runners or messenger chains.[64] A second change coming belatedly to the Russian camp was a tendency to employ cavalrymen as mobile infantry, using their horses for rapidity of movement and dismounting them to fight on foot.[65]

On 10 August the Russians occupied their forward defensive positions with the Eastern Group under the command of General A. A. Bil'derling deploying on the Russian left against Kuroki's 1st Army. The Southern Group, including Shtakel'-berg's corps, under General N. P. Zarubaev, deployed on the center and right against the advancing Japanese 2d and 4th Armies. These two groups were separated by a distance of twenty-three kilometers, but they remained under Kuropatkin's overall operational control. Kuropatkin retained one corps in general reserve and a second in army reserve at Mukden.

On 12 August Oyama attacked the Russian forward positions with his main effort designated in the 1st Army's sector. There, Kuroki delivered his own secondary attack in the center against III Siberian Corps to divert attention from his main attack on the Russian left against General K. K. Sluchevskii's X Corps. Following a penetration against X Corps, Kuroki planned to cut Kuropatkin's communications and then attack the Southern Group from the flank and rear. However, III Corps fought well enough to throw Kuroki's supporting attack into disorder, and Sluchevskii's X Corps held stubbornly to key terrain for most of the day. Only at nightfall did the Russians let go. Rather than launch a counterattack with his ample reserves, Kuropatkin gave the signal at midnight for his entire army to fall back to its second defensive position. Heavy rainfall for the next two days covered Kuropatkin's withdrawal.[66]

The second phase of the Liaoyang operation opened on 15 August with the

Manchurian Army occupying its second defensive position. I and III Siberian Corps and X Army Corps held the main line with II and IV Siberian and XVII Army Corps in reserve. This was an inordinately large force to keep in reserve, and the individual corps commanders reduced the frontline fighting complement even more by placing anywhere from 46 to 66 percent of their troops in local reserve! Evidently Kuropatkin's fear of envelopment had become contagious.[67]

In view of Oyama's plans, there was some cause for concern. As he moved his troops forward to engage the Russians in their new positions, he produced a maneuver scheme which called for a double envelopment. His attack would begin on Kuropatkin's right with a heavy blow against Shtakel'berg's I Siberian Corps. As Oku developed this success with his 2d Army, Kuroki's 1st Army would cross the Tai-tzu River to operate against the rail line running north from Liaoyang.

The action began at sunrise on the 17th with a diversionary attack against III Siberian Corps to obscure Oyama's main effort on the Russian right. After heavy fighting in the center that reduced some Russian battalions to half strength, at 1000 the 6th Japanese Division moved forward against the Russian right to slam into I Siberian Corps. The 6th easily occupied several villages, then had its advance stopped dead by the counterattacking Russian 12th Barnaul'skii Regiment. On the extreme right of I Corps the Japanese could approach no more closely than one and one-half to two kilometers, thanks to a machine-gun detachment which supported the right-hand battalion so effectively that no attackers penetrated its deadly fire curtain to turn the flank. Artillery, too, played a role in halting Oyama's advance against the Siberians. The 1st and 2d batteries of the 9th East Siberian Artillery Brigade each fired more than 1,300 rounds, while the 3rd battery fired 1,776. The majority of their fire was from covered positions. The day ended with the Russians holding all along the line.[68]

The key developments in the battle took place during the night of 17–18 August and the morning of the 18th. On the extreme Japanese right, Kuroki, evidently reacting to mistaken reports that Kuropatkin was again retreating, ordered his 12th

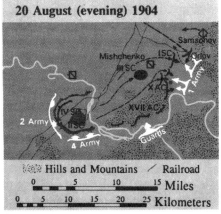

Division and elements of the 2d to ford the Tai-tzu to begin an enveloping movement against Kuropatkin's left. On receiving news of this crossing about 0600, Kuropatkin, operating under the assumption that Kuroki had six divisions and not just three and three-fourths, ordered into effect Directive No. 3, a plan which called for withdrawal to the main defensive position in the event of a Japanese crossing of the Tai-tzu. In reality, Kuroki's move had opened a wide gap in the Japanese line, and Kuropatkin might have won the battle that morning by exploiting this gap with the lightly engaged X Army Corps. Instead, the Russians withdrew to the main position on the night of 18–19 August.[69]

Kuropatkin's design was not merely defensive. By withdrawing he hoped to shorten his lines, then shift forces from his right to the threatened left to launch a sledgehammer blow against Kuroki's forces, the majority of whom now had turned left with the swollen Tai-tzu behind them. While A. V. Samsonov's cavalry screened on the far left in the vicinity of the Yentai Mines and XVII Corps held on the near left, General N. A. Orlov was to lead elements of V Corps (twelve battalions) in a drive past the Yentai Mines to roll up Kuroki's flank. Meanwhile, I and X Corps would strike a massive blow against Kuroki's thinly held front. According to Kuropatkin's plan—and in the best Leerist tradition—the enveloper would be smashed and himself enveloped.[70]

Unfortunately for the Russians, Kuroki was of no mind to fall into the trap, and Kuropatkin had underestimated the difficulty of executing the plan. The whole action turned on the possession of Manju-yama, a small elevation dominating the ground between the Yentai Mines and the Tai-tzu River. Unknown to Kuropatkin, while his commanders and their staffs were laboriously reshuffling their troops for the counterattack, Kuroki had occupied this prominence on the night of 19–20 August. The pivoting point for the Russian attack now lay in Japanese hands.[71]

Japanese seizure of the high ground caused Kuropatkin's plans to unravel on the 20th. Kuropatkin himself moved forward to oversee the attack, and in so doing lost control of the battle. As his troops moved to their jumping-off positions, they became disoriented while stumbling through fields of tall and seemingly endless millet. Orderlies and messengers became lost, and subordinate commanders received garbled or late communications. Unlike the Japanese, who used telephones extensively, the Russians relied primarily on messengers for communication. In the confusion that followed the Russian advance, X Corps did not immediately go to the aid of XVII Corps to recapture Manju-yama. V Corps became hopelessly lost in the millet, with some of Orlov's men actually firing by mistake on I Corps units before degenerating into a hopelessly disorganized state. Samsonov added to the confusion by suddenly withdrawing to uncover the Yentai Mines.[72]

Despite these reverses, at the eleventh hour of battle victory remained tantalizingly within Kuropatkin's grasp, only to vanish at the last moment. At 2030 on the 20th, reports came to headquarters that XVII Corps had managed to retake Manju-yama. On the southern front, II and IV Corps seemingly stood solid. Neither III Corps nor X Corps had been heavily engaged. Therefore, at 0230 on 21 August Kuropatkin prepared orders for General Bil'derling to launch an offensive that morning: "The Army Commander intends to attack the enemy on 21 August by our left wing." After dictating these orders the Commander lay down for a few hours' rest. Thirty minutes later, news began to trickle in that heavy pressure had caused General N. P. Zarubaev to use up all his reserves and nearly all his ammunition in the south, and still later reports described the effects of five days' continuous fighting on Shtakel'berg's troops. His men worn out, Shtakel'berg had begun a slow withdrawal northward from his present position. Finally, by 0530 complete word of Orlov's disaster also circulated at headquarters. Kuropatkin rose from a short sleep, digested the various reports, and ordered a full retreat. Although Kuropatkin later tried to put the best face on his action, Oyama had won a victory of sorts. Although his army had been fought to a standstill, it was Kuropatkin who had retreated.[73]

Because the Japanese were exhausted and short of supplies, including artillery ammunition, the Russians managed their three-day retreat to Mukden unharassed. In nearly nine days' fighting the Japanese had lost 23,000 troops, while the Russians had lost 16,000. Worse still for Kuropatkin, constant retreat was beginning to wear down his troops' morale.[74]

Sha-ho

Because exertions at Liaoyang had exhausted the Japanese, Oyama could neither pursue Kuropatkin nor disrupt arrangements for a new Russian offensive. During early September, I Army Corps and VI Siberian Corps arrived in Mukden to compensate for casualties at Liaoyang and end Kuropatkin's fears of Japanese numerical superiority. The Russians now had nine full corps (I-VI Siberian and I, X, XVII Army) in Manchuria for a total of 257.5 infantry battalions (195,000 troops), 143 cavalry squadrons, 678 guns, and 32 machine guns. Although short of some supplies, including clothing for the approaching winter and maps suitable for conducting a campaign north of Liaoyang, Kuropatkin began planning for a fall offensive. Thanks to the Liaoyang operation, Kuropatkin now had a more accurate picture of the Japanese and their limitations. He rightly concluded that the Port Arthur siege had aggravated the Japanese replacement problem and that Oyama's supply lines were becoming dangerously extended. Clear fall weather favored an offensive, and there was no telling how long the impasse at Port Arthur would assure him a favorable ratio of forces. St. Petersburg was also exerting considerable pressure for an offensive. Finally, Kuropatkin himself admitted that "an advance seemed more advantageous than waiting for the enemy to attack, for there seemed little chance of our being able to hold our ground on the Mukden positions."[75]

On 22 September Kuropatkin attacked south from Mukden across a front sixty-five kilometers wide. Against Oyama's 170 battalions the Russians advanced with 261 battalions divided into two large detachments, the Eastern under Shtakel'berg and the Western under Bil'derling. Thanks to the harvest, the Russians no longer had to contend with millet, although open-field stubble presented rough going for the poorly shod troops. As the Russians moved south, Kuropatkin's intention was to fix Oyama's main forces in the plains astride the Mukden–Liaoyang Railroad, while on the left (to the east) Shtakel'berg conducted an enveloping movement through hilly country. The object was to strike where the Japanese were the weakest, then swing from the east to the center in an effort to cut the rail line at Liaoyang and take Oyama's main forces from the flank and rear.

Despite the optimism evident in Kuropatkin's plans and orders, from the beginning the Russians operated at a serious self-inflicted disadvantage. Although Kuropatkin had created the Eastern and Western Detachments to endow his command with greater flexibility through decentralization, he retained veto power over dispositions affecting the employment of corps-size detachments. For units operating in unknown hilly country to the east, the issue of command initiative could become very important. If detachment or corps commanders lacked the freedom to employ their assets as they saw fit, then they needed near-instantaneous communication with Kuropatkin, something the Russians did not have. To complicate this sad state of affairs, as at Liaoyang, Kuropatkin would issue vague and incomplete orders.[76]

In addition to these difficulties, Kuropatkin's concept was seriously flawed—although perhaps not fatally if boldness became the watchword. He had designated a flanking movement over terrain which was hardly conducive to rapidity of movement. Indeed, almost from the beginning, the Eastern Detachment's artillery and infantry would find off-road travel very difficult. To compound this difficulty, Kuropatkin failed to take advantage of his superiority in cavalry to form a large mobile strike force which would have been invaluable—either east or west—in assisting with the flanking movement. Finally, and unfortunately for the Russians, as was almost always the case in Manchuria, Kuropatkin moved too slowly to forestall Japanese countermoves. The word *boldness* did not seem to occupy a prominent place in his operational vocabulary.

The Sha-ho operation actually unfolded in two distinct phases, Russian offensive and Japanese counteroffensive. During the first phase, extending from 22 through 27 September, General Shtakel'berg's Eastern Detachment assailed General Kuroki's 1st Army on Oyama's extreme right flank. After the Russians had recorded a series of inconclusive engagements and lost opportunities, the second phase, extending from 27 September through 4 October, opened with Oyama's riposte in the center and the west and concluded in bloody stalemate on the banks of the Sha-ho.[77]

In accordance with Kuropatkin's dispositions, Shtakel'berg (78 battalions, 50 squadrons, and 198 guns) slowly advanced in hilly country on the Russian left, his flank guard under P. K. Rennenkampf reaching the Tai-tzu river on 23 September. Although the extreme flank of Kuroki's 1st Army was held by only six battalions under General Umezawa, the Russians wasted the 24th conducting a reconnaissance, during which time the Japanese fell back to defend their base at Pensihu. By noon of the 26th, the Japanese began to receive reinforcements from General Inoue's 12th Division. Still, Shtakel'berg possessed overwhelming superiority—eighty-six battalions and squadrons against nineteen—and it remained only for him and his subordinates to fight their way through Kuroki's thin screen to begin an envelopment of Oyama's right.

For various reasons, however, Shtakel'berg failed to bring his overwhelming superiority into play, and his failure in turn decided the fate of Kuropatkin's offensive. Part of the failure stemmed from excess caution: Kuropatkin himself on 26 September instructed Shtakel'berg to conduct a full reconnaissance of Japanese positions while preparing for the attack. Meanwhile, the Japanese took advantage of the delay to shift reinforcements to the threatened area, and in hilly terrain only a few battalions could make the difference between success and failure, even against a greatly superior force. Once the Russians jumped off into the attack on 28 September, lack of coordination—stemming from the ruggedness of the terrain and clumsiness of command arrangements—deprived infantry and cavalry units of mutual support and adequate artillery fire. Despite Russian numbers, the consequence of these and other difficulties was defeat for Shtakel'berg and his lieutenants. In the most memorable of several failures recorded on the Russian left, on 28 September III Siberian Corps attacked six times and failed six times at a cost of 5,000 casualties to take Inoue's trench line on heights overlooking the Tai-tzu. The following day, despite additional heroic attacks, Shtakel'berg could make no further progress, and what little there had been of the Russian offensive in the east was overtaken by events in the Russian center and west.[78]

In the west, General Bil'derling's detachment (77 battalions, 56 squadrons, and 222 guns) had begun its southward movement on 22 September astride the Mukden–Liaoyang rail line. On the 23rd, after a march of only a dozen kilometers, Bil'derling reached the Sha-ho and dug in. His was only a supporting attack, and from the beginning the Western Detachment failed to press the initiative against Japanese forces arrayed in comparatively open terrain north of Liaoyang. In fact, Bil'derling's offensive technique was to advance unhurriedly through a series of defensive positions while waiting for news of Shtakel'berg's progress.

Once Marshal Oyama had determined that the main Russian effort was in the east, his own dispositions would not neglect the Western Detachment. As soon as he felt that Kuroki's defense in the east would hold, he characteristically decided to wrest the initiative from Kuropatkin by attacking in the center and west. His assertion was that "one ought never to allow oneself to be forced into the defensive, even by superior forces."[79] Therefore, on 27 September, he hurled Nodzu's 4th Army along the Mandarin Road and the rail line and struck out to the west with Oku's 2d Army against what he thought was Bil'derling's open right flank. Although there were few resources and little time for deep or elaborate maneuver, Oyama's intention was to envelop the Russians from the west. Consequently, contact between opposing forces would unfold in a series of meeting engagements fought for control of minor elevations and walled villages. Victory beckoned in the form of the unexpected gap in lines or the elusive open flank.

In the ensuing battles fought on 29 and 30 September within sight of the Shili-ho valley, the Russians suffered from lack of coordination, occasional lapses in judgment, and an inability to develop sound defensive dispositions. I Army and IV Siberian Corps, which were loosely strung out in Kuropatkin's center as a reserve, suddenly found themselves hastily occupying positions to ward off elements of the 1st and 4th Japanese armies. Because of broken terrain and the rapidity with which the Japanese developed their attacks, elements of the two Russian corps entered battle piecemeal and suffered a series of tactical defeats in detail. Counterattacks with cold steel did little to counter local Japanese superiority in artillery and rifle fire. Meanwhile, in the center of the Western Detachment, X Siberian Corps also gave

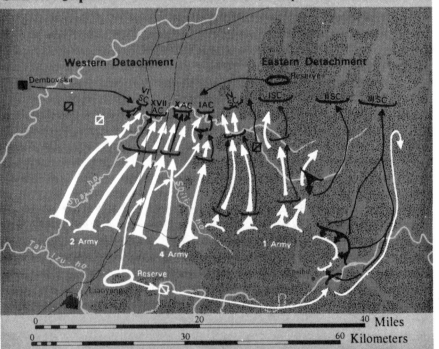

ground as its forward elements fell back to main positions along the Sha-ho. On the extreme right, XVII Corps grudgingly gave up terrain, then used nearly all its reserves to restore its original position, leaving nothing to ward off envelopment from the west. Until about noon on 30 September, VI Siberian Corps lay undiscovered and unengaged approximately eight kilometers to the rear and slightly to the west of the hard-pressed XVII Corps.[80]

The VI Corps commander, General L. N. Sobolev, justified inaction on the grounds that his troops, untested reservists, could better support the Western Detachment by remaining in a static defensive position that would eventually provide an anchor for Bil'derling's right. When XVII Corps began falling back under heavy Japanese pressure, Sobolev finally sent forward the 219th Iukhnov Regiment, which marched briskly to contact, swung onto line, dressed smartly in parade-ground order, and advanced to meet the enemy in a textbook assault formation of company half-columns. The astonished 8th Japanese Regiment promptly took the Russians under artillery, rifle, and machine-gun fire, inflicting 2,000 casualties upon the attackers within a matter of minutes. The Russian reservists performed bravely enough; they simply were not trained adequately to meet the test of contemporary ground combat. Their attack was a reversion to tactics which had been discredited nearly three decades before at Plevna.[81]

Despite this and other local successes, by 1 October, Oyama concluded that annihilation of the Russian army lay beyond his grasp. Mere existence of VI Siberian Corps, whatever its ability to assume the offensive, meant that the Japanese envelopment retained only tactical significance. While Russian dispositions appeared ragged in places, Kuropatkin's defenses offered little immediate promise for either avoidance or penetration and subsequent exploitation. Meanwhile, the best Japanese troops lay exhausted, thanks to their unrelenting attacks across the front. Losses were mounting, and artillery ammunition was running low. The time had arrived for a pause, and Oyama accordingly issued orders to limit pursuit of the Russians to the banks of the Sha-ho.[82]

While Oyama acted to dampen the flames of battle, Kuropatkin belatedly moved to fan them. Even as Shtakel'berg's Eastern Detachment prepared to withdraw, Kuropatkin ordered it both to conduct a limited counterattack to cover disengagement and to surrender its reserves for transfer to the west. On 1 October, II Siberian Corps marched into a local attack with mixed success against the Japanese 12th Division and the Guards Regiment. On 2 October, after a confusing exchange of dispatches, Shtakel'berg finally agreed to give up twenty-two battalions and four batteries to reinforce Kuropatkin's center and right.

With these reinforcements, the Russian commander heated up the battle in X Corps sector, where he believed that Nodzu's 4th Army now threatened to split the Russian defensive dispositions just east of the rail line. Two terrain features, One Tree Hill and another rise, which later came to be called Putilov Hill, dominated approaches to the Sha-ho. With his forces immobilized on the left and beginning to give way on the right, Kuropatkin frantically tried to reverse the tide in the center to retain control of his main route of possible withdrawal. On 2 and 3 October, Kuropatkin managed to pull together two brigades and a separate regiment to secure

One Tree and Putilov hills. By dawn on the 4th, after two days' heavy fighting and 3,000 Russian casualties, the center of the line on the Sha-ho remained in Russian hands.[83]

By this time both sides were exhausted, and the Sha-ho operation came to a novel close, with the adversaries facing each other south of Mukden in trenches across a no-man's-land which extended across a frontage of more than forty-five kilometers. Sha-ho had cost the Russians 41,351 casualties, of which about 11,000 died. Japanese losses were more than 20,000, of which nearly 4,000 died. Battles of varying intensity had raged for nine days across sixty kilometers of front. For the first time in modern combat, the Russians had engaged in a major regrouping of forces during the actual course of an operation, a departure heretofore considered impossible to implement with any degree of success.[84]

Inkou and Sandepu

Russian cavalrymen and military theorists had long dreamed of the large-scale raid which would yield strategic fruit and wrest the initiative from the enemy. In late December, while both the Japanese and the Russians were recovering from exertions on the Sha-ho and digesting the full implications of Port Arthur's fall, Kuropatkin devised a bold plan to launch his cavalry on a strike deep into Oyama's rear. The idea was "to play the trump card," as one staff officer put it, that is, to sweep far to the Japanese left, then cut back to the center both to sever rail and telegraph lines and to destroy supplies as far south as the port city of Inkou.

On 23 December 1904, Kuropatkin issued orders that General Mishchenko was to gather a force of about 7,500 cavalry and mounted scouts with sufficient supplies and demolition equipment to conduct the operation he had in mind. With only several days' notice and with forces hastily brought together for this operation, on 27 December Mishchenko plunged into the snow-swept plains west of Mukden to begin a wide detour to the left of Oyama's principle field dispositions.[85]

The challenge of commanding units brought together on short notice for a special operation was heightened by two additional difficulties: failure to bypass enemy dispositions and failure to effect surprise. Although Mishchenko had instructed his officers to avoid Japanese concentrations, several of his columns soon ran afoul of garrisons in small villages, and the Russians extricated themselves only after forfeiting both men and the element of surprise. Heavily laden carts and pack animals also did their share to slow the advance. Nevertheless, by noon on 30 December, Mishchenko's cavalry reached Takauzhen, where they awaited darkness to conduct a night attack against Inkou station and several nearby objectives.

Both the halt and Mishchenko's dispositions nearly proved fatal to the raid. Even as his men waited, a troop train carrying a battalion of Japanese infantry steamed into Inkou station. Despite this turn of events, Mishchenko allocated only a quarter of his force to attack the station, his primary objective. The remainder he assigned to security and a series of diversions and secondary objectives. Consequently, Mishchenko did not satisfactorily accomplish any of his tasks. His main attack on the station faltered when Russian light guns set fire to several storehouses, thus illuminating for Japanese defenders the Russian troops attacking on foot. After several fruitless assaults, Mishchenko's raiders withdrew after shelling the station and setting ablaze several additional outbuildings.[86]

The parties conducting diversionary and demolition assignments fared little better. Without heavy explosive charges, cavalrymen attacking rail lines and bridges did little more than superficial damage. When troopers conducting these subsidiary operations rejoined the main force, the entire detachment barely eluded pursuit and ambush as it recrossed the Hun-ho west of Newchuan. Altogether, Mishchenko's cavalry lost 341 officers and men while destroying two trains, taking nineteen prisoners, and briefly disrupting communications between Liaoyang and Inkou. The Japanese were able to restore communications within six hours of the raiders' departure, and one Russian commentator subsequently admitted that "the result attained by the detachment had not justified our hopes." In explaining the failure, he realistically cited three causes: organizational shortcomings, inability to effect greater speed, and failure to concentrate force in a timely manner at the principal objective.[87]

Even as Mishchenko's cavalry struggled across the frozen Manchurian plain, Kuropatkin summoned his commanders to consider the prospect of launching an offensive before Nogi's army could make its way north from Port Arthur to reinforce Oyama. Thanks to a command reorganization in the wake of the Sha-ho operation, Viceroy Alekseev had departed for St. Petersburg to leave Kuropatkin in overall control of Russian ground operations in the Far East. In addition, the arrival of fresh troops from European Russia and dissatisfaction with earlier ad hoc formations had prompted the creation of three distinct Russian field armies in Manchuria. Kuropatkin now commanded an army group, notching another first in Russian military history. He proposed to his army commanders, N. P. Linevich, O. K. Grippenberg, and A. V. Kaul'bars (respectively commanding the 1st, 2d, and 3rd Manchurian Armies), that they launch an enveloping operation with elements of the 2d Army (Russian right) moving across open terrain against Sandepu, an enclosed village which Kuropatkin reckoned to be the key to Oyama's dispositions just east of the Hun-ho, about twenty kilometers west of the railway and thirty-eight kilometers southwest of Mukden. Once the initial attack proved successful, offensive action would ripple across the entire front as the 3rd (Russian center) and 1st (Russian left) armies entered the fray. On 3 January, Kuropatkin issued orders for his attack, the object of which was to drive Oyama back across the Tai-tzu while inflicting as many casualties as possible.[88]

The engagement that followed took its name, Sandepu, from the village around which much of the action swirled. From the beginning of the offensive, which ran for four days, commencing on 11 January, the Russian advance lacked heart, preparation, and direction. Grippenberg—deaf, aged, and nervous—remained unconvinced of the wisdom of attacking with ill-equipped troops in temperatures that slipped lower than minus twenty degrees Celsius. He moved his attacking divisions forward too soon, thus giving up the element of surprise. He failed to coordinate his attack with Kaul'bars in the center. Although Russian guns lacked sufficiently powerful high-explosive shells to damage field fortifications of frozen earth and stone, the field artillery labored mightily to support the infantry attack—but against the wrong target. Instead of Sandepu, which the Russians had difficulty locating on their maps, most of the artillery preparation fell on nearby Heikoutai. The 14th Division, a unit earmarked for action against Sandepu, was sent at the last minute on a futile chase to attack a nonexistent adversary. At steep cost, I Siberian Corps finally drove two Japanese battalions from Heikoutai. A day later, the 14th Division returned to the scene tired and by mistake conducted an attack not on Sandepu but against Pao-taitzu. On 15 January, after quarreling openly with Grippenberg and reacting with horror to mounting casualty lists, Kuropatkin called off the attack.[89]

Kuropatkin had committed about a tenth of his resources against Sandepu. However, the increasing application of new technology, including especially the machine gun, meant that casualties were out of proportion with commitment. In addition, the weather proved nearly as deadly as the weaponry. In all, the Russians lost 12,000 killed and wounded, including over 7,000 from I Siberian Corps alone, against Japanese losses of about 9,000. Grippenberg declared himself ill and left in a huff for St. Petersburg, pausing long enough in Harbin to denounce Kuropatkin openly as a traitor. Meanwhile, the Russians had done little to spoil Oyama's prospects of pursuing his victory on the model of Sedan.[90]

Mukden

Oyama began planning for that victory during the last week of January, even as Nogi's troops were plodding north from Port Arthur. Russian dispositions actually favored a Japanese attack. Since the previous fall, reinforcements had continued to pour into Manchuria over the Trans-Siberian Railroad, and by February the Russians were strung out south of Mukden perpendicular to the railroad along a front over one hundred kilometers in length. On the Russian right (west) stood the 2d Manchurian Army (General Kaul'bars), in the center the 3rd Manchurian Army (General Bil'derling), and on the left (east) the 1st Manchurian Army (General Linevich). Altogether these forces represented more than a quarter million men under arms, but Kuropatkin's overall defensive positions lacked depth. True, the Russians had constructed a series of strong defensive positions south of Mukden, especially in the vicinity of One Tree and Putilov hills, but Kuropatkin's general reserves consisted only of XVI Army Corps, the 72d Division, and one separate brigade. In addition, with Japanese encouragement he had dissipated perhaps 50,000 men in a series of disparate and usually fruitless security operations.[91]

Before Kuropatkin could reconcentrate these widely dispersed forces for an effective offensive effort, Oyama intended to launch his own strike in yet another quest for his elusive Far Eastern equivalent of Sedan. To this end he planned to spring two surprises on the Russians, one organizational, the other operational. With the addition of Nogi's forces to Oyama's order of battle, Kuropatkin fully expected to face four separate Japanese field armies. Unknown to him, Oyama had created a fifth, that of General Kawamura, which included in its composition the veteran 11th Division of Port Arthur fame. Kawamura's mission was to concentrate in the mountains east of Mukden for an attack through Fushen against Linevich's 1st Army, thereby threatening the entire Russian defensive line with envelopment from the mountainous left flank.

However, this was only part of the larger operational design. Rather than rely for success on a pure single envelopment, Oyama planned to follow Kawamura's attack with both an energetic drive against the Russian center and a wide flanking movement by Nogi's army from the west. Thus, while holding constant pressure on the center, Oyama's flanks would clasp the Russians in the deadly embrace of a double envelopment. Because the ratio of Japanese-to-Russian forces stood close to one-to-one, Oyama had to count on deception and speed to tip the scales in his offensive favor. Therefore he planned to attack first in the east with Kawamura's 5th Army to draw reserves initially from the lightly engaged Russian right (Kaul'bars) and to divert Kuropatkin's attention to the Russian left. Once initial offensive dispositions had attained these objectives, energetic attacks in the center would unnerve and immobilize Bil'derling's army. While these parts of the puzzle were falling into place, Nogi's 3rd Army would undertake a wide detour in the plains around Kaul'bars's right to fall upon the rail line north of Mukden. Although Oyama lacked the cavalry to conduct an effective pursuit, he trusted to the speed of Nogi's flanking movement, when combined with Kawamura's renewed advance from the east, to close the ring, thus producing the complete encirclement—with attendant disruption and despair—necessary to bring about the surrender of Kuropatkin's entire army group. For attainment of this bold design, Oyama counted 270,000 troops with 1,000 guns and 254 machine guns. Against him Kuropatkin mustered 276,000 men with 1,200 guns and only 54 machine guns.[92]

To Oyama's credit, the Japanese marshal anticipated a partial Russian offensive, but not one of such magnitude that it might fatally disrupt his own plans. Kuropatkin had in fact provided for a renewal of the Russian advance on 12 February with a Sandepulike thrust from his right by Kaul'bars, but the attack was stillborn. It died amid reaction to what appeared to be alarming developments on the extreme left, where between 5 and 12 February detachments under P. K. Rennenkampf and Iu. N. Danilov recoiled under heavy pressure from Kawamura's 5th Army. Rather than push forward to develop the situation on his right in comparatively open terrain, Kuropatkin canceled the offensive and began laboriously shifting first his reserves and then entire formations from his right to the extreme left, thus playing into Oyama's hands. The poor quality of lateral communications plus the difficulty of hurriedly revising dispositions and command arrangements would militate strongly against rapid reversal of movement once the process gained momentum. This was precisely what the Japanese counted upon to make their operation a success. Oyama was in effect out to prove that battles are lost and won not only in the minds of generals but also in the capacities of modern staff organizations and subordinate commanders taxed with the necessity to react and adjust hurriedly to complex evolving situations.[93]

From 12 February, when Kuropatkin began shifting troops east, to the evening of 22 February, three discrete battles unfolded before Mukden. In the east Kawamura continued his advance, but as Kuropatkin's reinforcements arrived and screening elements fell back on prepared positions, the Japanese found the going

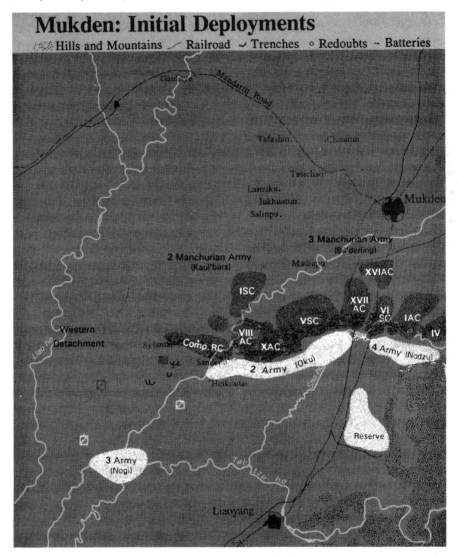

increasingly difficult. In the center, Kuroki and Nodzu attacked the heart of Kuropatkin's defenses, also making little headway. In the west, Nogi's flanking movement promised the greatest success, but not before he was forced to overcome a series of eleventh-hour Russian efforts to salvage the situation.

On 13 February, just after Kuropatkin had committed I Siberian Corps and additional reserves to reinforce his left, Nogi began his wide detour to the west. With the assistance of Oku's 2d Army, which commenced an attack to tie down Kaul'bars's front, Nogi set out in column of divisions, brushing the Liao-ho with his left, General Tamura's cavalry screen. Two days later, on 15 February, M. I. Grekov's

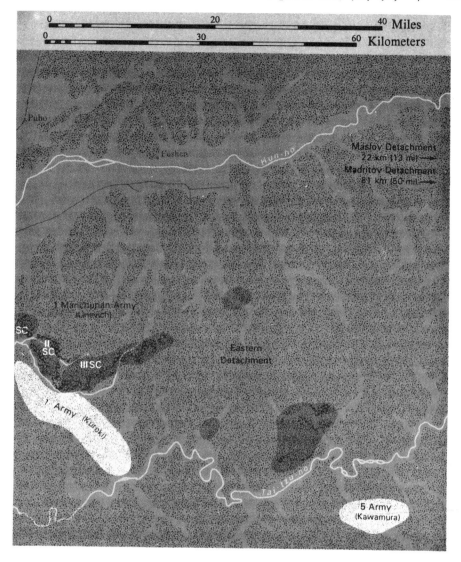

cavalry on the extreme Russian right engaged the lead elements of the Japanese 1st Division, then withdrew under pressure, leaving Kaul'bars's flank open. Although by noon of the 15th Nogi's troops had already reached the village of Syfantai—to the right and parallel with the 2d Manchurian Army's front—Kaul'bars was hard pressed to assist Grekov's cavalry. The 2d Army's dispositions lacked depth, and the only army reserves available would have to be stripped from units in the line. Without drastically altering his dispositions, Kaul'bars was no longer in a position to offer decisive resistance to Nogi's envelopment.[94]

By 15 February Kuropatkin had also become sufficiently alarmed to begin react-

190 / *Bayonets before Bullets*

Mukden Operation
5 - 12 February 1905

13 - 18 February 1905

ing to the threat unfolding on his right, but not to the degree that necessitated diminishing his commitment to the left. Drawing from XVI Army Corps, he hurriedly dispatched General A. K. Birger's reinforced brigade nearly forty kilometers northwest to Gaulityn with orders to stop what he believed was only a minor threat to Russian communications north of Mukden. Meanwhile, Kuropatkin had managed to concentrate 178 battalions on his left, a force that gave him a two-to-one local advantage over the attacking armies of Kawamura and Kuroki. To hold Kaul'-

Russo-Japanese War, 1904–1905 / 191

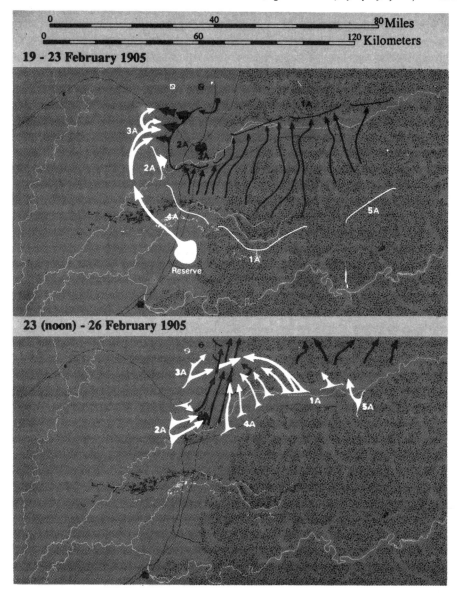

bars's attention, on 16 February Oku's 2d Army and Nodzu's 4th Army began a furious cannonade followed by a full-scale assault against the Russian 2d Army.

This attack and additional reports of 16 February from Grekov's cavalry finally brought home to Kuropatkin the realization that he was being pinned in the center while facing a full-scale envelopment from the west. With his line stretched to one hundred kilometers in length and with only the 25th Infantry Division in reserve, Kuropatkin now faced the moment of truth: neither his dispositions nor his assets would easily permit him to parry this threat descending inexorably from the west. In desperation he began once again to improvise, frantically throwing successive screens in Nogi's path and hoping that Kaul'bars could reposition enough of the 2d Army to eat into Nogi's flank.[95]

Accordingly, on 16 February, Kuropatkin took a series of measures—none of which were decisive—to counter Nogi. First, he made composites of two existing units, XVI Corps under General D. A. Topornin and a division under Major General N. A. Vasil'ev, both of which he dispatched northwest to reinforce General Birger's brigade. Next, he ordered General Kaul'bars himself to constitute from the 2d Manchurian Army a special counterattack force of thirty-two battalions, leaving General M. V. Launits in charge of a rump force to face the fury of Oku's 2d Army. Then, while Kaul'bars marched to strike Nogi in the flank, Launits was to wrestle with the tricky proposition of shortening his line and drawing back to his right flank while under attack and while retaining contact with the neighboring 3rd Manchurian Army. Finally, from the far left Kuropatkin recalled I Siberian Corps, which had not yet been committed to battle. Meanwhile, Marshal Oyama bade all his armies to renew the attack on the following day, the 17th.

Under pressure from renewed attack, Kaul'bars experienced difficulty in arraying his dispositions to face westward, extricating units, and assailing Nogi's envelopment. Early on 18 February, after delivering a series of local counterattacks and picking his way over roads cluttered with withdrawing baggage trains, Launits reported that his troops had reached their new defensive line. However, thanks to a lack of coordination between neighboring units of the 2d and 3rd Manchurian Armies, on 19 February a gap opened, and it was closed with resources and lives better spent elsewhere. Meanwhile, on 17 February, General Topornin's XVI Composite Corps had reached Salinpu, where it encountered Japanese cavalry. On the following day, Topornin's corps fought a meeting engagement with elements of two Japanese divisions, the 1st and 7th. All day on the 18th, the issue hung in the balance, while Nogi brought up units from two additional divisions and while the Russians also could have concentrated reinforcements. A misunderstanding of the situation—due to poor intelligence—caused Kuropatkin to squander his resources elsewhere. When Kaul'bars himself arrived on the scene, he also misread the situation, became alarmed when his dispositions appeared not to cover the western approaches to Mukden, then ordered Topornin to withdraw to a position which—unknown to Kaul'bars—was already covered by another unit. Because Kaul'bars had left the majority of his staff with Launits, he no longer possessed the ability to direct the battle.[96]

Ten kilometers to the northwest, Birger's brigade successfully engaged Tamura's cavalry in a daylong action along the old Mandarin Road at Tafashin, then failed to press its advantage both because of local uncertainties and because Kaul'bars had not established communications with his extreme right. Under the impression that he had been cut off, especially after Topornin's withdrawal, Birger stopped his attack with the onset of darkness. When he too attempted a withdrawal, his brigade broke up while moving in darkness in unfamiliar terrain. The withdrawal of Topornin and

Birger ended Kuropatkin's chances of striking Nogi's envelopment in its soft flank. Now the Russians faced the more difficult prospect of meeting it head on.

Both the Russians and the Japanese spent 19 February adjusting their dispositions. Despite the arrival of reinforcements, Kaul'bars failed to order an attack from either side of his newly occupied positions astride the Hun-ho. The Japanese took advantage of the lull to reinforce Oku and, more ominously, to deepen Nogi's envelopment maneuver by moving his striking tentacles laterally across the Russian front fifteen kilometers to the north. Kaul'bars's passivity assured the success of this difficult undertaking, but should Kaul'bars show signs of shaking off his lethargy, Oku planned energetic attacks for the 20th and 21st on the 2d Army's left.

After limited offensive action on 20 February, Kuropatkin ordered Kaul'bars to launch a full-scale counterattack on the 21st against Nogi. Accordingly, Kaul'bars formed three attacking columns (north to south: A. A. Gerngross, D. A. Topornin, and K. V. Tserpitskii) of about twelve separate detachments north of the Hun-ho. The mission of these columns was to launch coordinated attacks to establish a stabilized Russian position along a line running from Lamuku to Madiapu. The idea was that Gerngross with thirty-three battalions in the north would attack west and then swing south to crush Nogi's columns against Topornin in the center and Tserpitskii in the south. The problem was that the plan would have worked well two days earlier, when Nogi's columns were actually parallel with Topornin in the center. By the 21st, Nogi's columns were now even with Gerngross, a turn of events which would find Gerngross attacking Nogi head on rather than in the flank. Unaware of these developments, Gerngross went into the attack first against Japanese positions in and around the village of Chisatun. After initial success, his advance slowed, then stopped, thanks in part to lack of support and in part to strong opposition on the part of the Japanese 1st and 9th infantry divisions in the region of Tasichao. As Gerngross fought hard to avoid becoming enveloped himself, Kaul'bars was slow to come to his assistance. This prompted Kuropatkin's chief of staff, General Sakharov, to report, "It is necessary to ask the commander of the Second Army if he really fights with an army and not with a series of warriors for the rest of his troops to watch. . . ."[97] Kaul'bars simply had too little intelligence and too little control of his detachments to use them in concert where their strength would most count. As a result, at the end of the 21st, Kaul'bars called off the offensive, and one observer noted, "We shrank from the attack neither because of enemy superiority, nor because of moral weakness, nor because of losses, but only because of our inability to organize an attack on a large scale."[98]

The picture was brighter on the Russian center and left, where the 1st and 3rd Manchurian Armies stood firm against the Japanese onslaught. In fact, had either Kuropatkin or Linevich chosen to seize the initiative, they might have launched a general attack on the left to relieve pressure on the right. However, Kuropatkin preferred to confront Nogi's envelopment directly, and this meant that Russian reinforcements were needed on the right. On 22 February Kuropatkin warned his army commanders of the possibility of withdrawal to the Hun-ho to shorten lines and release troops.

On the evening of the 22d, Kuropatkin created still another detachment of twenty-four battalions, five squadrons, and fifty-two guns under General M. V. Launits, whose mission was to extend the right flank of the 2d Army's defensive line running north from Mukden along the rail line. Meanwhile, on the 22d Kaul'bars had spent more of his own precious manpower by throwing thirty-five battalions against several Japanese brigades in the vicinity of the village of Iukhuatun. Although a successful attack might have driven a wedge between the 2d and 3rd Japanese armies, the diversion of resources prohibited energetic countermeasures against Nogi, whose advance was temporarily slowing. As Launits's detachment made contact with Nogi's screen far to the north, a series of battles broke out along the Mukden rail line.[99]

Continued Japanese progress in the face of limited Russian resources on the right at last forced Kuropatkin to do what was now inevitable—shorten his front. On the evening of 22 February he issued orders for the 1st and 3rd Manchurian Armies to withdraw under cover of darkness behind the Hun-ho, thus reducing his front and freeing resources for commitment to the threatened northwest sector. But late on 23 February, as Kuropatkin siphoned off detachments for the west and as the 1st and 3rd armies settled into their new positions, the Japanese launched a series of probing attacks both to keep the Russians under pressure and to find weak places in the new defensive line. On 24 February the 1st Japanese Army broke the front of the 1st Manchurian Army east of Mukden at a point where a seven-kilometer segment of the front was held by only nine companies of IV Siberian Corps. As the Japanese developed this local success they were able to open a twelve-kilometer breach in the line, breaking communications between Linevich and Kuropatkin.[100]

Without knowing of this reverse, Kuropatkin had on the night of 25 February decided to abandon defense of Mukden in favor of establishing another defensive line south of Tiehling. To prevent the Russians from escaping, the following morning the 1st and 3rd Japanese armies broke through at Puho, cutting off the line of retreat for some units and baggage trains. The withdrawal of the 2d and 3rd Manchurian armies became especially difficult, in part because routes northward were clogged with fleeing baggage trains and in part because the Japanese breakthrough against the 1st Manchurian Army necessitated changing routes to avoid roving Japanese columns. The Russians incurred their worst losses when the 2d and 3rd Armies simultaneously fought a series of rearguard engagements and other actions required to keep open a narrow route of withdrawal north of Mukden between the rail line and the Mandarin Road.[101]

The Russians had been dealt another heavy—but not fatal—blow. The majority of Kuropatkin's troops managed to escape the Japanese grinding mill, although losses were serious. All the same, by 26 February his lead elements were approaching Tiehling, while on the 28th, the 1st and 2d Armies were behind defensive positions on the Chai-ho with elements of the 3rd Army in reserve. Kuropatkin's losses were 90,000 killed, wounded, and taken prisoner, of which the first two categories numbered about 59,000. The Japanese lost slightly over 70,000 killed and wounded. Not surprisingly, on 3 March 1905 Kuropatkin was summarily removed from command, although he successfully pleaded to be retained in a subordinate command capacity. During March the Russians would withdraw northward another ten days' march to a strong defensive position at Hsipingkai, where they continued to rebuild remnants of Kuropatkin's once mighty army and await the arrival in the Far East of Admiral Z. P. Rozhestvenskii's Baltic Fleet.[102] Only by summer would the threat of domestic revolution and Rozhestvenskii's catastrophic defeat at Tsushima Straits on 14 May cause the Tsar to accept American mediation that would end the war with the Peace of Portsmouth (5 September 1905 N.S.).

Rear Services

For several reasons, rear services during the Russo-Japanese War presented quite a different picture from the situation during the Russo-Turkish War. Varying circumstances notwithstanding, the results were the same: just as in 1877–78, Russian field armies in 1904–5 experienced shortages in key categories of supplies. Primary causes included the remoteness of the Far East, its poorly developed transportation infrastructure, and insufficient peacetime preparation for war.

The main culprit, according to the military historian L. G. Beskrovnyi, was the low carrying capacity of the Trans-Siberian Railroad. Although logistical planners assumed the capacity to transit as many as seven military echelons (rail transport serials allocated to military formations) per day, the actual Trans-Siberian figure for early February 1904 was five pairs of trains per day. Because of shortages in locomotives and rolling stock, the same figure for the Chinese Eastern Railway, which linked the Trans-Siberian with the supply hub at Harbin, was four pairs of trains per day.[103] During 1904, a combination of factors, including completion of a land link around Lake Baikal and improvements to sidings, roadbeds, and stations, yielded an increase in the transit capacity of the Trans-Siberian to twenty-two pairs of trains per day. In all, during the course of the war, some 2,698 troop and 2,529 freight echelons made their way by rail across Siberia. As A. A. Svechin opined, this total would have made a decisive difference only if it had arrived during the first five months of the war.[104]

Long distances and low carrying capacities meant greater reliance on local supplies accumulated in advance. Consequently, the intendance service had created thirty-seven supply magazines in the Far East, including twenty-four in the Priamur Military District, eight in Manchuria, and five at Port Arthur. Thus the majority of magazines was located in the closest military district rather than along the line Haicheng–Liaoyang–Mukden, where Kuropatkin's forces would actually concentrate. Just when combat-bound troops placed a premium on rail transit, the Priamur supplies would also have to be redeployed forward. To make matters worse, the intendance service was ill disposed and poorly prepared to avail itself of plentiful provisions available for purchase in the populous plains south of Mukden. On the positive side, Port Arthur's magazines at the outbreak of hostilities held flour, grain, and dry rations to last for more than 200 days.[105]

Despite various shortcomings, the army "was well fed," in the words of Svechin, who had witnessed events in the Far East at first hand. He attributed this fact to a well-organized supply system, a network based on possession of a secure rail head with steam and horse-drawn transport serving feeder lines linking the troops with their supply bases. Kuropatkin was so devoted to an orderly rear support system that he laid out more than 850 kilometers of narrow-gauge railroads, thereby simultaneously inhibiting his own operational flexibility while making his armies more vulnerable to disruption by envelopment.[106]

Medical services were more of an unalloyed success. Without taking into account losses suffered either at Port Arthur or in more isolated parts of the Far Eastern theater, the Russian army lost 40,000 dead from combat causes, another 140,000 wounded, and 13,000 dead from disease. This latter figure represented a respectable fraction of the total of 108,200 taken ill in the field during eighteen months of war. More than 50 percent of wounded soldiers were able to return to their units, a figure which in Svechin's estimation indicated that "medical facilities were equal to their task."[107]

Ammunition and armaments were another story. There were only two artillery depots in the Far East, at Chita and Khabarovsk. They held only sixteen quick-firing guns, sixteen mountain howitzers, four field mortars, and slightly more than 50,000 small arms. This meant that troops coming from European Russia and western Siberia would have to bring their own heavy armament. Even worse, as the conflict wore on, units remaining in European Russia would find their own stocks drawn down to support troops in the Far East. There was also an insufficient supply of artillery shells, especially at Port Arthur, where static defensive requirements would impose an especially high usage rate. While warehouses within the Priamur District held more than fifty million rounds of small-arms ammunition, this quantity fell short of army norms by some twenty-eight million rounds. Although there was sufficient clothing for winter operations, boots and outerwear were of poor quality, giving the troops a ragged appearance as they improvised substitutes. The realities of smokeless powder weaponry spontaneously encouraged officers and soldiers alike to dye their white summer tunics and forage caps a sickly yellow-green to achieve the same subdued appearance that newly adopted summer khaki uniforms now imparted to the Japanese adversary.[108]

With the exception of medical support, the rear service situation in the Far East appeared to reverse the terms of the Russo-Turkish War experience. That is, the provisioning of troops seemed at least adequate, while stores of munitions and weaponry fell short of requirements. Following 1904–5, regulations on army field administration would undergo alteration to eliminate the most glaring organizational discrepancies, while the continuing prospect of war in remote theaters would thrust into bold relief issues of peacetime stocks and their location. At the same time, officers who had served in the Far East reported a need for better coordination between agencies which oversaw the manufacture of military items and those charged with issue and repair.[109] Looming over everything was the problem of infrastructure. Without a flexible rail and road network and without the capacity to expand armaments and munitions production within a short span, much greater emphasis would fall upon detailed prewar planning for a short, intense war. Thanks in part to the Manchurian experience, systemic rigidities in support and sustainment would naturally reinforce prevailing expectations that any future military conflict must necessarily be brief.

Lessons Learned and Unlearned

Despite the changes of the intervening three decades, the Russo-Japanese War was curious for the way it repeated some of the fundamental lessons of the Russo-Turkish War. In 1904–5—just as in 1877–78—the entire shape of the land campaign was determined by the inability of the Imperial Navy to retain control of the sea. And again, although this time for different reasons, the preparation for the war revealed a serious—and nearly fatal—incongruity between policy and military capability. In this case, though, the Russians even more seriously underestimated their adversary than they had during the years of Alexander II and Miliutin.

At theater level, 1904–5 presented greater challenges for command and control. Increased use of the telegraph and the halting appearance of the telephone simply could not compensate completely for problems stemming from distances within theater, remoteness from the imperial capital, personality clashes, unsatisfactory command arrangements, and the prospect of high-level interference. Although the Tsar preferred to remain in St. Petersburg and occasionally intervene from afar, one might argue that his presence in the field could have resolved problems of conflicting and overlapping command.

In considering tactics and the conduct of operations, problems of personality were magnified by cumbersome command arrangements, lapses in coordination, and failures to employ new technology. As a result, the Russians never fully availed themselves of their inherent strengths, including possession of interior lines in Manchuria, to make up for weaknesses such as Kuropatkin's near-legendary indecision. To worsen matters, Kuropatkin lacked either the power or the resolve to rid himself of aging and incompetent officers in senior positions. An exaggerated deference to seniority, coupled with half-trained staff officers and overly rigid procedures and staff processes, further contributed to problems of command and coordination. Too late, Kuropatkin learned that he could not command from the front lines, and too late he came to the realization that hasty improvisation only aggravated fundamental deficiencies in command and tasking arrangements. Not without justification did A. A. Svechin, one of the foremost students of the conflict, conclude that Russian defeat stemmed in large part from various shortcomings in what he would later label "operational art."[110]

Because the conduct of operations is closely related to—and, indeed, reliant on—other aspects of war, Svechin's assertion should not blind the observer to other fundamental shortcomings in the Russian approach to war in the Far East. In an area distant from the heartland and bereft of supplies, logistical considerations tied Kuropatkin to the South Manchurian Railway and rendered his operational plans stereotypical and predictable. Port Arthur exerted a strong influence on Manchurian operations, when it should have been viewed as an economy of force operation tying down substantial Japanese resources. Neither the Main Staff nor governing authorities in St. Petersburg offered a sound strategic framework which would have guided the unfolding of theater-level operations. Improvisation—as in 1877–78—thus became the watchword, and improvisation was unable to compensate for naval failures, incomplete and unrealistic plans, difficulties of logistics, or shortcomings in command.

The war also revealed, but did not turn on, shortcomings in tactics and military organization. Although Dragomirov's emphasis on the offensive built around the infantry assault remained the centerpiece of Russian military art, troops in the field soon adjusted to the realities of ground combat in an era of smokeless powder weaponry. When weather and terrain permitted, they quickly learned to attack in open skirmish formation and to defend from field fortifications with overhead cover.[111] Except in the opening stages of the conflict, when the Japanese appeared better prepared to utilize the advantages of cover and concealment, the Russian infantry did not appear to suffer more than its Japanese counterpart from popular misapprehensions regarding the relationship between fire and shock action. Far too often after the initial stages of the war, Japanese and Russian infantry alike were recklessly thrown into the assault across open ground without adequate artillery support against concentrated rifle and machine-gun fire. Each side had its success stories with the bayonet and the bullet, and these stories required judicious analysis and assessment in order to understand what they meant for the future. Clearly the Japanese held an advantage in several areas, including rapidity of planning, communications, and execution of tactical designs. In addition, the Japanese benefited early from German instruction in the employment of artillery fire support from concealed positions.[112]

At the same time, Russian organization and tactics displayed a number of grave flaws which seem to have been peculiar to the tsarist system. Despite adjustments in cavalry organization and the addition of scouts since 1877–78, Kuropatkin was ill served in the realm of tactical intelligence. Despite emphasis during the interwar period on independent cavalry operations, Russian cavalry scored few successes in the Far East. Moreover, underestimation of the Japanese plus a European orientation meant that much of the burden of the Far Eastern conflict would fall on the shoulders of reserve formations. Yet these formations were often too ill equipped, ill trained, and ill led to confront the challenge of immediate combat against an experienced, determined, and disciplined adversary. Whether reserve or regular, an additional problem for Russian formations was a failure of coordination at the tactical level. Lapses in coordination not only meant that neighboring units failed to support one another but also that the three combat arms failed to support one another effectively. These failures were magnified by the appearance of new technology and by the greatly enlarged scale of battle and the battlefield.

Shortcomings in leadership—for several reasons—often aggravated lapses in coordination. Despite all the theoretical and doctrinal emphasis on offense, the Russian high command came to the Far East with a defensive mentality, and even after the required buildup for offensive operations, ingrained habits died slowly. In addition, if one of Svechin's major assertions is correct—and much of the evidence seems compelling—the Russian high command came to the Far East with a Napoleonic understanding of operations which had been outmoded by the pace of technological change and new methods for the conduct of battles and operations.[113] The propensity to prepare for set-piece climactic battle, a propensity reinforced by

1877–78 and Russian understanding of other recent wars, failed utterly to meet the challenge of the rapidly unfolding and often confusing series of engagements so eagerly sought by the Japanese. This Russian failure, in turn, was greatly reinforced by shortcomings in nerve, flexibility, planning, and tactical intelligence.

For all their inadequacies, the Russians adhered to the Peace of Portsmouth with an army group in Manchuria which the Japanese could not destroy. If it had not been for the growing prospect of revolution at home, the Russians might well have resumed the offensive. While the Russians were morally spent, the Japanese were physically and financially drained. So the war ended not with a ringing battle of annihilation but with the whimpering cries of exhaustion. Even in the opening years of the twentieth century, modern armies remained hostages to the home front.

6.

Theory and Structure, 1905–1914: Young Turks and Old Realities

> My objective insofar as possible is to help establish for all commanders both a *common* understanding of war and contemporary battle so necessary for our time and similar methods and procedures for the decision of combat missions (unity of doctrine).
>
> A. A. Neznamov[1]

> The departure point for all my measures was the goal of securing the creation of a Russian Army equivalent to the German Army.
>
> V. A. Sukhomlinov[2]

BETWEEN 1905 AND 1914, Russian military thought and action were fed by many streams, including lessons from the Far East, advances in speculative theory, foreign influences, and internal impulses of an evolutionary and revolutionary nature. The post-1905 climate of defeat and domestic revolution naturally gave rise to reappraisals and quests for new vistas. The Russian military press boiled over with contention as officers sought not only to rationalize defeat but also to prevent a recurrence. As ideas and programs gained adherents, cliques multiplied within and outside of the military establishment.[3] Perhaps not since the Crimean War had issues of military reform assumed such urgency for Russia. As various streams blended and separated, they sometimes flowed within familiar banks and other times carved new channels; many of the same streams—old and new—also came to water the intellectual and institutional development of a post-1917 regime that would become the inheritor of the modern Russian military tradition.

History and Theory

Military history remained an important source from which many theoreticians drank deeply. This was in part because officers were keen to use history as an instrument to understand what had happened in the war against Japan, and in part because military history continued to occupy a prominent place at the Nicholas Academy of the General Staff. Also, history in a grander sense still promised to provide the prism through which observers might refract experience to make sense of the larger philosophical and scientific issues confronting Russian military development. A vexing problem remained over how to relate military history narrowly interpreted as recent combat experience to military theory. Another problem was one of gaining a wider, serious audience. Or, as A. A. Neznamov, a 1900 graduate of the Academy and a veteran of 1904–5, confidently remarked in 1906, "now it is easier to study—military historical science has worked out everything and outlines everything in an accessible and understandable form; there is lacking only a desire and a serious regard for it."[4]

Whatever history's standing with professionals, one of the War Ministry's first priorities was to publish a comprehensive account of the recent war with Japan. Unlike the ill-starred official history of the Russo-Turkish War of 1877–78, the intent was that this project was to retain utility and timeliness. In September 1906, General Vasilii Iosifovich Gurko, son of the hero of 1877, assumed energetic chairmanship of a special military-historical commission laboring under the newly created Main Directorate of the General Staff.[5] Gurko assured the Tsar that he would produce a history "in three years," and the commission eventually assembled more than 10,000 volumes of archival material on the operations of Russian forces in the Far East, including memoirs and the observations and publications of foreign military attachés. Although a shortage of funds and the sheer mass of materials held up work, publication was actually begun in 1910 of a nine-volume official history (in sixteen books). Finished in 1913, the history treated most aspects of the war in excruciating detail and was subsequently translated into French, German, Italian, and Japanese.[6]

Energetic leadership assured the appearance of a new history in a more timely manner than volumes on the Russo-Turkish War, but in at least one respect Gurko failed to transcend the shortcomings of his predecessors: at crucial junctures the narrative neglected to provide the kind of critical analysis from which readers could profit in developing a realistic and sober view of "lessons learned" from wartime experience. Too many careers high and low remained at stake to subject the actions of commanders still on active duty to intense scrutiny in the glare of historically aided hindsight, and the trail of too many controversial issues led to high-ranking functionaries in and around the imperial court.

All this meant that more honest criticism would be relegated to the pages of unofficial publications less sensitive to prevalent political and military pressures. For example, Aleksandr Andreevich Svechin, a junior lieutenant colonel of the General Staff in 1910, published (apparently on the basis of materials borrowed from the historical commission) a survey history of ground operations in the Far East.[7] His book marked an important milestone in the career of a brilliant military theorist who would be ordained to serve both the tsarist regime and its successor. Svechin's treatment of Manchurian operations was remarkable for both its comprehensive nature and its sense of restrained candor. If so disposed, the discerning student of military affairs might find sufficient evidence in Svechin's account to assess blame, but at the same time the reader would have been subjected to what in many respects still remains the most incisive account of the Imperial Army's accomplishments and failures against the Japanese.

The Professional Perspective

Other and even more candid treatments appeared with private publishers, in unofficial sections of the military press, and in various publications of the staffs of military districts. While Svechin remained circumspect in his criticisms, allowing his narrative to uncover truth by degrees, other officers were less restrained. Kuropatkin was among the first to strike when in 1906 he rushed to press with his own apologia, a four-volume account of the war (*Otchet general-ad"iutanta Kuropatkina*) which tended to blame defeat on everyone except the author. Self-vindication aside, his memoirs (reduced to two volumes and translated as *The Russian Army and the Japanese War*) and the more balanced *Zadachi russkoi armii* (Tasks of the Russian Army, 1910) pointed out many faults that legitimately vitiated some of his culpability for Manchurian reverses. The same faults would require remedy before the army might be expected to fight successfully in any future war.[8]

Less self-serving and more creditable were the often bitter commentaries of junior staff officers who had witnessed the debacle from positions affording a tantalizing combination of perspective and scant power.[9] Particularly noteworthy were two books, *Posle voiny* (After the War, 1907) by A. V. Gerua and *Iz opyta russko-iaponskoi voiny* (From the Experience of the Russo-Japanese War, 1906) by Neznamov, who taught at the Nicholas Academy between 1907 and 1912. Together with similar works by E. I. Martynov, S. Mylov, and others, these books by Gerua and Neznamov gave voice to the indignation of military professionals who had seen their training and soldiers' lives washed down a bloody drain in the Far East. Professionals that they were, the young *genshtabisty* spared only the throne in their indictments of military incompetence and ill preparedness. When combined with evidence uncovered during the course of historical investigations, the assertions of this new breed of military professionals both explained military defeat and sought to give voice to a reform agenda.

They looked at the war systematically through the eyes of trained students and practitioners of the military art and saw some good and much that was bad. Despite lapses in training and the excesses caused by the Revolution of 1905, they were unstinting in their praise of the Russian soldier. "Our soldier does not merit reproach," Neznamov wrote, because "with inimitable energy he surmounted all difficulties of campaign in forty-degree [Celsius] heat and in impassable mud; he systematically went without sleep and emerged from ten to twelve days under fire without losing the ability to fight." Even when his officers were dead, this soldier voluntarily subordinated himself to others, and "he never gave up hope, never lost faith, never quit believing that he was stronger than the Japanese." Other officers echoed Neznamov's sentiments, noting especially the fighting qualities of the troops who made up the East Siberian contingents. What troubled the soldier most, what demoralized him more than anything else were orders to withdraw after he had fought well.[10]

Important items of armament also passed the test of combat. The Mosin rifle,

Model 1891, "revealed itself in the war to be completely satisfactory." Similarly, "the quick-firing field gun was wonderful," with accuracy, rate of fire, and range all exceeding comparable Japanese weapons. Experience did, however, reveal the need for modifications in shell fusing and explosive effect. Neznamov and others called for a more powerful high-explosive round to attack field fortifications and a shrapnel round with greater bursting effect against troops in the open.[11]

Other armaments and pieces of equipment had not fared so well. The Engel'-gardt mortars proved themselves nearly useless because of unreliability and insufficient range. At Sandepu, A. A. Ignat'ev watched the mortars brought up to attack Japanese infantry sniping from behind stone walls. After one or two rounds, the mortars' carriage wheels shattered: rubber recoil suppressors had frozen in the subzero cold, thus transmitting to the wheels the full impact of recoil.[12] Similarly, the diaphragms of field telephones had frozen, rendering them inoperable. The call went out for more and better equipment. Observation balloons had demonstrated usefulness, but they required adequate maintenance and support so that they would not be stranded on some rail siding with no hydrogen gas, as Neznamov had once noted during the Sha-ho operation. Searchlights would also retain utility in the future, but their location and subordination had to be worked out in advance to assure successful employment. Neznamov noted that "the Japanese searchlights operated on almost all dark nights and very much disturbed Russian work on positions and scout detachment reconnaissance."[13]

The litany of failures continued until it encompassed most perceived shortages and shortcomings. If mortars were a problem, shortages of field howitzers and mountain pack howitzers presented even greater difficulties. At the same time, the battlefield application of Sir Hiram Maxim's automatic death made a deep impression on many Russian officers, including Neznamov, who wrote that "machine guns have assumed great significance, and their absence (while the enemy had them) was sorely felt."[14] In addition, the Russians needed better field equipment, including boots, entrenching tools, and field transport.[15]

Even more telling in the struggle to resume the initiative in future war were problems of preparation, perception, time, space, and control. Those pre-1904 graduates of the Nicholas Academy who bothered to ponder deeply the events they had witnessed concluded that their staff academy education had been inadequate in more than one respect. In practical terms, the interminably long mapping and sketching exercises were of marginal utility. The course in astronomy was nearly worthless—unless as Ignat'ev noted, an officer happened to be cutting across country at night and needed to locate the exact position of a locality unmarked on any map.[16] Not nearly enough time had been devoted to applied tactics. Military history, while valuable, had to be firmly related to contemporary war fighting concerns. Graduates had to be educated to size up situations quickly, make accurate judgments, and capture their commanders' tactical concepts in understandable and workable field orders.[17] In other words, commanders and their staff officers must learn to grasp the essence of situations, react, then shape them to Russian advantage on the basis of common conceptions, procedures, and vocabulary.

New Departures

For these and related skills to make sense, they needed to be embedded within an intellectual framework which facilitated a clearer understanding of contemporary war. Even as updated in the 1890s, General Leer's understanding of strategy as a science with its own immutable laws—demonstrable through a close study of military history and especially the campaigns of Napoleon—remained too rigid, remote, and unimaginative to convey a sense of the complexities of contemporary battles, operations, and campaigns. Modern mass armies stubbornly resisted defeat in the single climactic battle which during the previous century had often decided the fate of an entire campaign, or even an entire war. The nature of battle itself was changing from a deadly affair mercifully lasting only several days to protracted struggle dragging on for several weeks. The railroad and the telegraph—and more recently the telephone and wireless—continued to play havoc with traditional notions of time, space, and timing. The same changes prompted a renewed call for a reevaluation of fundamental conceptions of envelopment and operation on interior and exterior lines. In a word, Jomini was out, and Clausewitz and the elder Moltke (as modified by experience and observation) were in. However, Moltke had to be understood not so much in the way he related to Napoleon as in the way that he and his disciples had revolutionized modern warfare in a new age of industrialism. More important, Russian officers needed time and encouragement to read systematically and ponder deeply.

In the realm of tactics, General Dragomirov's principles cried out for rigorous updating in light of new weaponry and attendant requirements for more flexible application and a new emphasis on combining the effects of the combat arms, especially infantry and artillery. The lethality of smokeless powder weaponry begged for a fundamental reevaluation of the relationship between fire and shock action in both offensive and defensive battle.[18] Greater dispersion seemed inevitable, but the problem was how to achieve mass and retain control with manpower and firepower spread over larger areas. Manchurian battles delivered new experience and new data into the hands of those who would update the lessons which A. M. Zaionchkovskii and others had seen in the conflict of 1877–78.

Preliminary conclusions offered scant comfort to tacticians. The Russo-Japanese War seemed to indicate that the modern tactical headache, the meeting engagement, was to remain a standard feature of military operations. To escape set-piece battles with their steep casualty rates, the sensible commander now sought both to avoid the assault of fortifications and to retain the initiative by attacking his adversary from the march while both sides were still moving to contact. Conventional wisdom now decreed that commanders also attempt to catch each other in the rear or on the flank to avoid costly confrontation with frontal firepower. Only the offensive promised both decision and all-important retention of the initiative in warfare. One of the ironies of the period was that renewed emphasis on offensive battle did not occur in utter disregard of changing military technology; rather, it evolved as a way of minimizing the lethality of the new technology.[19]

Whether or not stress on the offensive proved sound, conflict rattling across space and time had to conform with some kind of overall design. New means and methods of deploying mass armies within theater had led to engagements and battles unfolding seemingly helter-skelter across vast distances for days and even weeks, until physical, moral, and material exhaustion called a temporary halt. But how to make sense out of chaos, how to weld apparent confusion into a coherent whole? In *Posle voiny*, Colonel Aleksandr Gerua reflected on the teachings of Russian military thinkers and the writings of the German strategist Blume and called for a new concept to bridge the intellectual gap between Dragomirov's elementary tactics and Leer's undying (and ethereal) principles of strategy. Already in 1907, Gerua labeled his version of the bridge "applied strategy" (*prikladnaia strategiia*). Of emphatically modern significance, its function would be "to afford a series of firm rules for moving armies along contemporary routes of communication, securing these routes, equipping bases, maneuvering large armies toward the field of engagement, and organizing reconnaissance and so forth in the field."[20] Somewhat later, perhaps under the influence of the German *operativ*, he would interpose between strategy and tactics something that he called *operatika*.[21] Its function was to provide an intellectual perspective from which commanders and their staffs could envision and plan for the sum of disparate activities and actions over time and space that went into the makeup of a modern military operation. Gerua's term never gained currency. In the 1920s, however, Svechin and other Soviet military writers would supplant it with the less elegant term "operational art" (*operativnoe iskusstvo*).[22]

Perhaps Gerua failed to introduce new terminology because traditional conceptions of strategy retained sufficient flexibility to be hauled back to earth from Leer's ether. Theorists might differ with each other in their definitions of strategy, but there was common agreement that Leer's legacy lacked practicality. A new generation of officers extended the criticisms which Leer and his disciples had already witnessed in the 1890s, with the result that old terms and concepts were subjected to rigorous reexamination in the light of new urgencies. Thus there was more to the anti-intellectual grumbling of V. A. Sukhomlinov, future War Minister (1909–1915), and other critics of theory than the innate conservatism of the old guard, although, to be sure, reaction had its share of advocates. Regardless of persuasion, officers often subscribed to a false dichotomy, seeing themselves and their contemporaries as either "theorists" or "practitioners." Leer himself had once complained that his opponents had equated "theorist" with "scoundrel" (*negodiai*), and had not the great Suvorov once referred to men of ideas as "poor academics"?[23] With all its dangers, there was an understandable and growing impatience with theories that did not promise immediate application. For the time being, few saw the inherent danger in emphasizing practice over theory that Svechin—paraphrasing a French commentator—would warn against years later in a different context: "theory strives always to go hand in hand with experience, and sooner or later avenges itself if it is ignored too much."[24] After 1905, then, for a variety of reasons Leer's idealism was carried away in a new wave emphasizing theater and battlefield application.

Theory and Application at the Staff Academy

Military history shared many of theoretical strategy's vulnerabilities. Between 1905 and 1908, several commissions reviewed the curriculum of the General Staff Academy, with the last under General Sukhomlinov, then Commander of the Kiev Military District, recommending abolition of the Chair of the History of Russian Military Art. Part of this chair's teaching functions would revert to the more conventional Chair of the History of Military Art and part to a new Chair of the History of the Russian Army. The assumption was that instruction in the military art at the Academy should stress the common heritage and mutual relations of Russian and European military affairs. Or, except perhaps in a peripheral sense, the urgencies of modern military reconstruction seemingly outweighed concerns for the uniqueness of Russian military institutions. Not that the two were mutually exclusive—the point was that Russian military history had to demonstrate relevance to contemporary European military issues.[25]

The way in which the Academy faced this and related dilemmas in no small part determined how the army would meet the challenges of Russia's next war. In truth, Sukhomlinov's commission represented only part of a larger onslaught. P. A. Geisman, who remained an influential member of the faculty, wrote that "the failures of the last war evoked in the press and among the reading public rumbles and assaults against military science and the highest military school, the Academy, and against a whole group of its professors."[26] In response, the General Staff Academy created its own investigative commission and formally circulated a questionnaire among its graduates to determine what areas they thought needed improvement. Of the 300 addressees, sixty responded, with a majority grading the Academy on overall favorable terms. On a less positive note, some graduates noted that the Academy no longer held a leading role in military science, in large part because the courses in strategy, tactics, fortification, and artillery had become outmoded. Theory must be related more closely to practice, and something had to be done to revitalize key courses.[27] Consequently, between 1906 and 1909, the professors themselves drafted a new Academy regulation and took steps to make the program of instruction more relevant to the requirements of modern warfare. Similar steps were taken at the various service academies and within those institutions providing initial exposure of Russian officers to their branches of the ground forces.

The question was how far to push change, and the response was—not too far. The General Staff Academy rose to meet post-1905 challenges with a combination of half-hearted reform and inertia. The Commandant, N. P. Mikhnevich, combined with the powerful Academy Senate to frustrate Sukhomlinov's initiative by appointing A. K. Baiov, a well-regarded military historian, to a revitalized Chair of Russian Military Art. Meanwhile, several veterans of the Far East, including the redoubtable Lieutenant Colonel Neznamov, were recruited to lend the perspective of more recent combat experience to the teaching of tactics and strategy. At the same time, the majority of the faculty and its methods escaped major alteration. The result was that the old and new coexisted, partly for mutual benefit and partly to mutual detriment. Courses were updated and syllabi revised. Yet the lecture approach and many of the examination procedures remained the same. So did the dreary map-sketching exercises. Clausewitz found only an indirect impact on the curriculum. There were no war games, only exercises with stereotypical "school solutions." B. M. Shaposhnikov, an officer-student in 1907–10 and later Chief of the Soviet General Staff, recalled that one of the major changes came with the introduction of a mandatory first-year course in equitation. At Liaoyang, when Lieutenant Colonel N. I. Globachev, chief of staff of the 54th Infantry Division, had been summoned to replace his wounded commander, Globachev's horse bolted out of control, carrying off its rider and leaving the already badly mauled and confused 54th to disintegrate without capable leadership. The lesson learned was that competent troop direction required horsemanship, and the Academy did not overlook this fact in revamping its curriculum.[28]

Despite the always threatening weight of tradition and the ever-present danger of confusing superficial with substantive change, the atmosphere at the Academy proved conducive in a limited sense both to reexamining older verities and searching for new ones. Examinations and student projects often focused on comparisons across time, and faculty members such as Neznamov made their presence felt in field exercises and tactical problems. In addition, many of the students themselves were veterans of the Russo-Japanese War. As Ordinary Professor of Strategy, Neznamov brought a combination of background from the field and theoretical insights to his instruction. He was a brilliant tactician whose Manchurian experience and reading of German military theory prompted him to link tactical and operational conceptions with broader issues of strategy. Indeed, without using the terminology, Neznamov's course in strategy probably did a reasonable job of bridging the gap which Gerua had pointed out in 1907. Shaposhnikov later recalled that Neznamov's lectures were "something like instruction about operational art, neither grand tactics according to Napoleon's definition nor Leer's strategy of the theater of military action."[29] Students were at first captivated, then put off when they discovered that many of Neznamov's ideas came from a German military theorist, General Sigismund Wilhelm von Schlichting, whose works were first translated into Russian in 1908.

Reaction to Schlichting's influence indicated the degree to which segments of the military consciousness remained captive to "we–they" notions of indigenous evolution and foreign military domination. By the end of the nineteenth century, the appearance of contending nationalist and academic schools in Russian historiography had further aggravated the xenophobia of those officers who felt it was their duty to defend the Russian military patrimony from pernicious foreign influences. Thus, although Neznamov went out of his way to cite Russian military authorities with great frequency in his works, he was branded a "Westernizer," as were many reform-minded *genshtabisty*, who soon earned for themselves and their adherents the sobriquet "Young Turks." Against these Westernizers were arrayed latter-day descendants of the "Russian nationalist" school, which now championed the development of "a national military doctrine." One side preached the merits of military modernization, whatever the inspiration; the other trumpeted the necessity to search the immediate and more remote past to retain harmony with Russia's true national lines of military development. The issue, of course, was one of degree. While Neznamov considered military history an indispensable adjunct to theoretical development, the nationalist school saw historical understanding as the key to theoretical advances. In effect, the old feud between the Russian national and academic schools was now rekindled under different terms with different participants. Lines between the camps often blurred, but their discourses, definitions, and debates helped establish a framework for the continued development of Russian military theory.[30]

The Mature N. P. Mikhnevich

Nikolai Petrovich Mikhnevich, the strategic thinker who inherited Leer's mantle at the Academy, actually stood with one foot in either camp, but his publishing record and deep regard for historical studies meant that he was usually identified with the nationalist school. A disciple of the positivist philosopher Auguste Comte, Mikhnevich firmly believed in the evolution of both human institutions and knowledge from simple to more complex forms. This conviction simultaneously put him at odds with Leer's unchanging laws of military science and endeared him to historians wedded to an approach which stressed studying change within context over time.

Neither Mikhnevich's attraction to history nor his dislike for closed systems seem to have impeded his career: in 1904, he relinquished the Chair of Russian Military Art to become Academy Commandant; then in 1907 he left St. Petersburg successively for division and corps command, returning in 1911 to become Chief of the Main Staff, a position he held until 1917. All the while he managed to continue his theoretical work, producing a third edition of his landmark *Strategiia* (Strategy) in 1911. Despite his own sympathies, Mikhnevich no doubt hoped to place the weight of his reputation behind reconciliation of warring factions so that teaching and publication on military art would proceed under more harmonious circumstances. He was unsuccessful, and the subsequent and highly contentious interpretation and reinterpretation of historical and contemporary military realities and their implications for the conduct of war helped account for an outburst of originality in Russian military thought and doctrine before 1914.[31]

Individual and group allegiances aside, a common point of departure for those who seriously pondered military issues was speculation over the nature and character of future war. Would future European conflict be a "lightning war" in the manner of 1870–71? Or would it follow the lines of protracted struggle in the manner of the wars of the French Revolution and Napoleon? Answers to these questions drew upon insights and experiences gained from a variety of sources and experiences. The same answers also determined issues and emphases across a range of war-related considerations. By now, nearly everyone acknowledged the impact of technology—although perhaps in varying degrees. Likewise, everyone acknowledged the likelihood of a future coalition war waged by multimillion-man armies. However, here the similarities in outlook ended. If war was to be brief and violent, stress would fall upon immediate preparation, speedy deployment, a spirited offensive, and firm tactical and operational linkages during the initial period of conflict. If war was to be protracted, then stress would fall on strategic depth, full mobilization capacity, measured responses to operational challenges, and maintenance of internal unity and firmness of purpose.

Mikhnevich's *Strategy* touched on nearly all these issues with its far-reaching and integrative inquiry into the nature of military science, strategy, and tactics in an age of mass armies. In accordance with his own positivist views and in contrast with Leer's penchant to look for many laws, Mikhnevich saw only two: the law of evolution and the law of struggle. Both military institutions and knowledge about military affairs were evolving from simpler to more complex forms. For him, such a thing as military science existed, but only in so far as it was "an objectively verifiable and systematic knowledge about real phenomena from the perspective of their regular recurrence [*zakonomernost'*] and unchanging order." Within Mikhnevich's dynamic scheme, the search for laws gave way to a search for principles with an emphasis on the need to seek unity of theory and practice. The main objective of a theory of military art, he wrote, was "to establish firmly its fundamental principles, to study the most fundamental elements of a situation, and to indicate in light of the situation how principles are to be applied in war."[32]

Mikhnevich agreed with apostles of the Russian national school that the individual remained the center of war, but his understanding was more complex than simple emphasis on the role of individuals. In the past, the human element had been manifested in war through strategy and tactics. Now, the evolving complexity of human society meant that emphasis fell upon the manifold aspects of humanity in the aggregate. Or, to put it another way, the human element remained, but now manifested itself through more sophisticated institutions in a new, mass form. There was no romantic wistfulness for times gone by, only hardheaded acknowledgment of an emerging new order. For Mikhnevich, then, the laws of war were embodied in those social characteristics (numerical, physical, economic, intellectual, and moral superiority) which in sum determined the outcome of armed conflict. In a more limited military sense, the principles of war governed application of mass against the main objective and the attainment of moral superiority over the factors of materiel, accident, and surprise.[33]

In arguments reminiscent of the German theorist von der Goltz, Mikhnevich went on to assert that the competitive stakes were now so great that nations would go to war only on the basis of all their resources, physical and moral. With everything committed, modern war held the distinct possibility of transforming itself into protracted struggle that would involve the total resources of the state, a concept already advanced by Gulevich and Bloch. This vision of deliberate and calculated engagement explicitly called for a new kind of preparation and mobilization for war. It also called for a firmer link between domestic and foreign policies, a position that at least implicitly criticized the tsarist government's conduct of the Russo-Japanese War.[34]

Mikhnevich also held that Russia possessed some distinct advantages in waging modern war. One was the strong monarchy, which he saw as the best form of government for waging modern war. Another was the combative spirit (*voinskii dukh*) of the population, which promised persistent moral superiority. At the same time, Russia's comparative backwardness meant that its society was proof against the kind of wartime dislocations which would quickly imperil more complex western European societies. In different terms, his ideas were reminiscent of the nineteenth-century Slavophile conviction that Russia's backwardness was actually virtue when viewed from a different perspective. For Mikhnevich, durability and inherent spiritual strength meant that there was no need for Russia to be stampeded into a "lightning war." If need be, the Russians could revert to a Scythian strategy, calling upon depth and the resources of their land to outlast the enemy in protracted conflict. "Time is the best ally of our armed forces," he wrote; "therefore, it is not dangerous for us to adopt 'a strategy of attrition and exhaustion,' at first avoiding decisive combat with the enemy on the very borders, when superiority of forces might be on his side."[35]

Yet this was no excuse for deliberately embarking on a defensive war. A theorist of strong convictions and perhaps even stronger perceptions, Mikhnevich remained enough of a historian to know that the political price could be steep when trading land and lives for time. He encouraged the monarchy to increase military expenditures and to double the size of the army "in order not to fall behind the other states." Otherwise, "in a future general European war without allies, Russia would be forced to begin on the defense as against Charles XII [of Sweden] and Napoleon, which of course is *undesirable and disadvantageous*."[36]

Although very much a traditionalist in cultural terms, Mikhnevich saw changing technology exerting a profound impact on war. Indeed, since the 1890s, his evolutionary model of military reality owed much of its dynamism to an acknowledgment that technology was changing the very nature of battles and operations. He saw, for example, that the new smokeless powder weaponry imposed new battlefield requirements for calculating distances, intervals, and depths. These requirements, in turn, called for new tactical and organizational structures. At the same time, other technologies, including steam propulsion and telegraphic communication, imposed still more new requirements in planning for and conduct of mobilization, deployments, and operations.[37]

Mikhnevich's emphasis on planning not only called attention to the pressing need for rational economic development but also laid stress on the purely military aspects of the initial period of war. He held that strategic deployments should not occur in close proximity to the enemy so that concentrating forces would not be subject to attack before an army was fully capable of conducting operations. It seems likely that he also borrowed from the Germans and expanded upon Leer's teaching to evolve a suitable terminology to describe what occurred in war. Just as in the 1890s, he continued to write that "every war consists of one or several campaigns, and every campaign of one or several operations."[38] However, the understanding now was that this conception underlay a more modern understanding of operations and encouraged the kind of conceptual linkages across the war-fighting spectrum which Gerua had found so lacking in the pre-1905 intellectual environment.

The National School Redux

Mikhnevich's emphasis on the war plan as an embodiment of the human element in war found a sympathetic audience among military historians of the Russian school. From his position as Ordinary Professor of Russian Military Art, the historian A. K. Baiov stressed that military art was the product of military thought interacting over time with other human activity in all its manifestations. The specific emphasis on context—time, place, and people—meant that evolving military art bore a profoundly national character. All people might hold theory in common, but theory was translated into practice only within a specific national context. In pitting man (spirit, intellect, art, nationality) against material (matter, means, numbers, technology), Baiov stressed the role of the individual and said that "the main weapon of war has been and always will remain the human being." For him, the "indisputable national characteristic of our army is the unity between the army mass and its leaders. . . ."[39]

Emphasis on unity of theory and practice went to the heart of the utility of military history. Baiov retained access to and credibility with various significant audiences, including officers at the Nicholas Academy and members of military-historical societies. While he acknowledged the need for occasional borrowing from the West, he inveighed against pursuing "an alien view with doctrinaire obstinacy." He warned that "we are possessed by foreign views, the results of which have invariably produced among us a lack of ties between theory and practice." If Russian officers had done their historical homework better, the question of shock versus firepower, of bayonet versus the bullet and other false dichotomies never would have occurred. If the Russians did not know their past, it could not perform the function of unifying theory and practice. These were powerful and probably true assertions, but they lacked the substance and immediate application which contemporary urgencies seemed to require. For those who sought immediate answers to pressing tactical and operational exigencies, Baiov's ideas were simply not palpable enough.[40]

Neznamov and the War Plan

More palpable were the views of Aleksandr Aleksandrovich Neznamov, who shared some of Baiov's interest in history and more of Mihknevich's preoccupation with the war plan. Neznamov was one of the most outspoken of the "Young Turks" whose views were frequently interpreted as diametrically opposed to the nationalists, although differences were often more of degree and approach rather than substance and program. Neznamov well knew the value of history but chose to use it only as a point of departure, for 1904–5 had convinced him that the Russians simply did not understand the nature of contemporary war.[41] For Neznamov, the most pressing task confronting the Imperial Russian Army was not a generalization of past experience but an analysis of the probable means and methods of waging future war. "Even the past does not provide a full idea of the present, especially in our fast-moving century," he wrote. Therefore "past military thought cannot be ignored, but [military thought] must constantly make corrections because of present technical advances and, where possible, also peer ahead."[42]

As if in answer to Gerua's pleading, Neznamov extended Mikhnevich's thought to evolve a modern theory of military operations which joined planning and preparation to the actual conduct of operations and battles. Central to his thought—just as to Mikhnevich's—was the war plan. Like Mikhnevich, Neznamov believed that modern war would no longer be decided by the outcome of a single climactic engagement (Russian *srazhenie*). Rather, modern war consisted of a series of engagements and operations linked to one another by the overall concept of the war plan. The plan guided the fulfillment of discrete but related tasks; therefore the accomplishment of general strategic objectives occurred during the actual course of operations.[43] Neznamov owed his concept not only to a close and original study of the Russian experience in the Far East but also to a reading of contemporary European, especially German, military theory. He quoted von Falkenhausen on preparedness, paraphrased von Schlichting on the meeting engagement and modern battle, and believed in the relevance of von der Goltz's notion of the nation in arms.[44]

N. P. Mikhnevich (1849–1927) Strategist, Commandant of the Nicholas Academy of the General Staff 1904–1907, Chief of the Main Staff 1911–1917
Photo: Courtesy of A. G. Kavtaradze

A. A. Neznamov (1872–1928) General Staff Officer, Tactician, and Strategist
Photo: Courtesy of A. G. Kavtaradze

Neznamov's war plan was an integrated concept calling for a nation's total involvement in modern conflict. Implicit was a fundamental devotion to Clausewitz's definition of war as politics by other means, with all the attendant implications for unity of civil-military will. The necessity for a truly singleminded effort meant that before embarking on modern war, a nation had to take into account a number of considerations other than purely military factors, including economics, politics, morale, and culture. Neznamov's intent was not merely to emphasize method in war planning but also to underscore the importance of preparing the entire body politic for future conflicts, which would probably not resemble past wars. In actual war preparations, he parted ways with those who emphasized the importance of great Russian captains. Leadership was no doubt important, but the war plan itself was less the province of supreme authority than it was a function of a relatively constant set of objective factors: geography, climate, communications, strategic objectives, and centers of political and economic concentration.

The idea behind the war plan was to translate preparations into military realities which would allow one state to impose its will on another through offensive operations. This was the essence of strategy. The army's strategic deployment remained the clearest expression of a nation's war plan and its determination to seek decision. In the past, Neznamov declared that faulty dispositions had been "a chronic Russian weakness." In contrast now with Mikhnevich, who emphasized the inherent advantages of depth and the ability to trade space for time, Neznamov asserted that dispositions must be governed by a requirement to achieve speedy and superior concentration against the main threat while lesser threats were held at arm's length. Security of concentration was an absolute necessity, but distances were to be calculated not by historical rules of thumb but in accordance with knowledge of actual rates of deployment, concentration, and advance. Above all, in determining courses of action, Neznamov repeatedly intoned that "we must know what we want."[45]

Whether the Russians wanted it or not, Neznamov read the combined lessons of the past and present theoretical projections to emphasize maneuver warfare. Along with his contemporaries, A. G. Elchaninov and V. A. Cheremisov, Neznamov pondered the nature of contemporary and future battle to emerge with a vision that called for new attention to the application of mass through combined fire-and-maneuver tactics. The old combination of skirmish line and closed ranks in the assault had to give way to new forms of organization and attack. In addition, new ways had to be devised to concentrate all forms of firepower, for in Neznamov's view "fire was the primary factor in contemporary battle."[46] The appearance of various kinds of air assets both promised new forms of reconnaissance and attack and created the problem of air defense and active and passive security measures against air power. Despite the importance of mass, the lethality of modern weaponry opened distances and added depth at all organizational levels in the field.[47]

Battles, he believed, would be integral components of future operations conducted not only by a single army but by groups of two or three armies, a development that would create the need for additional organizational and intellectual linkages. Under the pressure of modern combat, success beckoned to commanders at all levels who displayed confidence and mutual trust in their own and other commanders' dispositions and decisions. Such confidence flowed from a common understanding of the nature of contemporary war and from adherence to a common plan. "Only battles are decisive in war; everything else serves only to prepare for them," Neznamov asserted. Therefore "it is understood that each troop unit, each column must press into battle with all it has . . . [and] under conditions in which units enter battle in the normal organizational structure."[48] Kuropatkin's Manchurian muddle had made a strong impression on Neznamov.

Manchuria also influenced the way that Neznamov viewed seemingly discrete aspects of combat within theater. He saw armed confrontation both as physical struggle and as a struggle for information and time. Speed enabled a commander to win these struggles, thus assuring retention of the initiative and constantly forcing an adversary to react. At the same time, Neznamov perceived that "just as all of a war is broken down into a whole series of operations, so is each operation broken down into a whole series of immediate tasks, in which the preceding ones condition the following ones, and all of them are combined into a single operation just as all operations are joined with one another."[49] Just as contemporary war could not be fought with older methods, neither could contemporary armies be defeated in a single engagement. Future war might well assume a protracted character. Manchuria had demonstrated that war was now a series of "separate offensive leaps forward and defensive leaps backward." Thus appeared in embryonic form a theory of successive operations.[50]

One of Neznamov's lasting contributions to military theory was to ascribe a central place to the operation as a phenomenon of contemporary war. In contrast with Leer's more abstract categories of operations (main, preparatory, and supplementary) Neznamov offered a down-to-earth classification of operations as either offensive, defensive, meeting, or delaying (*vyzhidatel'nye*), with the latter two being variants respectively of the first two. He also emphasized preparation and conduct, asserting that these aspects of operations were evolving in complexity from the concepts of individual commanders to "purely scientific" requirements which involved not only the art of army commanders but also the precise work of their staffs.[51]

From these and related ideas flowed conceptions which visualized modern war unfolding across a broad strategic front in which envelopments and breakthroughs became major operational objectives. While envelopment (shallow and deep) and encirclement operations had long held central stage in German military thought and teaching, there was increasing evidence that the breakthrough was gaining its share of adherents, both Russian and German. The objective of the breakthrough was to drive a wedge into the enemy's strategic front, then to develop success in depth and outward, thus threatening at their core an enemy's communications and organizational coherence. Individual successes during the course of the breakthrough would assure larger successes within the theater of operations. Overall success depended upon superiority in forces and means, particularly in the realm of artillery support. It was emphasized that the breakthrough would enjoy success only under conditions of the cooperative action of all arms.[52]

How to conduct Neznamov's vision of future military operations? In rejoinder to the nationalists, he asserted that the traditional Russian virtues of bravery, self-sacrifice, stolidity, and self-sufficiency—although still necessary—no longer sufficed. Now, more than ever, the army needed knowledge, training, and correct utilization of national assets, and it needed to apply them in accordance with mutually understood principles and methods. Schooling and training in advance of war were the keys to releasing the moral potential of the Russian soldier and elevating the competence of his leaders.[53]

Unified Military Doctrine

In 1911 and 1912, these assertions placed Neznamov in the center of a storm of controversy over the concept of a "unified military doctrine." The uncertainties of modern battle and the new emphasis on "maneuver-mass-fire" in war meant that unity of outlook was vital in the preparation for and conduct of contemporary battles and operations. Such a singleminded outlook could flow only from common adherence to a unified doctrinal vision before the onset of hostilities. Thanks to the imminent appearance of a new field regulation and the prodding of Neznamov and others, soon a lively discussion about the need for such a doctrine appeared in the pages of the Russian army daily, *Russkii invalid*.[54] As Neznamov's notion of doctrinal unity garnered adherents and generated enemies, resulting debate fed additional fuel to the fires of the nationalist-Westernizer controversy.

The debate must be understood from several perspectives. One problem was that conscious use of the term *doctrine* itself was new to Russian military thinking. For historians such as A. M. Zaionchkovskii, who had delved deeply into the Russian military past, the translation of any sort of accepted ideas into practice would invariably lead to "dogma" and stereotypical execution (*shablon*). For Zaionchkovskii, who had just completed a history of the Crimean War, the imposition of a common doctrine would quite naturally reinforce some of the worst organizational tendencies he had seen within the Russian military of an earlier era.[55] Other commentators, including N. Chertyrkin and A. Adaridi, believed that doctrine smacked of something foreign, something that distinguished rigid Japanese and German approaches from truly flexible Russian military art.[56] Thus, varying perceptions lay at the heart of the dispute: one side saw promise; the other saw peril. On one hand, advocates of a unified doctrine saw the necessity for unity of concept, language, preparation, and action. This meant more than commonality of command language. Neznamov argued that each division commander in Manchuria had fought with sixty-four fine companies, but that the whole rarely generated sufficient energy to exceed the sum of its parts.[57] He might have added that the same was true—even more so—at successively higher levels. In future war, doctrine would be the key that unlocked the full potential of Russian forces. On the other hand, the enemies of doctrine saw it as an intellectual straitjacket which inhibited initiative, limited independence of command, and stifled military creativity.[58]

Muddying the water were extraneous "we–they" issues and genuine misperceptions about roles played by institutions, history, and ideas in doctrinal development.

Because Neznamov was vitally involved in writing the new field regulation as Professor of Strategy at the Nicholas Academy, some observers saw in his activity an ominous trend toward intellectual centralization. To be sure, there were observers and commentators who genuinely believed that doctrinal research and instruction legitimately remained in the province of the Staff Academy and the General Staff. This was anathema for those who saw the genius of Russian military art flowing upward from the company and the regiment where the will of officer and soldier combined to achieve that much-sought-after unity of theory and military practice.[59] All that was required was to update Suvorov's *Art of Victory* to account for changed technology. Others, especially Zaionchkovskii, correctly saw that Neznamov had failed to perceive the difference between German doctrinal prescription and the realities of Japanese combat application in Manchuria. To understand the limits of application, the historian must carefully sift the evidence, then interpret his findings to the theorist. Finally, others saw that doctrine simply promised more than it could ever deliver under conditions of genuine combat.

The controversy raged until the late summer of 1912, when the Emperor himself intervened to end the dispute. General N. N. Ianushkevich, Commandant of the Nicholas Academy, was summoned to Tsarskoe selo, where Nicholas II slashed the Gordian knot of definition by proclaiming, "Military doctrine consists of doing everything which I order." Lest the Emperor be misunderstood, he instructed Ianushkevich to order Neznamov "not to appear any more in print about this." *Russkii invalid* was also to be informed of the Tsar's decision.[60] Not until after the Tsar's death in 1918 would the issue of unified military doctrine reappear in the pages of the Russian military press.

The salvo against Neznamov was only part of a continuing general offensive against the Young Turks. Their links to the Tsar's uncle, Grand Duke Nicholas Nikolaevich, and their growing political influence had made them, in the words of War Minister Sukhomlinov, "a co-government."[61] Between 1909 and 1912, suppression came in the guise of reassignment, with the more prominent Young Turks, including V. I. Gurko, A. S. Lukomskii, and N. N. Golovin, suddenly receiving postings to responsible positions outside the imperial capital. Too late they would return to prominence—after the onset of war had vindicated many of their views.[62]

Against the background of personnel turmoil and intellectual ferment, Lieutenant Colonel Svechin remained a voice of sober calculation. He understood the contemporary emphasis on the offensive from the beginning but was careful to look ahead in case initial operations failed to produce decision. In 1913, he assessed the significance of potential coalition operations both west and east and concluded that the strategic center of gravity was slowly shifting to the east, thanks to demography, distance, and improved Russian military preparedness. In the event that French and tsarist armies failed to deliver rapid decision in any future conflict, the two nations would be well served to seek a balance between offense and defense. Svechin did his calculations and concluded that combatants might plan for an early victory but must be prepared for protracted conflict.[63] This was a theme to which he would return in the 1920s with lamentable personal consequences. Before 1914, his voice was lost in the whirlwind accompanying overcommitment to the French and overconfidence in the decisive effect of initial operations.

New Directions, Old and New Constraints

Both during and after the Russo-Japanese War, the same sense of shock and soul-searching which rocked the theoretical community also shook the foundations of government and high military administration. Gradually, as reverberations heightened, then subsided, St. Petersburg responded to real and perceived lessons of Far Eastern debacle. The monarchy's work was complicated as it wrestled with a rising wave of revolutionary unrest and contended with limited constitutional reform. After 1908, crisis in the Balkans and a heightened sense of alliance obligations added new urgency to the chorus of voices calling for renewed attention to military preparedness. As civil and military changes wound their tortuous courses, new structures and procedures emerged to confront old monarchical and military realities. As in the realm of theory and doctrine, the mixture of traditional and novel both established the context for reform, set its limits, and shaped its development.

As an armed force, the Russian army came away from the Far East spent, better suited to quell revolutionary disturbances than to face modern war. Even as civil authorities pressed the army into reluctant action against the internal foe, the army itself was shaken by mutiny.[64] Once the tide of mutiny subsided, the prospects for military reconstruction faded before the magnitude of the task. The Russo-Japanese War had cost the tsarist government more than 2.5 billion rubles directly and another four or five billion in indirect losses to the economy. Hostilities had eaten deeply into the wartime stocks of nearly all military districts. Many shortages, ranging from machine guns to heavy artillery, required immediate attention. The combat arms cried out for structural modifications on the basis of recent combat experience. Other branches such as transport required reorganization from the ground up. The War Ministry had to begin the task of identifying shortages and remedying wartime dislocations which would outlive the fighting to threaten future mobilization allocations and timetables. Some army corps had no engineering troops, while others were at a loss to match their infantry with appropriate artillery units. While the lessons of defeat remained fresh, the Russian army had to look at itself from the top down, then, while time remained, retrain, reequip, and reorganize itself to face the possibility of future war.[65]

Already in the spring of 1905, the Emperor conceded the need for greater integration of military direction at the top, from which flowed two significant organizational changes, neither of which fulfilled its initial promise. The first involved the convocation in May 1905 of the State Defense Council under the chairmanship of Grand Duke Nicholas Nikolaevich. With six permanent and five changing members, the Council's mandate was both to impose unified leadership on Russia's ground and naval forces and to determine the Empire's overall military policy. Already in the 1890s, Obruchev had envisioned something like this, but the post-1905 reality of shifting priorities and allegiances, coupled with the complexities of court and state politics, were to make Obruchev's dream of unified direction unattainable, even under drastically changed circumstances.[66]

The second important—and contending—organizational initiative was the creation in June 1905 of the Main Directorate of the General Staff, or the GUGSh (for *Glavnoe Upravlenie General'nogo Shtaba*), as an entity independent of both the old Main Staff and the War Ministry. Although many commentators had cautioned against the institution of a Prussian-style General Staff, the chief of which reported directly to the Emperor, the desperate military mood of 1905 seemed to call for drastic measures. Under its first chief, General F. F. Palitsyn, the GUGSh assumed undisputed direction of imperial military planning and mobilization.[67] The GUGSh was divided into three directorates, of which the most important was that of the Quartermaster General. The latter's directorate consisted of four subdivisions: First Over Quartermaster (for war plans, fortress, movement, organizational-mobilization, and intelligence issues); Second and Third Over Quartermaster (operational-statistical data for potential European and Asian theaters); Fourth Over Quartermaster (for military-historical, administrative, and general staff cadre issues). Two other directorates, military communications and military topography, rounded out the GUGSh organization. Following the GUGSh's appearance, the Main Staff retained authority in matters related to military personnel assignments and pensions, and after 1909–10, in civil-military government and police-type affairs for selected provinces.[68]

Like the State Defense Council, the GUGSh was to run afoul of old and new political and financial realities before 1909, when it was subsumed into the War Ministry. General Iu. N. Danilov, who served its successor as Quartermaster General between 1909 and 1914, remembered that the anomalous position of the GUGSh outside the War Ministry tended to estrange it from troop-related concerns. More seriously, he saw the War Minister and the Chief of the GUGSh competing with one another for attention at court and within government. Worse, the GUGSh lacked the War Ministry's ability directly to match money with programs.[69]

A significant part of the difficulty in the post-1905 period lay in the changed political environment and its impact on tsarist political conduct. Although the Fundamental Law of 23 April 1906 created something like a cabinet-style government and ceded some of the autocrat's power to the State Duma, as the new Russian parliament was called, the Tsar retained considerable discretionary power in matters of budget and overall state policy, especially in foreign and military affairs. In the event the Tsar failed to support a given program, elements within the Council of Ministers and the ministries could appeal to various committees and deputies within the Duma. Even if the Duma failed to support ministerial initiatives, exposure to public debate brought notoriety in and pressure from the press. Although the Tsar remained largely immune to public opinion, he continued to play off against each other factions and personalities at court and within government, at times displaying uncharacteristic resolve and stubbornness in advancing pet programs such as naval reconstruction. Meanwhile, more adroit ministers, including the Minister of Finance, A. A. Kokovtsov, utilized this more complex governing situation to manipulate outcomes corresponding with their own political and ministerial priorities.[70]

The logical result, as William C. Fuller has noted, was interest-group politics run amok. Ministries and political figures pursued parochial objectives with scant thought devoted to overall coordination and integration of state policy. The War Ministry and the Ministry of Foreign Affairs withheld information not only from each other but also from the State Defense Council and the Prime Minister. The Tsar clung to his role as final arbiter, retaining the last word in those matters to which he ascribed vital importance, while the Duma and the government often worked at cross-purposes. Meanwhile, the government increasingly lost the confidence of Duma members and groupings, including the influential A. I. Guchkov and his Octobrist party, which were often disposed to support increased expenditures for defense.[71]

A major obstacle to all parties, whatever their persuasion, was a lack of unified vision necessary to mobilize the requisite political, moral, and fiscal resources. For a time, Grand Duke Nicholas Nikolaevich and his loose network of like-minded Young Turks appeared well situated to orchestrate a reform program and bring it to fruition amid a plethora of competing personalities, opinions, and agencies, high and low. However, it soon became apparent that the Grand Duke's followers lacked the cohesiveness and political clout to overcome the mixture of inertia, animosities, and contending priorities afflicting postwar tsarist political-military development. Thus, despite the army's needs, through 1906 and 1907, naval reconstruction held center stage at the imperial court.[72] Moreover, in 1906–7, an army going through the painful process of demobilization was dragged much against its will into rural and urban pacification.[73] Under these circumstances, the army soon found itself unable to train, let alone change itself. Or, as A. F. Rediger, the angry Minister of War, put it to P. A. Stolypin, the Prime Minister, "the army is not training, it is serving you." In 1906 alone, pacification required the efforts of 15,298 companies, 3,666 squadrons, and a host of lesser units.[74]

Even worse for long-term military development was the realization that the state treasury stood nearly empty. In the immediate postwar period, the state's ability to raise funds, whether by increased taxation or foreign subscription, failed to meet all but the barest necessities of military and naval reconstruction. By borrowing beyond its means between 1907 and 1909, Russia increased its foreign debt by three billion gold rubles and its annual debt service by 150 million rubles. After 1910, thanks to better harvests and limited industrial expansion, annual income rose to one billion rubles, of which about one fourth went to military and naval expenditures. With the help of revenues from an improved economy and still more loans, defense expenditures rose from 573.8 million rubles in 1910 to 716.21 million rubles in 1913.[75]

Against the background of fiscal stringency, the Tsar intervened in the planning process to support the so-called Small Program of naval reconstruction calling for a limited resurrection of the Baltic Fleet to fulfill requirements for coastal defense. In April 1907, the State Defense Council refused to approve even this modest plan, holding that imperial defense required a restoration of Russian armed forces beginning with the army, not the fleet. When the Grand Duke reported the Council's decision to the throne, the Tsar bypassed the Council to approve a reduced variant

(126.7 million rubles) of the already modest Small Program that would support construction of four line vessels, three submarines, and a floating dock for the Baltic and fourteen destroyers and three submarines for the Black Sea.[76]

Meanwhile, as State Defense Council deliberations indicated, representatives of the army stood just as prepared as the navy to advocate their own programs. As early as August 1906, when the tide of revolutionary disturbances had subsided sufficiently to enable civil authorities to look ahead, Prime Minister Stolypin had asked the War Ministry to prepare preliminary proposals for the reorganization of the army. On 9 December 1906, War Minister Rediger responded through the State Defense Council with a series of proposals to reduce the composition of the army and improve its combat readiness. Next day he laid out plans for extensive improvements which would require a one-time expenditure of more than two billion rubles and a decade-long series of annual 145-million ruble appropriations. Realizing that there was little real hope for achieving this program, Rediger then proposed a more realistic immediate expenditure of 425 million rubles and a 75-million-ruble annual expansion of the War Ministry's budget. Apparently, neither Nicholas Nikolaevich nor his retinue of Young Turks agreed with Rediger's priorities. To these impatient reformers, Rediger appeared cut from the same cloth as Vannovskii-Kuropatkin rather than Miliutin-Obruchev. Nonetheless, the Grand Duke could not work financial miracles: early in 1907 he delivered the sad news that little money would be forthcoming, even for the army's most pressing needs. Instead, he recommended reducing the size of the army to save expenditures and using the savings to improve what remained. Rediger muttered that such a policy would have dire strategic and political repercussions.[77]

With the treasury empty, much of 1907 and 1908 witnessed endless—but sometimes fruitful—wrangling within the State Defense Council as the GUGSh, rump elements of the Main Staff, and the War Ministry all either put forth or supported several versions of pet programs for reorganizing and modernizing the army. A special commission headed by General M. A. Gazenkampf advocated one of the variants, a proposal drawn up by General Palitsyn and General M. V. Alekseev, Over Quartermaster of the GUGSh, that called for fundamental transformation. The GUGSh design urged simplification of infantry organization, immediate transition to three-battalion regiments, and adoption of a scheme wherein reserve cadres would train with regular units until mobilization forced creation of separate reserve units. In time, Palitysn and Alekseev hoped to reorganize the other combat arms; meanwhile, they expected some reduction in actual army strength but an increased level of combat readiness. Less costly was a proposal made by the Main Staff which stressed working within the existing organizational framework to find solutions to vexing problems through minor alterations in the status quo. In March 1908, the State Defense Council, while meeting in the presence of the Prime Minister and the Ministers of Finance and Foreign Affairs, in essence approved the Main Staff's proposal. However, the argument came to a boil once again in the fall of 1908, when the GUGSh's Mobilization Committee under General A. Z. Myshlaevskii recommended a program which substantially repeated the Palitsyn-Alekseev initiative.[78]

At this point Nicholas II intervened to reduce dissonance and perhaps also to diminish the influence of the Young Turks and the Grand Duke. On 26 July 1908, the Tsar removed his uncle from the State Defense Council "to free" him for exclusive assignment to command the St. Petersburg Military District and the Guards.[79] In November 1908, the Tsar dismissed Palitsyn as Chief of the GUGSh, replacing him with V. A. Sukhomlinov, Governor-General and Commander of the Kiev Military District, and henceforth depriving the chief of independent access to the throne. The GUGSh reverted to the status of a directorate within the War Ministry. The other shoe dropped in early 1909, when the Tsar abolished the State Defense Council and replaced Rediger with Sukhomlinov as War Minister. Sukhomlinov's emergence at the top meant that the reconsolidation of military power was now complete, although some vestiges of the 1905–8 system persisted until World War I.[80]

It was the 1909–14 version of the GUGSh which would shepherd a reformed Russian army into World War I. The reintegration of the GUGSh into the War Ministry both resolved the anomalous position of the chief and reduced the amount of duplication between the two institutions. As the revamped GUGSh gradually assumed shape, its organizational structure came to correspond more closely with the requirements of centralized military planning and coordination. It consisted of five sections: Quartermaster General, Structure and Service of Troops, Mobilization, Military Communications, and Military-Topographical. The First (or Quartermaster General's) Section now held undivided responsibility for war planning, military intelligence, and control over all officers "of the General Staff." The Third (or Mobilization) Section oversaw and coordinated all issues related to mobilization. While these and other changes represented an advance over previous structures, one problem which the design could not overcome was politics: between 1908 and 1914, five different generals occupied the post of the chief of the GUGSh, thus undermining organizational continuity.[81]

Changing Distribution and Composition of Forces

With all its advantages and shortcomings, the Sukhomlinov-dominated military order would bear ultimate responsibility for effecting military reconstruction where other ventures had failed. A cavalry officer and close student of the Dragomirov system, Vladimir Aleksandrovich Sukhomlinov was intelligent but not an intellectual. The death of his first wife and remarriage to the free-spending widow of a civil engineer stretched the limits of his state compensation. The desire to retain power made him a political creature, but he was both aware of the need for change and sensitive to the Tsar's aversion to fundamental reform. There was no doubt that he remained committed to building Russia's defensive and offensive military power.[82] These considerations, plus Sukhomlinov's political style—he was not above venality—and his own feelings of personal vulnerability, dictated the War Minister's method and policies. To guard against political opponents he built spy networks; to retain power he manipulated the military promotion and assignment machinery; to prevent dilution of power he reduced the GUGSh to the status of a directorate within the War Ministry and insisted on frequent rotation of staff chiefs; and to bring about change he sponsored reform in the guise of limited initiatives to revitalize the army.[83]

In the spring of 1909, after a renewed Balkan crisis during the previous year had underscored a desperate need for strengthening the army, Sukhomlinov gratuitously reported to the Tsar that "in view of the State Defense Council's failure to work out any final instructions for the desired reorganization of the army, it was necessary once again to set to work on the general question of the restructuring of the army." Although Sukhomlinov might have disliked his predecessors, he was not above rummaging for profit among their leftovers. Therefore he promptly dusted off the well-worn Palitsyn-Alekseev plan and adopted it as the basis for his own report to the Tsar, which, in turn, became the foundation for the 1910 reforms of the army's composition, organization, and training. The rejected prophets had thus laid the groundwork for a reform program that would take the Imperial Russian Army into World War I.[84]

On the most fundamental level, Sukhomlinov's program sought a better system of matching soldiers with their units. Although the foundation for universal military service remained the conscription law of 1 January 1874, demographic change coupled with evolving force deployments and training requirements now called for adjustments in the recruitment system. Miliutin's legislation had originally divided the districts (*uezdy*) of the Empire into three categories, Russian, Ukrainian, and non-Slavic, or *inorodcheskie*. For purposes of conscription, Ukrainians and Belorussians were classified as "Russian." Within the framework of the territorial system, each infantry regiment and artillery unit possessed a specified area of recruitment, while cavalry and engineer units were drawn from the imperial population as a whole. In matters of ethnic composition, the rule was that no more than one-fourth of any unit was to consist of non-Russian minorities. Since the 1880s, however, significant distortions had gradually crept into the territorial system as an increasing number of active formations were deployed farther west in the frontier military districts. By 1910, the general situation was such that the physical location of the forward-deployed peacetime army no longer corresponded with the districts assigned to its component units for territorial recruitment. Consequently, more than 87 percent of the army's new recruits had to travel outside their home military districts to discharge initial tours of active duty. The Warsaw Military District, for example, drew more than 99 percent of its annual recruit contingent from other districts, while a comparable figure for the St. Petersburg Military District was about 58 percent. In the aggregate these and other mismatches between population base and deployed units engendered substantial expenditures for transportation and, worse, made for added complications in mobilization and training schedules. Those reservists assigned to fill out the wartime troop strength of active formations had to be shuffled over long distances simply to discharge periodic training obligations.[85]

Actual mobilization would further aggravate asymmetries inherent in the lopsided relationship between deployments and population base. The three western military districts (Vilnius, Warsaw, and Kiev) had to field sixteen corps, while the remainder of European Russia accounted for only fifteen more. Therefore the three military districts bearing the brunt of western frontier defense counted in their

A. F. Rediger (1853–1918) General Staff
Officer, Specialist in Military Administration,
War Minister 1905–1909
Photo: Skalon, *Stoletie Voennogo Ministerstva*

V. A. Sukhomlinov (1848–1926)
Commander and Governor-General of the
Kiev Military District 1904–1908, Chief of the
Main Directorate of the General Staff
1908–1909, War Minister 1909–1915
Photo: Skalon, *Stoletie Voennogo Ministerstva*

M. V. Alekseev (1857–1918) General Staff
Officer, Military Planner, Chief of Staff at
Supreme Headquarters (Stavka) 1915–1917
Photo: Courtesy of A. G. Kavtaradze

Iu. N. Danilov (1866–1937) Military
Planner, Quartermaster General, Main
Directorate of the General Staff 1910–1914
Photo: Danilov, *Rossiia v mirovoi voine
1914–1915 g.g.*

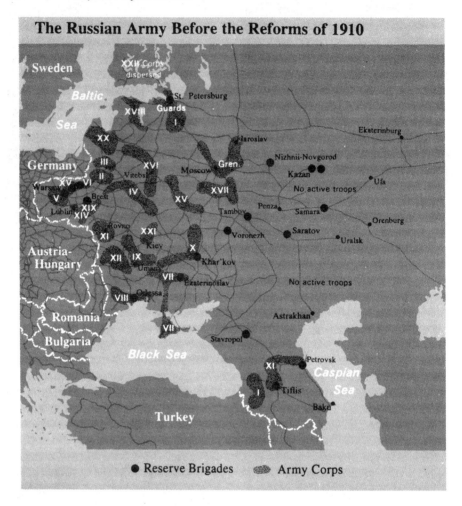

peacetime tables of organization 42.5 percent of the Empire's military forces, while the far more populous interior districts accounted for much smaller percentages. This peacetime distribution of forces, termed *dislokatsiia* by the Russians, immensely complicated planning both for the wartime augmentation of active units and for the mobilization of reserve units. Just to join their units in the event of war, more than 220,000 reserves would have to be laboriously shuttled from district to district. For example, plans called for 82,000 reserves to be shifted from other areas to the Warsaw district, while 40,000 went to the Vilnius district.[86]

At the same time, the lessons of recent Far Eastern conflict supported the contention that the sparsely populated areas of Siberia badly needed augmentation from European Russia in the event of renewed war with Japan. Western-oriented deployments only aggravated the situation. Sukhomlinov summed up the challenge by observing that "we cannot any longer concentrate our attention on the west: we

The Russian Army After the Reforms of 1910

must be prepared for serious struggle also on our entire broad eastern borders to which the density of our forces in the west does not at all respond."[87]

Based on the Palitsyn-Alekseev projections, Sukhomlinov's answer to dilemmas inherent in east-west threat calculations and distorted peacetime deployments was to reorganize the army's active and reserve forces and redistribute them more evenly across the populated regions of the Empire. Sukhomlinov withdrew 128 battalions from the Warsaw and Vilnius Military Districts and relocated the majority of them (seventy-nine) in the Moscow and Kazan Military Districts. By 1913, the peacetime distribution of army corps within the major military districts was as follows: Warsaw—five; Vilnius—four; Kiev—five; Odessa—two; St. Petersburg—four; Moscow—five; and Kazan—two.[88] At the same time, the War Ministry reorganized and simplified the induction and training of new recruits so that within practical limits the army reverted to something at least partially resembling a territorial system.

Divisions stationed in the interior districts now drew about 45 to 55 percent of their recruits from their primary and secondary territorial bases. This meant that greater numbers of recruits and reserves were now subject to active duty stints in areas closer to home. In matters of composition, the army continued to observe the principle that no unit be composed of more than one-fourth ethnic minorities.[89]

Redeployments were only one step in the direction of resolving dilemmas of money and manpower. One of the more pressing post-1905 problems was a shortage of trained personnel to fill out active and reserve formations. Hasty demobilization and stringent budgets initially discouraged maintenance of the army at full peacetime strength. More insidious over the long term were various exemptions to compulsory service which resulted in a reduced flow of recruits. As of 1908 about 45 percent of males subject to service received assorted deferments and exemptions. Although annual recruit contingents numbered approximately 450,000 men, the active army on 1 April 1909 was understrength by 41,076 soldiers, or 3.3 percent. Measures were needed to reduce the number of exemptions and spread the military burden more evenly, but it was not until 1912 that the Duma produced legislation designed to step up the pace of army recruitment.[90]

Meanwhile, the War Ministry employed another expedient to reduce the burden of active service and to enlarge and revitalize the pool of trained military manpower. Because 1904–5 had revealed a shortage of trained reservists, in 1906 the span of active service was reduced to three years for the infantry and artillery and to four years for the other arms. This change, when coupled with a level rate of recruitment, produced nearly 25 percent more trained reservists.[91] At the same time—also because of observations drawn from 1904–5—the nonunit reserve manpower pool (*zapas*) attracted renewed attention. It was divided into two categories (*razriady*). Category I consisted of the younger and more recently trained men (twenty-four to thirty-one years of age), who were now assigned in the event of war to fill out active army formations. Category II consisted of older males (thirty-two to thirty-nine years of age) assigned solely to reserve formations and rear support organizations. The Manchurian experience had revealed that older males withstood the rigors of field service less satisfactorily than their younger counterparts, and altered regulations for reserve service reflected this understanding. In addition, the traditional category of the state militia (*gosudarstvennoe opolchenie*), which included all physically fit males between twenty-one and forty-three years of age not subsumed into other categories, was endowed with greater structure and a range of assignments supporting regular and reserve forces within the general mobilization schedules.[92]

These and related alterations occurred within the broader framework of change in the army's organizational structure. In January 1908, General Palitsyn had presented the Emperor with a report which surveyed Russian military organization from the Crimean War through the Russo-Japanese War. Palitsyn's principal conclusion was that the peacetime size of the army had continued to grow despite creation of the very reserve system which was intended to augment wartime strength while reducing peacetime strength and expenditures. On 1 April 1909, the peacetime strength of Russian ground forces, including the corps of gendarmes and border forces, which were recruited on the same basis as the army, numbered 1,348,769 officers and men. It is interesting to note that this figure represented 1.8 percent of the male population of the Empire (including Finland), then calculated at 78.9 million. Estimated expenditures for the ground forces in 1909 were 470,620,239 rubles, or 19 percent of the entire projected state budget, a figure which represented about three rubles expenditure per individual in the entire population. Each active-duty soldier cost the government about 350 rubles per year.[93]

Expenditures rose not only because of anomalies in deployment and recruitment but also because of the very nature of the relationship between reserve and active forces. In reality, Palitsyn noted that projected wartime field forces relied very little on designated cadre reserve contingents, the 196 infantry battalions of which accounted for only 15 percent of Russian infantry in the event of hostilities. In addition, both unit and nonunit reserve forces displayed low states of readiness and inadequate training. These shortcomings were also characteristic of the fortress troops, who were generally lumped together with the reserves. In a word, Russia received too little from its cadre and noncadre reserve forces to justify the levels of commitment and expenditures which they required.[94]

Changing Organization and Structure

To remedy these shortcomings, Sukhomlinov continued to borrow from Palitsyn and Alekseev to propose a general overhaul of the system which would neither impose a greater service burden nor increase expenditures. The Emperor approved the War Minister's recommendations, which went into effect during the second half of 1910. In view of the low level of training within cadre reserve formations, Sukhomlinov simply abolished the twenty-six reserve brigades along with the fortress troops. They were replaced by "secret cadres" of officers and men who trained in peacetime with regular field formations but who, in wartime, would be detailed to form the cadres of mobilized reserve units. In the event of mobilization, these cadres formed the nucleus for 560 battalions which, in turn, made up thirty-five reserve divisions (those numbered 53 through 84, and the 12th, 13th and 14th Siberian divisions). In 1914, these newer cadre reserve units comprised about one-third of the mobilized Russian field armies. During wartime, the field forces themselves would provide garrisons for fortresses. The new design reduced peacetime expenditures by almost nineteen million rubles annually. It also improved the training of reserve cadres and made any potential adversary's task of discovering cadre deployments much more difficult.[95] The War Ministry used savings from this scheme both to enlarge all existing rifle brigades to eight-battalion size and to create seven more field divisions and a rife brigade, thus enlarging the peacetime army by 13 percent for a total of 1,252 infantry battalions in seventy-four divisions and seventeen rifle brigades.[96]

Thanks to Sukhomlinov's reforms, the peacetime strength of the Imperial Russian Army on the eve of World War I reached 1,423,000 officers and men. They were spread among thirty-seven army corps, including twenty-seven in European Russia, three in the Caucasus, two in Turkestan, and five in Siberia. The corps was the largest permanent all-arms formation, numbering two divisions, each of which was composed of two infantry brigades with two regiments each. Each field army corps also counted in its complement a Cossack regiment, a field artillery howitzer *divizion* (twelve 122-millimeter guns), a sapper battalion, an aviation detachment (six planes), and telegraph and searchlight companies. In addition to the normal infantry complement, each division carried on its table of organization an artillery brigade (forty-eight 76-millimeter guns), a Cossack "hundred" (*sotnia*), a cavalry *divizion* (half-regiment), a sapper company with 10 percent of a bridge park, and a convoy security detachment.[97]

The army's seventy-plus divisions were made up of 236 infantry regiments (twelve Guards, sixteen grenadier, and 208 line), and the regiments were sequentially numbered throughout the army. This meant, for example, that the 16th Infantry Division (M. D. Skobelev's old command) of VI Army Corps consisted of the 61st (Vladimir), 62d (Suzdal), 63rd (Uglich), and 64th (Kazan) regiments. As was customary, each regiment also bore the name of a city, province, or river of the Russian Empire, and sometimes also the name of an honorary colonel of the regiment. Between 1910 and 1914, the typical line infantry regiment had four battalions with four infantry companies each. The fundamental tactical unit was the infantry company, the wartime complement of which was four officers, twenty noncommissioned officers, and 202 rank and file. Each company was further broken down into four platoons of four sections each. In addition to its infantry complement, each regiment usually had one hundred mounted and sixty-four dismounted scouts, a machine-gun detachment of eight guns in four platoons, a detachment of thirteen mounted orderlies for messenger communications, a noncombatant service company, and a telephone detachment suitable for serving nine posts with twelve kilometers of wire to connect them.[98]

The army's organizational table aptly illustrated a supplementary benefit of Sukhomlinov's reform: simplification. The disappearance of traditional categories of reserve and fortress troops simplified infantry organization so that the various types of infantry regiments could be reduced to two, the four-battalion line infantry regiment and the two-battalion rifle (or "sharpshooter") regiment. In turn, simplification reduced the categories of various types of companies and troops within larger units. While additional "secret cadre" officers and noncommissioned officers increased the peacetime complement of field units, their presence helped spread the instructional and training loads. On the surface, at least, Sukhomlinov's reorganization promised much and seemed to cost little.

This was because the costs of reform were more often measured in rubles than in impairments to mobilization, readiness, and combat efficiency. Although the Far Eastern experience had indicated that four-sided, or rectangular, formations lacked flexibility and were often clumsy to control and apportion between attack and reserve missions without violating unit integrity, the post-1910 Russian army at division level and below retained many aspects of pre-1905 organization by multiples of four. The logic of command, reinforcement, and support argued for simplification in the form of triangular, or three-sided, organization. However, previous patterns persisted, perhaps because Sukhomlinov and his adherents found it less difficult to increase size than to face the financial, political, and staffing challenges of fundamental restructur-

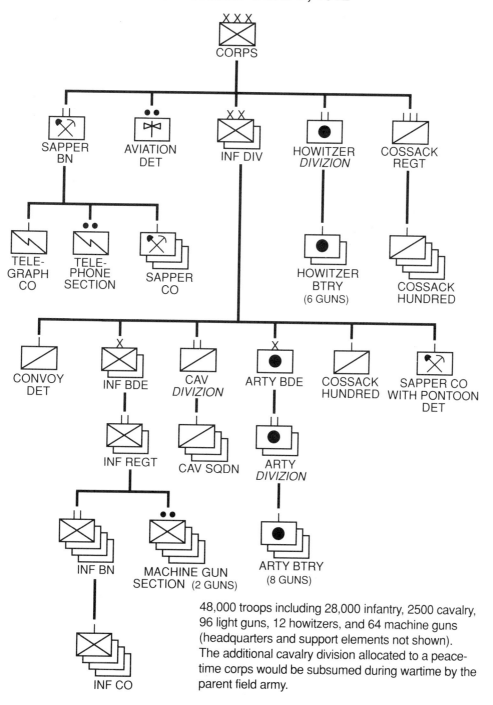

ing for greater tactical efficiency. Consequently, retention of old structures imposed a heavy—perhaps impossible—command-and-control burden on battalion and regimental commanders and staffs.[99]

Difficulties of Mass and Concentration

Dislokatsiia remained another serious hidden cost. The changed deployments of 1910 simply exchanged one set of complications for another. Because field army units and their reserve augmentation forces were now more evenly distributed across European Russia, mobilization in the event of European war would require additional time and altered transportation schedules. Opposite the Germans on the western frontier, time-distance calculations and the inevitable mobilization gap meant that Russia would have to accept designating its initial concentration areas deeper within friendly territory to assure absolute security of concentration and initial deployments. Concentration farther from the border simultaneously imposed additional requirements for careful mobilization planning and greater attention to the preparation for initial operations, including planning for reconnaissance and march-maneuver from field deployments to initial meeting battles. Whether planners subscribed to Mikhnevich or Neznamov, their work had suddenly become more difficult than ever before in Russian military history.[100]

Greater depth of concentration aggravated problems of security (screening and reconnaissance) and deep attack, both of which were traditional cavalry functions. During concentration, horsemen formed a mobile screen to cover the detraining, assembly, and field deployment of mass infantry forces. Once the infantry began moving forward, cavalry patrols served as the eyes of commanders seeking gaps and open flanks in their adversaries' advancing formations. Cavalry eyes and ears also helped guard against surprise and entrapment. Meanwhile, even larger cavalry formations were to strike the enemy's rear in attempts to disrupt and delay his mobilization and deployment schedules. All these requirements lobbied for numerous large and small cavalry units in highly mobile formations either in the field or about to take the field in the event of war.

Cavalryman that he was, Sukhomlinov tried valiantly to forge the mobile instrument Russia needed to support offensive operations in the early twentieth century. The total active cavalry force, after dropping to 74,300 in 1905, gradually rose to 83,517 in 1908. Under Sukhomlinov's tutelage in 1909–10, twenty-six dragoon regiments were added to the sixty-seven existing regiments (ten Guards, twenty-one dragoon, seventeen uhlan, eighteen hussar, and one Cossack), for an active force total of 112 regular and Cossack regiments formed into twenty-two divisions and two separate brigades. This figure leveled out between 1911 and 1913 at twenty-four divisions and eight brigades, including eighty-nine regiments in the western military districts.[101]

Even these considerable numbers did not meet the diversity of visions and requirements. This was because even Sukhomlinov's creative reorganization schemes failed to stretch scarce money and assets far enough. The events of 1904–5 had demonstrated the need both to go beyond mounted scout detachments for reconnaissance and to create something more mission-oriented and flexible than composite detachments for deep raiding. However, despite Russia's cavalry riches, there simply were not enough trained horsemen immediately available at the outset of mobilization to fulfill diverse requirements at field army and corps levels. Despite talk of massive raiding to disrupt Prussian and Austrian mobilization schedules, the Russians possessed no permanent cavalry force larger than a division (3,500 sabers and twelve guns).[102] Worse yet, in the event that movement to contact began before completion of mobilization, the Cossack regiments designated for corps-support missions would not yet have detrained, which meant that the corps would undertake half-blind their initial march-maneuvers to contact. This was indeed what occurred with XIII Corps during Samsonov's ill-fated offensive of August 1914 into East Prussia.[103]

Like the cavalry, Russian field artillery both suffered and benefited from changes within army organization between 1910 and 1914. Shortly after the cessation of hostilities in the Far East, the Main Artillery Directorate (GAU) had concluded on the basis of wartime experience that ease of control and maneuverability favored organization of field batteries at six guns each. However, since reorganization would entail substantial outlays, the Finance Ministry successfully moved to block implementation of GAU's decision. Only in 1910 did reformers turn their attention to the field artillery, then not in connection with the size of firing batteries but with the larger issue of force structure. In accordance with requirements stemming from an expanded infantry force, each field and reserve infantry division received an artillery brigade made up of six eight-gun batteries. Each rifle brigade received an artillery division (*divizion*) of three eight-gun batteries. Altogether, these changes necessitated the creation of nine new artillery brigades and nine new artillery *diviziony*.[104] What these changes meant was that between 1910 and 1914 the field artillery would continue to reflect prevailing views that emphasis should fall on fire support coming from light and highly maneuverable field guns. If Russian threat handbooks on German field forces correctly reflected St. Petersburg's thinking, then Russian artillerists believed that German heavy guns inhibited agility and maneuver in the field. Conventional Russian wisdom explained German emphasis on heavier calibers as an attempt to compensate for shortcomings in maneuverability of fire support.[105]

Technology and the Large Program

This does not imply that the Russians ignored experiences which lobbied in favor of additional types of artillery, including heavy, only that in some respects the 1904–5 precedent was either misread or underestimated. The absence of suitable mountain artillery received attention as early as 1907, and two years later GAU placed orders with the Putilov works for 214 three-inch mountain guns, Model 1909. These were built on the Schneider-Creusot model, and their suitability for pack and draft transport far outstripped older models. In contrast, the issue of heavy artillery did not lend itself to easy resolution. The War Ministry allocated scarce funds for purchase in 1907, but withheld decision until 1910, when it ordered 122 152-millimeter guns of Schneider manufacture. The Russians later ordered additional 4.8-inch howitzers of Krupp manufacture.[106] The purchase of these guns, together with organizational changes, enabled reformers to create twenty-four new peacetime and sixty new wartime batteries of heavy artillery, for a total of 620 guns.[107]

The Russians did somewhat better with machine guns than with heavy artillery. Machine-gun detachments of eight guns each were allocated to infantry regiments and cavalry divisions. Normally, the Russians employed two-gun sections in support of each battalion, although the machine-gun section commander retained considerable "liberty of action within the limits of his task."[108] This allocation of weaponry, though not generous by later standards, was considerable for an army whose adaptive capacities have never attracted wide attention. However, as had been the case with heavy artillery, Russia lacked sufficient manufacturing capacity to satisfy demand for weapons. Therefore, although the Russians produced many of their own Maxim Model 1910s, the army called upon foreign manufacturers, including the American Colt and the French Hotchkiss, to provide additional guns. At the beginning of World War I, the Imperial Russian Army counted in its inventory 4,157 machine guns.[109] The majority of these were the standard Vickers-Maxim model, modified in 1910 to incorporate a shield and lightened mount. Although Russian inventors had produced a few experimental automatic and semi-automatic rifles, the Imperial Army of 1914 had no light and easily portable automatic firepower.[110]

The issue that always loomed large in the background for weapons procurement was that of the western fortresses. To retain even limited effectiveness, these seven aging dowager sisters once again cried out for reconstruction, reequipment, and, in some cases, relocation. Proponents of modernization invoked the precedent of Port Arthur, while their opponents argued that fixed defenses were obsolescent in a new age of maneuver warfare. Meanwhile, apostles of military reform understood prevailing budgetary constraints, correctly perceiving that fortress modernization would siphon off funds desperately needed to enlarge and reorganize the field army and reequip it with modern weaponry. Sukhomlinov came down on the side of those who would raze the fortresses, but he eventually had to settle for a compromise. He was able to do away with the old category of fortress troops and reorganize them into reserves, but he was unable to fend off repeated calls for modernization from various quarters, including the Duma, the military districts, and the artillery lobby. Therefore, between 1911 and 1914 several millions of rubles were expended in purchasing ordnance for the fortresses, which were imperfectly integrated into war plans, to provide the screen behind which Russian forces could supposedly concentrate and selectively choose their main offensive directions.[111]

Evolving technology posed additional challenges on the ground and in the air. As early as the Krasnoe selo maneuvers of 1906, the army had tested various types of auto transport with an eye toward military application. Although a special commission had concluded that armored cars might find use in various roles, including reconnaissance, anticavalry combat, and exploitation, the War Ministry was slow to obtain funds for acquisition and experimentation. Only between 1910 and 1914 did the army finally acquire 259 automobiles and 418 trucks as its share of 21,000 vehicles purchased mostly abroad by the Ministry of Trade and Industry. On the basis of this acquisition, the War Ministry created five automobile companies, one training company, and six separate detachments. This was essentially the vehicular complement with which the Russian Army entered World War I.[112]

Thanks to a small group of visionaries and possibly also to a slightly longer tradition, the Russian Army did better in the air. Although observation balloons had been poorly supported and utilized in Manchuria, their unquestioned utility prompted the army in 1907 to order an additional fifty-two balloons of several types sufficient to form an additional twelve fixed balloon companies for observation purposes. By 1910, successful experimentation with dirigibles provided justification for purchasing nine long-range and eight short-range airships.[113]

Just when lighter-than-air flight appeared to have won unqualified acceptance, heavier-than-air craft entered the competition for scarce resources. Following Louis Bleriot's cross-channel flight in July 1909, the Russian War Ministry allocated funds for aircraft purchase abroad and established at Gatchina an aviation branch of the St. Petersburg Aeronautics School. With patronage assistance from Grand Duke Alexander Mikhailovich, aircraft advocates gradually won grudging private and public support for aviation, sponsoring air races, publicizing their efforts, and winning governmental funds. By 1911, airplanes successfully participated in army maneuvers, and in 1912 a special aviation section of the GUGSh assumed control of Russian military aviation. Thanks to growing recognition of the military utility of heavier-than-air aviation, by August 1914 the Russian Army counted 244 airplanes in its inventory. This number was sufficient to form training companies and aviation detachments for the majority of army corps and fortresses located in the western military districts. When the First Balkan War of 1912–13 underscored the growing importance of aircraft, the War Ministry found itself confronted with additional requirements for aircraft procurement and production.[114]

The idea apparently was that these and other urgent needs would be remedied by the next round of organizational reform and procurement. That round began in the fall of 1912, when the outbreak of the First Balkan War focused renewed attention on Russia's continuing lack of military preparedness. For nearly a year, the Main Directorate of the General Staff labored in great secrecy to devise a coherent plan for strengthening imperial defenses during 1914–17. On 22 October 1913, the Tsar approved the resulting draft of the "Large Program for Strengthening the Army." After some debate within the Duma, the Large Program became law on 24 June 1914, allocating more than 433 million rubles to support additional military expenditures.[115]

The intent behind the Large Program was both to remedy those deficiencies which the reforms of 1910 had ignored and to address requirements inherent in the changing international situation. In view of the growing German and Austrian threat, the legislation of 1914 made provisions to expand the peacetime strength of the army by 11,592 officers and 468,200 men.

The Large Program additionally made provisions for strengthening the army's chief components. With selective reinforcement, more infantry battalions, especially those in frontier military districts, would be maintained at personnel levels close to wartime strength. Throughout the army's cadre units, minimum manning levels for infantry companies would be raised from forty-eight to sixty men, while legislation would provide for the selective creation of new units amounting to about three additional army corps.[116]

The Large Program also addressed those cavalry and artillery requirements which were so close to War Minister Sukhomlinov's heart. Because Russian cavalry was insufficient to serve both tactical and strategic purposes, the new legislation envisioned creation of twenty-six new cavalry regiments, or a 39 percent increase over the 1913 figure. The size of troop (*voiskovye*) cavalry divisions would be raised from four to six regiments, while army-level (*armeiskie*) cavalry divisions would be concentrated in frontier regions for fulfillment of immediate strategic missions. The Large Program further stipulated that Russian field artillery would undergo thoroughgoing reorganization, with primary effort devoted to reducing batteries from eight- to four- and six-gun composition. Each army corps would receive additional 122- and 152-millimeter howitzer batteries and 107-millimeter cannon batteries. These alterations and additions would raise corps artillery assets to 108 light guns and thirty-six heavy guns, a number sufficient to bring about rough firepower parity with German corps-size formations.[117]

While the Large Program devoted a lion's share of attention to the requirements of field formations, the Main Directorate of the General Staff additionally sought to strengthen selected support, training, and technical means. Thus the legislation of 1914 expanded the officer education system at entry level, increased the number of the army's sapper, field engineering, and railroad troop units, and expanded support facilities for the army's fledgling aviation detachments. The number of aviation companies was raised from five to seven, while each corps and fortress was to have its own aviation detachment. In addition, the legislation of 1914 called for the formation of twenty-eight additional aviation detachments, five companies, and two schools.[118]

The Large Program was just as significant for what it neglected as for what it envisioned. Although the program's authors were successful in addressing the army's most immediate requirements, key items of support and infrastructure went begging. Although reformers sought to strengthen corps logistics, almost nothing was done for the construction of strategic rail lines and still less for the accumulation of munitions and supplies.[119] Thus priorities continued to reflect that which was financially possible and that which assumed a short war. In this, despite the warnings of Mikhnevich, Neznamov, and Svechin, the Russians differed little from their western European counterparts.

Command and Control

The 1910–14 period also witnessed considerable improvement in the sphere of command and control of the armed forces in the event of war. The experience of 1904–5 had clearly shown the need to modify the field regulation of 1890. A commission under the direction of General N. S. Ermolov concluded that the Far East had shown the highest command echelons ill suited to cope with the challenge of preparing for and conducting war in a theater so far removed from the center.

The field regulation itself required revision, while staff officer preparation had fallen short of the task. By 1911 a new draft regulation was submitted for review to corps and division staffs and also to regimental commanders. It was approved in 1912 and published on the eve of the war, on 3 June 1914.[120]

While the regulation had much to say about the nature of modern war and battle, its more practical provisions related to army organization in the field. It introduced new forms of field army organization, noting inter alia that "forces assigned to joint action are formed into armies, separate corps, and detachments—of various types of troops." In addition, the regulation stated that "several armies assigned to the attainment of a single strategic objective and operating on a definitive front may be combined into a still higher troop unit, forming the armies of a given front." An army not belonging to a front was designated a separate army with its own independent directing apparatus.[121]

Significant shortcomings of the new regulation included its failure both to do more than prescribe the relationships between new formations and to touch more than perfunctorily on issues of high command. Fronts and field armies now incorporated into their structures immense material and human resources. New higher-level formations required time and practice not only to assimilate these resources but also to familiarize themselves with new means of direction and communication. The nature of contemporary battle and the evolution of concepts such as the modern all-arms operation required both flexibility and experience in all aspects of staff operations. Staffs had to have the ability to acquire, process, and understand the complex and often confusing information flowing in from various sources during field operations. Then they had to make decisions on the basis of that information and communicate the essence of their decisions to actors in the field.[122]

Field forces also required unity of command, especially at the very apex of the strategic-operational structure. The field regulation provided for a Supreme Commander who would be "the high commander of all ground and naval forces designated for military action." He was endowed with extraordinary powers, and without exception retained in subordinate capacity representatives from all services and organizations within the theater of military action. The Supreme Commander was accountable only to the Tsar. Under Russian circumstances, these qualifications meant that the Supreme Commander would be either be the Tsar himself or a member of the imperial family. The regulation provided for a headquarters, or *Stavka*, of the Supreme Commander and a staff suitable to direct military and naval operations, including a quartermaster general (for operational control), an adjutant general (for communication with the War Ministry and troop structure), chiefs of military communications and naval forces, and a headquarters commandant. Significantly, the regulation did not provide for *Stavka* control of material and technical support services, the functions of which would be separately discharged by each of the fronts communicating directly with the War Ministry. Ominously, then, in 1914 there was initially no unity of operational and support structures at the highest instance of field command.[123]

What unity existed for the conduct of operations depended to a large degree on the competence and professional preparation of higher-level command and staff cadres. Here the pre-1914 Russian experience was ambiguous. Because the Russo-Japanese War had highlighted the issue of command fitness, in 1906 the War Ministry had created the Higher Examination Board, an organization subordinate until 1909 to the Tsar (thereafter to the War Minister) and designed to review qualifications of senior officers for high command. Between 1906 and 1908, the board retired from service some 4,307 officers, including 337 generals, 711 colonels, and 1,206 lieutenant colonels. Although the board no doubt weeded out the most incompetent and superannuated officers, it had little influence on the preparation of senior commanders, most of whom would still have completed their higher military education before 1905. In addition, once the board fell under Sukhomlinov's jurisdiction, there was no guarantee that it could not be used for political purposes.[124]

At the same time, just as before 1905, the surest inoculation against damaging political appointments was to insure that qualified officers "of the General Staff" occupied significant staff positions at all levels from the War Ministry at least down to brigade. General Staff service lists for 18 July 1914 reveal that slightly more than 80 percent (602 of 733) of positions throughout the army staff system calling for officers of the General Staff were filled by qualified officers. While this stood in undeniably positive contrast with the pre-1904 situation, there remained grounds for concern: numbers were greatest within district staffs, then thinned out appreciably at corps and division levels, where the assignment system provided respectively for only three and two officers of the General Staff.[125] This meant that much of the maneuver army would find itself relying on officers without a general staff background for its intellectual and procedural linkages.

Consequently, despite improvements in various kinds of linkages, their overall effectiveness remained suspect. In the end, too much of the burden of command and staff work still lay on the shoulders of officers lacking rigorous and consistent preparation. The military historian A. M. Zaionchkovskii, who in the 1920s shifted his research focus to World War I, would write that "in general the Russian Army went to war with good regiments, average divisions and corps, and poor armies and fronts, basing this evaluation in the widest sense on preparation and not on personnel qualities." Further, Zaionchkovskii noted that it was impossible not to agree with a 1913 conclusion of the German General Staff about the Russian Army's lack of operational effectiveness: "Thanks to slow-witted staffs and poor communications, to expect the Russian command rapidly to avail itself of a favorable operational situation would be as difficult as the rapid and exact execution by troops of a prescribed maneuver on order." Therefore, "in contact with the Russians, the German command will enjoy the possibility of accomplishing the kind of maneuvers which it would not be permitted by another equivalent enemy."[126]

Zaionchkovskii laid much of the blame for this state of affairs on the Russian General Staff, which he believed had failed to imbue the Russian army and its officer corps with a common understanding of military phenomena and a common approach to the resolution of military problems. Although Zaionchkovskii stopped short of decrying the absence of a unified military doctrine, his mention of alien French and German influences implied that by the late 1920s he had modified his own position, even if he stolidly continued to ascribe significance to Russian nationalist lines of military-intellectual development. Tenacity of sentiment aside, Zaionchkovskii's conclusion was that lack of intellectual unity at the upper reaches of the military hierarchy had inevitably produced lack of coordination and implementation at the lower reaches.[127]

The same lack of unity was apparent in civil-military direction. On the eve of the war, when queried about Russia's preparedness for conflict, War Minister Sukhomlinov had unequivocally pronounced Russia ready for war. Ten years later, Quartermaster General Danilov would rhetorically answer the same question with a resounding "No!" Too many organizations were incomplete, too many programs remained unrealized, too many expenditures were yet to be made, and too many concepts required testing before implementation. Eight years of serious thought, only part of which found its way into regulations and plans, and three years of incomplete reform were precious little preparation for the Russian army's entry into World War I.[128]

7.

Dilemmas of Design and Application, 1905–1914

> We must know what we want.
> A. A. Neznamov[1]

> I warn you: you're standing on the edge of a bottomless pool—and not a pool of water either, but pitch.
> A. A. Svechin to Colonel Vorotyntsev, General Staff, in *August 1914* by Alexander Solzhenitsyn

JUST HOW WELL theory and structure meshed in practice would depend in large part on the existence of firm planning and doctrinal linkages to join concept and organization. Both Mikhnevich and Neznamov had argued for a single war plan which would serve as the expression of what Russia intended to accomplish in any future war. Once the plan fell into place, execution would rely on the ability of commanders and staffs to incorporate their units into a single integrated design. Success would derive from a common understanding of contemporary war and the procedures and techniques required to wage and support it. Russian peacetime preparation for war between 1905 and 1914 revealed some advances, many lapses, and not a few lost opportunities.

Russian Mobilization Schedules and War Planning, 1906–1914

Four major determinants on Russian war planning after 1905 included the recuperative capacities of the army, the internal redistribution of forces, the demands of the military alliance with the French, and St. Petersburg's own shifting strategic priorities. For the first four years after the Russo-Japanese War, considerations of money, internal unrest, and the requirements of reorganization prevented any abrupt changes in the overall emphasis on a defensive posture imposed by necessity since 1904–5. However, by 1910 the situation had changed sufficiently so that planners edged perceptibly toward recalculating the traditional balance between defense and offense in anticipation of future war. The French also began to levy heavier demands on St. Petersburg, with the result that by 1912 the Russians were to abandon their former prudent approach to war planning with disastrous consequences.

Until 1909, the Russians were incapable of much more than speculation about how they might readdress the issue of readiness and war plans. Because treasury and army lay exhausted, Russia might respond to the outbreak of war in the west by assuming the strategic defensive while laboriously mobilizing reserves covered by a screen of troops deployed behind aging fortress barriers in the border military districts. Essentially, stopgap measures adopted in 1904–5 governed Russian planning until army and state could recover from the shock of military defeat and revolution.[2]

This state of affairs did not preclude strategic reassessments. Already in 1906, Major General M. V. Alekseev (the GUGSh Over Quartermaster) and Colonel Sergei Dobrorol'skii presented to F. F. Palitsyn, Chief of the GUGSh, a report which reviewed Russian strategic priorities in light of changed circumstances in the Far East. The two officers argued that Russia now confronted potentially hostile combinations in both the west (the Triple Alliance) and the Far East (Japan and Great Britain). Of the two, the Triple Alliance presented the graver threat. After concluding a strategic survey of the western border regions, Alekseev and Dobrorol'skii asserted that in the event of war Russian forces could no longer accomplish their strategic concentration forward in the border regions and must instead complete their assembly deeper within Imperial Russian territory. Forces within the Polish salient would be reinforced sufficiently to threaten the flank of any German attack toward St. Petersburg, while Russia completed its mobilization, safely assembled its forces, and finally shifted from the defensive to the offensive. "In this way," they wrote, "the strictly defensive idea of our plan for the first period of a war" would permit assembly of forces under secure circumstances, then facilitate their "transfer to the offensive and rapid closure with the enemy upon completion of concentration."[3]

In view of the army's condition and pressing internal requirements, the Alekseev-Dobrorol'skii scheme made good sense. The only potentially discordant note lay in their startling reassertion of the late General M. I. Dragomirov's strategic priorities. That is, Alekseev and Dobrorol'skii now argued that the main Russian effort should be directed against Germany because on the western frontier that nation was viewed as "the soul and unifying link" of the hostile coalition. In the event that Germany's primary objective was France, something which would become evident during the second week of mobilization, then the Russian army might advance to conduct initial operations against Germany without waiting for full mobilization and completion of strategic deployments.[4] Thus, under favorable circumstances, Russian war planners even before 1910 assumed that they might risk offensive operations earlier than dictated by the normal requirements of prudent planning.

Against this backdrop, the Emperor on 3 February 1909 signed a resurrected version of Schedule 18 which accepted as probable adversaries Germany, Austria-Hungary, and Romania, while acknowledging that each potential adversary would be able to mobilize and concentrate his armies faster than Russia. Therefore the opposing coalition might initiate hostilities with an attack on Russia, even in the event that Germany directed its main blow against France. Indeed, despite numerical inferiority, the Germans might open hostilities with a spoiling attack designed both to exercise limited initiative and to disrupt Russian concentration and deployments. Following from these assumptions, the mission of the Russian army initially was to determine the enemy's grouping and objectives, cover mobilization, and concentrate in the appointed regions. Once preparations were complete, the Russians would shift to the offensive in accordance with directives coming from the high command. Five armies and the Odessa separate corps were designated for operations in the west.[5]

More than a year later, on 26 June 1910, the Emperor approved Mobilization Schedule 19, first version. Although Schedule 19 had much in common with its predecessor, the updated plan differed from Schedule 18 in three important respects: it called for an "active defense," it revealed a new preoccupation with mass and depth, and it focused more clearly on Germany. In accordance with General Alekseev's earlier concern (sanctioned by War Minister Sukhomlinov) that preponderant Russian weight must be brought to bear in the west, Schedule 19 called for marshaling additional forces against the Triple Alliance at the expense of other nonwestern strategic priorities.[6] Thus the design called for seven armies (instead of five), with additional forces coming to the west from the Omsk, Irkutsk, and Priamur Military Districts of Siberia and from the Turkestan and Caucasus Military Districts. Responsibility for the Black Sea shoreline was shifted to the Caucasus Military District, and both the Black and Baltic Fleets were subordinated to the high command. Here the Alekseev-Dobrorol'skii influence clearly manifested itself: "If we take into account that on the western frontier the fate and future of Russia will be decided, then is explained the evident necessity of counting on these forces in spite of whatever complications can occur in our Asian frontier regions."[7] By denuding the eastern military districts, Schedule 19 would put at the disposal of western commanders an additional three army corps numbering seven infantry divisions, 1.5 cavalry divisions, and 386 guns. Deployment planning for these and other forces reflected current thinking about the most dangerous threat. And General Iu. N. Danilov, Quartermaster General and chief architect of the plan, left little doubt on this score: he allocated fifty-three divisions against Germany and only nineteen against Austria-Hungary.[8]

Mobilization Schedule 19 was thus a blend of traditional caution and novel means and ends. Or, as the Soviet military historian I. I. Rostunov has observed about the Danilov plan, "in general the strategic outlook of 1910 reflected the same excessive caution which was characteristic of the military leadership of the time."[9] Accordingly, Schedule 19 emphasized depth and the attainment of greater numerical superiority, whatever the turn of events or expected hostile combinations. Reflecting the novel, in turn, were two recent departures which made mass and depth possible— even necessary. These included Sukhomlinov's reorganization of the army on a more clearly territorial basis and the abandonment for planning purposes of much of Poland, including Warsaw, the forward fortresses, and the Narew defensive line. Russian armies would concentrate farther to the east, along a line extending from Vilnius through Belostok to Brest. Danilov justified fixation on Germany and concentration deeper in the interior on grounds of Russian military weakness, the distinct possibility of German attack, and the uncertainty of adequate French support.[10]

The original version of Schedule 19 remained in effect for only two years, thanks in part to changing threat perceptions, the reassertion of alliance interests, and a partial revision of strategic priorities. All of these influences made themselves felt within the context of conventional and bureaucratic politics. Although Russian war planners had already conceded in 1902 that they might temporarily give up as much as two-thirds of Poland in the event of war, Danilov's 1910 plan aroused a storm of controversy within higher government and military circles.[11] When the full military implications of Schedule 19 hit home in the border military districts, a number of influential officers, including General Alekseev and General N. A. Kliuev (the former was now chief of staff in the Kiev Military District and the latter his counterpart in Warsaw), agitated on various grounds for at least partial restoration of their districts' traditional anti-Austrian missions. Supporting their arguments were changing intelligence assessments, which revealed both growing Austrian strength and firmer evidence that Germany would turn west before dealing with Russia in any future coalition war. Linked to these assumptions were other considerations, including improved Russian combat readiness and a persistent unwillingness to cede whatever offensive advantage the Russians might derive from possession of forward positions in the Polish salient.[12]

Further resistance to the original Schedule 19 strongly manifested itself in Moscow during a February 1912 conference of the military district chiefs of staff, where adherents of a more concrete and decisive plan of concentration and deployment temporarily gained the upper hand over a more cautious Danilov and the GUGSh. In brief, the district chiefs of staff proposed concentrating two powerful armies (the 2d and 4th) in the Polish salient with the objective of attacking across the Vistula to threaten the flanks and rear of enemy concentrations in both East Prussia and Galicia. In essence, the proponents of forward concentration proposed simultaneously to deprive the adversaries' most threatening formations of security of concentration and to drive immediately for a coup de main.[13] However, in the event that Germany unexpectedly turned east, the Russians would find themselves much more vulnerable to counterstroke.

For this reason and because of intervention from the center, the district chiefs of staff found their views reflected only partially in the next evolution of Russian mobilization schedules and deployment planning. Thanks to the intervention of War Minister Sukhomlinov, who strongly supported Danilov's resistance to "collegial planning," the revised Schedule 19 (May 1912) incorporated the district staff position only imperfectly. The 4th Army had been reduced in strength by three divisions, and the area of concentration for the 2d Army had been shifted from the lower to the middle Narew. Subsequently, the GUGSh would reduce by 144 the number of battalions designated for deployment in the Polish salient and permit front staffs to withdraw of their own accord to the deployment line established in 1910. The military historian A. M. Zaionchkovskii has drily noted that the GUGSh "persistently drew deployment back to the east, and, with the same personnel composition, would probably have returned it to the deployment of 1910 had the beginning of the war not intervened."[14]

Under these conditions, the Russian approach to mobilization and war planning hardly bore the unified character for which Mikhnevich and Neznamov had so ardently pressed. Not surprisingly, the mobilization schedule approved by the Tsar in May 1912 was a compromise, as Jack Snyder has recently put it, among three competing sets of concerns: operational logic and Russia's historic objectives in southeast Europe, security of deployments, and alliance objectives.[15]

Since 1910, alliance considerations had steadily reasserted themselves to play an increasingly important role in Russian planning calculations. Following Sukhomlinov's territorial reconfiguration of the army, the French had urged the Russians to take whatever measures were necessary to convince Germany of the inevitability of a Russian offensive between $M+15$ and $M+30$. In return, the French promised already in 1910 "an immediate and decisive offensive against the German armies with all forces." In staff conferences held at Krasnoe selo in August 1911, General Auguste Dubail, the French Chief of Staff, asserted that the French would attack at $M+12$ with the British Army on their left flank. When he asked what the Russians would do in return, General Zhilinskii, the new Chief of the GUGSh, responded lamely that the Russian Army would not have completed its present cycle of reequipping until 1913–16. Nevertheless, Zhilinskii did promise that Russia would complete much of its concentration by $M+15$, thereby serving French purposes by forcing the Germans to retain five or six corps in East Prussia.[16] By the time that Zhilinskii and a new Chief of the French General Staff, General Joseph Joffre, met in Paris in August 1912, the Russians would have grounds to offer something more concrete.

Unfortunately, the concrete was grounded in compromise. The GUGSh's mobilization schedule of May 1912 was a refined version of Schedule 19 in two variants, one designated A for a main effort against Austria-Hungary and the second designated G for a main effort against Germany. Even as watered down by the GUGSh, the A variant represented an important triumph for the Alekseev-inspired district planning staffs over the Danilov-dominated center. The general intent of A was "to go over to the offensive against the armed forces of Germany and Austria-Hungary with the objective of taking the war into their territory." The mission of the armies which would be arrayed against the Germans was "the destruction of German forces remaining in East Prussia and the conquest of the latter with the goal of creating an advantageous assembly area for further operations." In accordance with A, forty-five infantry divisions in three—later four—armies (4th, 5th, and 3rd, followed by the 8th) would comprise the Southwest Front for offensive operations against Austria-Hungary. This left twenty-nine infantry divisions in two armies (1st and 2d) available for deployment as the Northwest Front against Germany. Two additional separate armies, the 6th and the 7th, would cover the approaches respectively to St. Petersburg and Bendery.[17]

In contrast with A, the intent explicit in variant G was to "go over to the attack against German forces threatening us from East Prussia, [while] paralyzing the enemy on the remaining fronts." The mission of the Russian forces confronting

Germany was the same as in Schedule A. Meanwhile, the armies arrayed against Austria were not to permit the enemy to break into the rear of Russian forces operating against Germany. The forces deployed against Germany would include three armies, the 4th, 1st, and 2d, totaling forty-three divisions. This left thirty-one divisions in two armies, the 5th and 3rd, for deployment against Austria-Hungary. As in variant A, the 6th and 7th Armies covered the northern and southern flanks.[18]

The Soviet historian Rostunov has noted that these plans have often been characterized as "offensive" in contrast with earlier plans, an evaluation which he frankly characterizes as "exaggerated." Nearly all earlier plans had envisioned a transition to the offensive. The question was not so much one of ultimate intent as it was one of timing, means, and method. In the absence of an improved rail and communications net, Russian planners still had to contend with a comparative mobilization and concentration lag. If revised strategic priorities and objectives called for earlier offensive operations, then the appropriate formations would have to concentrate farther forward and begin their march-maneuver to initial contact without waiting for either the attacking armies or their component divisions to attain full strength.[19]

Deployment plans were based on the most probable turn of events, which the Russians calculated would catapult the Germans into a two-front war, with the preponderance of German forces mobilized for an attack on France. Unless the Russian high command perceived that German concentrations exceeded a certain unspecified level, Russian deployments would automatically occur in accordance with Schedule A. Although by 25 September 1913 the Russians had completed a new design, Schedule 20, it was not to take full effect until the autumn of 1914. Therefore, a modified version of Schedule 19 (A), which incorporated elements of Schedule 20, actually governed Russian mobilization in July 1914.[20]

Opposite the Austrians on what would become the Southwest Front (General Ivanov), modified plan A envisioned the concentration of two powerful army groups, one consisting of the 4th and 5th Armies (seven corps) poised to strike south from the Polish salient, and the second consisting of the 3rd and 8th Armies (nine corps) poised to strike west from the Ukraine. The objective of Southwest Front was to encircle and destroy Austro-Hungarian forces concentrating in Galicia, then to seize the Carpathian passes to facilitate subsequent operations across the Hungarian plain. The 4th Army (General Zal'tsa) would concentrate in the area of Lublin, the 5th (General Pleve) in the area of Kholm-Kovel', the 3rd (General Ruzskii) along the line Lutksk-Dubno-Liakhovtsy, and the 8th (General Brusilov) along the line Nikolaev-Chernyi Ostrov-Dunaevtsy. In the aggregate, Southwest Front was to deploy sixteen corps with supplementary and supporting troops along a line describing a 400-kilometer-long semicircle.[21]

Opposite the Germans on what would become the Northwest Front (General Zhilinskii), plan A envisioned the concentration of two armies, the 1st (General Rennenkampf) with four corps and the 2d (General Samsonov) with five corps. The 1st Army would concentrate west of the middle flow of the Nieman, while the 2d would concentrate along the line Grodno-Lomzha-Belostok. The general mission of the two armies was to destroy German forces concentrating in East Prussia, then

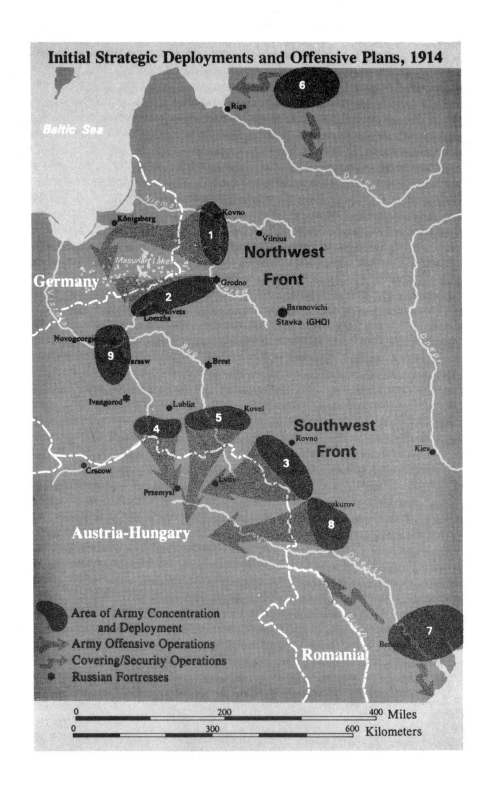

drive for the lower Vistula. Rennenkampf's 1st Army would pass north of the Masurian Lakes, while Samsonov's 2d Army passed south. In the aggregate, Northwest Front was to deploy nine corps along a broken line extending 250 kilometers.[22]

Two key aspects of Schedule 19 (A) which would return to haunt the Russians were time and mass. In conversations between the French and Russian general staffs which occurred in August 1912, General Zhilinskii, then Chief of the GUGSh, assured the French that the Russians would assume the offensive against Germany "after M + 15" with forces in excess of 800,000 men. Although the historian A. M. Zaionchkovskii has stated that the phrase "after M + 15" could have been construed to mean anything, including M + 30, the understanding was that the Russians would attack Germany early enough in sufficient numbers to fulfill their alliance obligations in a meaningful way. By September 1913, Article III of the Franco-Russian Military Convention declared unequivocally that Russia would begin offensive operations against Germany with 800,000 men on the fifteenth day of mobilization.[23] Thus the Russians chose to interpret Zhilinskii's assurances literally, with the result that less than a year later Zhilinskii himself, in an ironic twist of fate, would find his own Northwest Front advancing to the attack before completion of concentration. Last-minute improvisation would wreak additional privation as Supreme Headquarters, or *Stavka*, weakened 1st and 2d Armies to create an additional army, the 9th, to press the attack against Germany from the vicinity of Warsaw.[24] The situation was even more precarious on the Southwest Front, where greater distances and complex troop movement schedules imposed even more serious time-distance lags.

Failure to complete concentration held significant operational implications. It disrupted organizational integrity, deprived advancing units of key assets, and reduced all-important force ratios for numerical superiority along separate axes of advance. If Northwest Front attacked at M + 15, General Zhilinskii would find himself pitting 304 battalions, 196 squadrons, and 1,144 guns (12 heavy) against German forces of 196 battalions, 89 squadrons, and 1,044 guns (39 heavy). Although the preponderance of force appeared on Zhilinskii's side, the reality would be far different. Geographical necessity meant that 1st and 2d Armies would have to march separately on either side of the Masurian Lakes until they could join forces deep within East Prussia. When subtractions were made to account for uneven deployments, the exclusion of second-echelon divisions, and miscellaneous security missions, Rennenkampf's 1st Army would attack with six and one-half divisions and Samsonov's 2d Army would attack with nine and one-half divisions.[25] Against them the German General Max von Prittwitz would initially count about thirteen divisions and good lateral rail communications. In view of the terrain and the widely separated Russian axes of advance, little wonder that General Joffre had once labeled East Prussia "an ambush."[26]

On an even larger scale, Schedule 19 (A) also failed to provide the Russians with sufficient mass to deal an immediate knock-out blow to the primary adversary, Austria-Hungary. Quartermaster General Danilov retrospectively observed that overall Russia would be able to dispatch 1,224 battalions, 664 squadrons, and more than 5,000 guns to the west during the first period of war with the Triple Alliance. By his calculations Germany and Austria-Hungary would be able to oppose Russia with 850–950 battalions. His conclusion was that "we calculated in this instance at the war's beginning to have over our enemies a sufficiently significant superiority in forces (higher than 20 percent)."[27] Danilov did not say how the 20 percent rule of thumb was derived, but the precedent of 1904–5 suggested that 20 percent superiority was probably too slim to assure a safe margin of victory over similarly trained and equipped foes.

Worse, staff planners allocated only 60 percent of the total combat strength of the Russian army to the effort against Austria. David R. Jones has recently indicated that the total forces deployed by Russia and Austria on the Southwest Front in August 1914 would be nearly equal (for the Russians, 34.5 infantry divisions, 12.5 cavalry divisions, and 2,099 guns; for the Austrians, 37 infantry divisions, 11 cavalry divisions, and 1,854 guns).[28] Surprising as these figures were, raw numbers failed to reveal an additional important anomaly: those Russian armies with the most significant offensive missions (the 3rd and the 8th) were precisely the ones left with the farthest march-maneuver distances from strategic deployment to initial contact. What these calculations meant in the aggregate was that adherence to Schedule 19 (A) would fatally condemn the Russians to dissipate their offensive effort, allocating too many troops to operations in East Prussia while failing to provide sufficient mass to assure decision against Austria-Hungary during the early weeks of any future war.[29]

The subsequent debacle of August 1914 has generated endless debate over Russian motives in supporting an initial offensive design which proved wildly optimistic. Historians have offered a variety of rationales, ranging from selfless devotion to the alliance to professional and planning incompetence to offensive mindsets which contemporary social scientists label "ideologies." In the absence of insight based on new primary materials, it seems reasonably safe to surmise a combination of at least four explanations for Russian behavior: overconfidence, poor intelligence, alliance pressures, and strategic miscalculation.

The first explanation was overconfidence. Thanks to the Sukhomlinov reforms, by 1912 the army was in undeniably better condition to support offensive operations than at any time since perhaps the era of Miliutin. Even discounting Sukhomlinov-inspired euphoria, there were some grounds to justify taking the offensive with a newly reorganized army in anticipation of a short war. The alternative was to rely on national resilience and the empire's deeper reserves in protracted conflict—a course for which the recent Russo-Japanese War scarcely provided grounds for optimism.

A second explanation lay with a new willingness to accept risk, thanks to important (and misleading) intelligence assumptions. Although the Russians presided over a fragmented and underfinanced espionage effort, their most important agent, Colonel Alfred Redl, had served between 1905 and 1913 as chief of staff for the Austrian Army's VIII Corps in Prague.[30] Before committing suicide in 1913, Redl succeeded in selling his patrons in St. Petersburg the general Austrian mobilization plan against Russia, possession of which made Russian staff planners overconfident, especially in regard to potential operations on the Southwest Front. The GUGSh based its estimates for Austrian dispositions on intelligence derived from Redl. In reality, the Austrians understood that a leak had occurred, and unknown to the

Russians, they adopted countermeasures in the form of changed dispositions. Thus the Russians would play their war games and press their first offensive into Galicia on the basis of outdated and misleading intelligence.[31] Reality would find the Austrians deployed farther west than the Russians expected, with the result that initial combat actions would assume the form of envelopments rather than the anticipated frontal attacks.[32]

Against the Germans there was an intelligence failure of another sort. As early as 1910, the Russians had become convinced that Germany would send the bulk of its forces west against France, leaving only minimal forces in East Prussia, possibly to conduct a spoiling attack, then more surely to delay any Russian advance. By 1912, the Russians considered it probable that Germany might leave only three active corps and ten reserve divisions in East Prussia. While Russian intelligence sources had obtained information that German war games projected fighting the Russians separately on either side of the Masurian Lakes, Russian planners apparently dismissed the advantages which a dense rail net might confer on even residual German forces in East Prussia. Therefore the overriding Russian impulse would be to discount intelligence-based warnings, press forward with all due haste to brush aside the German screen, then reorganize for subsequent offensive operations west of the Vistula. Intelligence estimates failed to correspond with the prevailing operational winds, so they were apparently dismissed.[33]

Consequently, in contrast with their pre-1910 caution, the Russians were prepared to accept greater risk after 1912 because their army displayed improved combat readiness, because they possessed the luxury of heeding or disregarding intelligence, and because the odds which they faced—at least against the Germans—were arguably lower than earlier worst-case scenarios had led them to anticipate. Interestingly, the GUGSh appeared neither to overestimate or underestimate their potential adversaries' numbers but rather to dismiss possibilities and circumstances which might either impede a Russian advance or confront field commanders with unexpected combinations.[34]

A third explanation for Russian conduct was the weight of the alliance. Although altruism often provides scant motivation for behavior in affairs of state, there is good reason to believe that self-interest prompted the Russians to accept their alliance obligations seriously. If France readily fell to the Germans, the Russians had little confidence they could hold against the combined hordes of Germany and Austria-Hungary. Therefore the governing assumption was that Russia and France either rose or fell together and that Russia should strain to the utmost in meeting its alliance obligations, even to the point of conducting initial offensive operations at M + 15, and even to the point of dividing the offensive effort.[35]

Against this background, a fourth explanation came into play: a failure through miscalculation to resolve complex strategic dilemmas of mass, concentration, and movement. In their prewar writings, Neznamov and Mikhnevich had proceeded from many common assumptions to justify two distinct strategies, each designed to produce victory, but with each ascribing varying degrees of importance to offensive and defensive action. On one hand, if Russian planners sought to wage a short war of annihilation (*sokrushenie*), then they needed to de-emphasize the defensive while accomplishing significant strategic aims (destruction of main enemy forces or seizure of key objectives) during the flawless unfolding of a seamless design embracing mobilization, deployments, and initial offensive operations. On the other hand, if Russian planners sought to wage a protracted war of attrition (*izmor*), then they needed a more cautious, phased approach to the same separate tasks, including a temporary and deliberate reversion to the defensive, a calculated buildup of deployments to attain mass and concentration, and a dynamic shift to the offensive as prelude to a series of successive strategic operations. As the situation stood on the eve of World War I, Schedule 19 (A) mobilized and deployed the Russian army for *sokrushenie* but left it with time-space-mass anomalies that would inexorably lead to *izmor*. Thus if Russian planners and military administrators could be held culpable for serious miscalculations, then their culpability was rooted in a failure to understand the complex and shifting relationship between offense and defense over time in the initial period of any future war.[36]

Related to these miscalculations was the issue of static defenses covering forward positions in the Polish salient. With even modest investment and integrated planning, some kind of modified defensive barrier would have won additional time for the Russians to complete strategic deployment and concentration for offensive operations. However, the emotionally charged atmosphere of 1910–13 afforded little opportunity for a sober-sided examination of this and other alternatives, including a calculated reassessment of static and dynamic defensive requirements. In the end, a combination of scarce resources and compromise conditioned Russian planning and preparations for future war. Although there is evidence that the Large Program would have rectified some of the more obvious anomalies, it was not calculated to take full effect until 1917.[37]

It is difficult to escape the conclusion that failure to translate theory into meaningful action was in large part a function of institutional drift. Beyond his own traditional and narrowly conceived notions of state and monarchy, Nicholas II failed to imbue the military hierarchy with a sense of overarching purpose that would have served as the foundation for a "unified military doctrine" (even as the Tsar himself understood that term). And, as Schedule 19 (A) so aptly demonstrated, compromise and collegiality were poor substitutes for clear, singleminded direction from the top.

Command and Communications

Part of the grounds for Russian planning optimism no doubt stemmed from a perception of improved competence on the part of the army's field command. Although the popular Western conception is one of sloth and incompetence, the Russians had made significant strides in comparison with the situation in the Far East at the outset of the Russo-Japanese War. Or, as General Danilov noted in his discussion of the high command in 1914, "all measures were taken to assure the appropriate selection of personnel in the designation of wartime high command appointments and in the establishment of order in the formation of front and army staffs." General Ia. G. Zhilinskii (b. 1853), Northwest Front Commander, had been Chief of the GUGSh for three years, before which he had commanded the Warsaw

Military District. His chief of staff was General V. A. Oranovskii, who had served as Linevich's quartermaster general in Manchuria, then as chief of staff of the Warsaw Military District. Oranovskii's peacetime specialty had been a close study of the potential German front. General N. Iu. Ivanov (b. 1851), Southwest Front Commander, had been Commander of the Kiev Military District since 1908 and had seen combat in both the Russo-Turkish and Russo-Japanese Wars. His chief of staff was the extraordinarily capable M. V. Alekseev (b. 1857).[38]

Similarly, the subordinate army commanders were far from an incompetent or inexperienced group. General P. K. Rennenkampf (b. 1854) had commanded cavalry formations in Manchuria and since 1911 had served as Commander of the Vilnius Military District, where his 1st Army would begin its deployments. The 2d Army's General A. V. Samsonov (b. 1859) had commanded a brigade in the Russo-Japanese War and served before World War I as chief of staff of the Warsaw Military District. Although some observers traced the later inability of Rennenkampf and Samsonov to coordinate their armies to personal conflicts dating to 1904, little or no evidence exists to support this contention. On the Southwest Front, General N. V. Ruzskii (b. 1854), who would command the 3rd Army, had served as chief of staff of the 2d Manchurian Army in 1904–5, then had served in various staff capacities in St. Petersburg, including chairmanship of the commission which revised Russian field regulations. Generals P. A. Pleve (Commander of the Moscow Military District) and A. E. Zal'tsa (Commander of the Caucasus Military District) appeared no less experienced. General A. A. Brusilov (b. 1853), future commander of 8th Army, was a cavalryman with division and corps command experience. In 1912–13, he had served in the Warsaw Military District as assistant troop commander.[39]

Within the context of projected assignments and varied career patterns, several sources of potential difficulty revealed themselves. One was the fact that nearly all the potential commanders on the two Russian fronts had been born in the 1850s. This meant that they would typically have received their baptism of fire in 1877–78 and completed their military education before 1900, that is, before the winds of change had begun to alter traditional higher-level military thinking. The second difficulty was implicit in an assignment system which earmarked too few younger general staff officers to serve as the military-intellectual glue binding together various levels of the staff hierarchy. A third problem was the Emperor's failure to designate a commander in chief in advance of actual mobilization. Law and the new wartime field regulation made the Emperor himself commander in chief, provided that he chose to exercise personal leadership over the armed forces. In the event, Nicholas chose to retain flexibility by keeping the appointment open until 19 July 1914, when he bowed to political pressure and named his uncle, Grand Duke Nicholas Nikolaevich, to the high command.[40]

The price of flexibility was lack of immediate integration at the highest instance of field command. It was not for nothing that in 1903 War Minister Kuropatkin had insisted on convoking a meeting of his potential subordinate commanders to establish face-to-face familiarity and test the mettle of staff officers. Although in 1914 few observers questioned the Grand Duke's ability to inspire confidence, he had been involved only on the fringes of war planning and had little opportunity before the outbreak of hostilities to take the measure of the commanders and staff members with whom he would be working in the first hectic months of the war. In fact, the assumption was that the Grand Duke would command 6th Army covering approaches to the imperial capital.

However, the high command calculus was sufficiently worked out so that at least *Stavka*, as Supreme Headquarters was called, would have a chief of staff in the person of General N. N. Ianushkevich and a quartermaster general in the person of General Iu. N. Danilov. Unfortunately, their qualifications presented a mixed picture. Ianushkevich had only recently been appointed Chief of the GUGSh, and his expertise in the field of military administration was more academic than practical. In contrast, Danilov possessed vast planning experience, having served since 1908 as Quartermaster General of the GUGSh. Beyond these two officers, the situation remained unclear, with likely assignments to the rump staff at Stavka subject to prevailing organizational and political winds.[41]

Added to the uncertainties of staff arrangements were problems of communication, personal and practical. Norman Stone has recently emphasized the partisan split between adherents of the Grand Duke and those of Sukhomlinov. Stone insists that this division affected command and staff appointments down to army level, so that every headquarters possessed representatives of the two camps who would find it difficult to communicate with one another. If this was so, then divided personal loyalties worsened an already potentially disastrous communications picture: even in peacetime the Russian army had little experience with wireless communication, and there was insufficient telegraph equipment. Samsonov's 2d Army had only twenty-five telephones, several Morse-coding machines, and a single slow and cranky Hughes teleprinter. Under conditions of offensive operations, the communications picture was likely to get worse, not better.[42]

Rehearsals

Perhaps nowhere was the gap between peacetime potential and wartime reality more clearly revealed than in tsarist war games. In theory, war games were a valid device for linking mobilization planning with operational concept. They resembled the contemporary command exercise without troops. That is, commanders and coordinating staffs were assembled to conduct extensive map exercises, the purpose of which was to test insofar as possible the feasibility of plans and the conduct of initial operations in any future war. In view of the GUGSh's shifting fortunes between 1906 and 1914, it was no surprise that Russian war games failed to display consistent rigor and purpose.

B. V. Gerua, the brother of A. V. Gerua, has recounted how games were sometimes preceded by staff rides in which officers of the general staff toured areas of potential operations to familiarize themselves with local terrain and the nature of possible theaters of operations. While assigned to the Kiev Military District, Gerua participated in two such rides, one to Volynia and the other to Podolsk. He remembered the problems as stereotypical, the impressions of picturesque localities as pleasant, the instructional results as useless. In contrast, A. G. Vineken recalled that comparable French exercises under the watchful eye of General Ferdinand Foch cast participants in the role of skeleton staffs engaged in genuine operations. Each officer's day was accounted for to the minute, and the directing officers laid out problems which required accuracy in calculating time and distances and "colossal preparatory work."[43]

Evidently, F. F. Palitsyn hoped to introduce games on the Prussian model to the military districts, but the GUGSh Chief's influence was neither consistent nor persistent. Early in 1907, he proclaimed a major war game in the Kiev Military District to show local officers "how it was done." Every district general staff officer was forced to memorize detailed materials on the Austro-Hungarian army and its order of battle. A "Grand Inquisitor" then appeared to test the officers' knowledge, using a confessional-like box in the staff communications center to quiz staff members on classified materials. When Palitsyn arrived, the situation room was in full readiness, with maps laid out on tables and officers assembled to represent various staffs. Both Palitsyn and Sukhomlinov, new commander of the Kiev District, were cavalrymen who possessed strong opinions about the use of cavalry for security, screening, and reconnaissance operations. While supporting staffs assiduously assembled and quartered the main warring forces, the two prime actors wasted three exercise days playing nostalgically and interminably through the initial cavalry encounters. Soon they exhausted the entire time scheduled for the game without accomplishing three-fourths of the tasks allotted, including the movement and engagement of the principal combatant forces. Gerua summed up the game's effect by dryly observing that "the mountain had given birth to a mouse." After Palitsyn's departure, the Kiev staff lapsed into its normal routine.[44]

As War Minister, Sukhomlinov initially accomplished little more than Palitsyn. In the spring of 1911, Sukhomlinov received permission from the Emperor to occupy an entire wing of the Winter Palace with games that would involve army-level commanders and staffs designated for possible action against Germany and Austria-Hungary. One objective was to test conceptions of future war since "the situational data on hand in the Main Directorate of the General Staff gave us the ability to create a situation and pattern of possible actions with a great degree of probability regarding the likely military operations of foreign armies."[45] Another objective was to observe how commanders might respond to real situations, with the understanding that those who were found wanting would find other employment. When Grand Duke Nicholas Nikolaevich got wind of the exercise's possible effect, he took exception to "giving an exam to army commanders" and successfully importuned the Tsar to have the game canceled at the last minute. The Grand Duke thus managed to call into question not only the War Minister's credibility but also the Emperor's loyalty to his War Minister. A minor scandal ensued before Sukhomlinov arranged a face-saving formula which enabled him to retain his portfolio while replacing the exercise with an imperially sanctioned "conference."[46]

In 1914, exercises produced a more successful variation on the themes of 1907 and 1911. Although Gerua labeled the exercise results of 20–24 April 1914 "inconclusive," they were informative not only because of what was assumed and what happened, but because of what was left out. To avoid "intrigue" (that is, the Grand Duke's interference), Sukhomlinov arranged to hold the 1914 games in Kiev. The assumption was that Austria-Hungary and Germany would be the primary adversaries and that the mobilization schedule of 25 September 1913 would govern the exercises. It was also assumed that Germany would launch its heaviest blow against France, after leaving small forces deployed in East Prussia. Although the Russians did not categorically rule out a change that would pit Germany's main forces against Russia, in any case "the primary objective of actions was the destruction of the Austro-Hungarian Army." Whatever the situation, operations against Germany would support those against Austria, although the Russians conceded the importance of supporting France "in order to facilitate success in the West."[47]

In comparing forces and time available, the Russians assumed that they would retain absolute superiority in forces, although in the event that Germany mobilized first against Russia, continued tsarist superiority would have to be assured by bringing supplementary forces from the interior, including Asiatic Russia. Germany would complete its mobilization at $M+13$ and Austria at $M+16$. At $M+16$, Russia would have assembled only half its forces, and complete assembly would occur only on $M+26$, and that only if the Far Eastern contingents were not counted. Their inclusion extended the time requirement from $M+26$ to $M+41$. Without counting contingents from non-European Russia, the tsarist General Staff understood that its mobilization lagged at least thirteen days behind that of Germany and ten days behind that of Austria-Hungary.[48]

The 1914 games anticipated groupings of forces on the western frontier in much the same fashion as they would actually occur. Mobilization and assembly would produce eight field armies, of which the 1st and 2d combined to form the Northwest Front for action against East Prussia. The 4th, 5th, 3rd, and 8th Armies comprised the Southwest Front for action against Austria-Hungary. The 7th Army was to concentrate in the region of Bendery for possible action against Romania, while the 6th Army covered St. Petersburg and its sea approaches.

The 1914 games also anticipated many of the later campaign's major actors. General Ianushkevich acted as chief of staff, while General Danilov acted as quartermaster general. The exercises were devised by the GUGSh, signed by General Zhilinskii, and approved by the War Minister, General Sukhomlinov. Zhilinskii served as commander, Northwest Front, while the commander, 1st Army, was General-Adjutant Rennenkampf, and the commander, 2d Army, was General Raush von Traubenberg. General-Adjutant Ivanov commanded Southwest Front, while his army commanders were General Zal'tsa (4th), General Pleve (5th), General Churin (3rd), and General Ruzskii (8th).[49]

There were also some important lacunae. Neither the 6th nor the 7th Armies took part in the game. More significant, transportation schedules were not played, and rear services were not depicted "in order not to complicate the play."[50]

The game predicated a situation in which France and Russia had gone to war against the Triple Alliance. It was assumed that Germany would direct its primary offensive blow against France after deploying only ten regular and eleven reserve divisions in East Prussia. Despite the disparity of forces in the east, the Germans were to conduct a spoiling offensive along the middle flow of the Nieman on the front

Grodno-Olita both to disrupt Russian mobilization and concentration and to divert attention and Russian forces away from the main Austrian effort in Galicia. There, after concentrating their primary forces (forty infantry divisions and ten cavalry divisions) in Galicia, the Austrians were to assume the offensive with seven corps (twenty-one divisions) against the forces of the Kiev Military District along the front Sedlets-Brest-Kobrin. The game's first turn occurred on M + 10, one day after the Germans were able to assume the offensive in the north and at the critical moment when the Russian high command would have to decide whether to bring the preponderance of its forces to bear against the Germans or the Austrians.[51]

Northwest Front seized the initiative by attacking at M + 12 with its two armies before they had completed their concentration and before 2d Army could bring effective forces to bear against the German right flank. Against the Austrians, Southwest Front responded that its armies could not assume the offensive before M + 20 (8th, 3rd, and 5th) and M + 21 (4th). M + 12 witnessed the first serious battles in the north between German forces and the 1st Army, while 2d Army lagged two days behind in arraying its lead corps for an attack on the German rear and right. Consequently, Northwest Front used its second turn to slow 1st Army's attack to permit 2d Army to coordinate its attack with Rennenkampf's offensive against the German left.[52]

In anticipation of the third turn, Northwest Front received information that a British expeditionary force had landed on the continent to reinforce the French, thereby tipping the military balance in the west against Germany. The Germans responded both to this news and the advance of the 1st and 2d Russian Armies by ordering a withdrawal in East Prussia behind the Angerapp River. As the Germans gradually gave ground, they dispatched three corps to the west to reinforce the German offensive against France. This news reached the Russians on the evening of hypothetical M + 21. Northwest Front responded by ordering an energetic offensive against both German flanks north and west of the Masurian Lakes. At the same time, the high command confidently dispatched one Russian corps from Northwest Front south to assist the beleaguered Southwest Front, where time was fast approaching to deal the Austrians a decisive blow.[53]

Southwest Front used its second turn on M + 19 to advance by one day the attack of 4th Army south to relieve growing Austrian pressure on 3rd and 8th Armies. On M + 21, 8th Army's scheduled offensive finally began to relieve pressure on 3rd Army, whose situation now became "completely satisfactory." At the same time, 5th Army prepared to operate in conjunction with 4th to strike the main Austrian advance in the flank. At this point, on the eve of the Southwest Front's third turn, the game drew to a close for lack of time, and the outcome of the envelopment against the Austrian left remained unclear. What was clear, however, was that the four additional corps scheduled to reach the front by M + 25 would endow the Russians with a six-and-one-half division superiority over the Austrians, a situation deemed sufficiently advantageous to assure Russian success.[54]

What lessons did the Russians draw from the game? Not many on the official level, if Sukhomlinov's report of 22 May 1914 can be accepted as the definitive pronouncement from the War Ministry. His account consisted primarily of a recitation of the game's course with almost no critical commentary. Perceptions of "lessons learned" must be found elsewhere, in the form of the participants' disparate notes and minor reports. As the postwar commentator A. N. Suvorov noted, a "common thread" running through nearly all their commentary was the necessity not to hurry the offensive, to complete concentration, to establish logistical order, and only then to press forward. General Ivanov, who would later command Southwest Front, held that since the Austrians could mobilize faster than the Russians, in order to defeat them decisively the Russians would first have to put everything in order, then go over to the offensive "to avoid isolated failures which would affect the spirit of the troops and the nation." General V. I. Dragomirov, who commanded the 3rd Army for purposes of the game, echoed these sentiments, saying that a substantial supply buildup would be a necessary precondition for serious operations at any depth in Galicia. Meanwhile, Suvorov dryly commented that "problems of the rear were evidently forgotten."[55]

General Dragomirov also seriously questioned the adequacy of ammunition reserves and replacement rates. Although the Russo-Japanese War might have lent credence to those who believed that 300–400 shells per field gun were adequate, Dragomirov felt that the figure would "hardly be adequate for the first serious battles, after which army supplies will have been dried up." The same held true for casualty replacements, and Dragomirov demanded to know why the accepted norm for those killed in action was set "at 3 percent when the expected mortality rate was 15–20 percent?" Of course, his conclusion was that the accepted 5 percent replacement figure was much too low. This question, together with vexing issues of rear service support, raised serious doubts about the country's ability to support not just a prolonged war but even a brief war of movement and maneuver in depth for decision.[56]

The obsession with speed in the offensive also raised a serious question about the Russian ability to gain sufficient mass to bring about decision. Rennenkampf apparently found the order for his 1st Army to take to the offensive before completion of concentration "hardly correct." His army's hasty advance would launch it across the Nieman to do battle with three German corps at a time when his III and IV Corps had only three and one-fourth of their divisions! The same problem of incomplete concentration meant that 2d Army would be pressed into the offensive before it received a full complement of army-level cavalry units which would serve as its eyes in entering foreign territory.[57]

A key issue which none of the commentators seems to have raised was that of operational direction. Despite the avowed emphasis since 1908 on Alekseev's vision of the separate fronts pursuing common objectives, little thought appears to have been devoted to problems of coordinating the actions of fronts and armies. In particular, the physical gap between 1st and 2d Armies caused by geography and different rates of concentration and forward movement ought to have become a source of genuine alarm. The difficulties—if not the implications—inherent in this situation were very evident in the game of April 1914. What remained less evident— because actual army-level maneuvers were not conducted—was the ability of army commanders to control and coordinate the actions of their constituent corps commanders. There was no evidence other than the concerns of theorists that the operational-level issues raised seven years earlier by Colonel A. V. Gerua had been addressed more than on paper. Troop exercises involved formations up to corps size. Map exercises involved fronts and armies. There was nothing to link the two. Therefore officers (except those who had fought in the Far East) had no real or vicarious experience in leading corps- and army-level formations. Not surprisingly, in August 1914, when General Samsonov encountered serious difficulty, he would obey his commander's natural instinct to abandon his command post and move to the sound of the guns.

Logistics

Recent observers have viewed the army's failure to account for logistical problems as one more symptom of a fatal disease born of venality and incompetence and destined to produce catastrophe, stalemate, and scandal. The reality was more complex. In truth, the Russians had studied the lessons of 1904–5 and had emerged with what they felt were sound logistical projections and structural calculations for the conduct of a brief contemporary war of maneuver. It was in many respects their elementary understanding of combat experience that would play them false.

Despite the misgivings and projections of theorists such as Mikhnevich, a short-war mentality governed Imperial Russia's material preparations for future European conflict. The prophets of a short war argued that Russia would fight on the basis of war stocks built up in advance during peacetime preparations. Indeed, the Chief of the Main Artillery Directorate, in testimony offered during hearings in May 1915 on the munitions shortage, asserted that "previous wars provided clear evidence that an army got along on that reserve of munitions which existed in peacetime. Everything ordered with the declaration of war was caught up with only after the war's conclusion and served to make good on expended reserves."[58]

When a European war came, reliance on precedent created two significant limitations, material and structural. On the basis of 1904–5, Russian artillery specialists calculated that 1,000 rounds per gun would suffice in any European conflict. This figure actually surpassed the usage rate in Manchuria by more than 300 rounds per gun, a figure which most planners assumed would provide an adequate hedge against excessive consumption. Similar indices were applied to required stockage rates for field artillery pieces, machine guns, and small arms.[59] In contrast, an outstanding instance of constrained budgets causing a shortfall was in the area of small-arms munitions, where a shortage of funds caused the Mobilization Committee of the GUGSh to revise its stockage target downward by about 500 million rounds. Thus the Russian army would enter World War I with only two-and-a-half billion rounds of small-arms ammunition.[60] Although partial retention of the Polish fortresses introduced additional distortions and self-imposed limitations into the supply picture, the Russians were not overly concerned with the possibility of shortages. Thus David R. Jones has recently noted that "what deficiencies existed between the quantities on hand and those stipulated as necessary in 1914 were minimal."[61] The problem was that the realities of combat were to reveal the hopeless inadequacy of these prewar calculations.

Less apparent and perhaps more damaging in the short term was the structural aspect of logistics. The Far Eastern experience had caused the Russians to evolve a dedicated supply structure which would serve front-line units working back up the hierarchical chain through division, corps, army, and front. In the best Kuropatkin tradition, army corps would be joined by a network of roads and narrow-gauge field railroads with a major railhead at front level.[62] The assumption was that as initial deployments and initial operations wound their course, units would draw on this structure for personnel replacements, provisions, munitions, and supplies. This well-defined structure presented two significant drawbacks: rigidity and finite supply limits. The structure was such that it could not accommodate the rapid shifting of supplies from one axis to another in accordance with a command impulse to support tactical and operational success. Therefore units advancing in areas with a poor transport network would quickly find themselves outracing their logistical support means. The advance would then come to a halt while the entire network was laboriously displaced forward and reconstructed.

The second difficulty involved rear area supply limitations. Each supply echelon drew on its immediate rear services until the geographical limits of a front (a transformed military district) were reached, at which point there was no fully integrated system to draw from the empire as a whole. This meant that as each military district utilized its supplies, there would be a lag until deficits could be made good on the basis of the national economy, which, in turn, must have been mobilized. Until linkages were worked out between the various fronts and Stavka, there would be no more supplies in the pipeline. This occurred very quickly in the Russian instance and explained why Russia experienced a mobilization crisis in 1915.[63]

The blame, therefore, for logistical discrepancies lay not not so much with those who accepted the probability of a short war as with those who failed to endow the system with sufficient flexibility to support a maneuver war that would severely tax the infrastructure already in place. The problem was that the Manchurian experience had to be understood in its full context, not as an ironclad precedent but only as an indicator of what might be possible in future war. In this respect, the evolution of norms and indices based on combat experience provided an incomplete data base for Russian planners to project usages in a much different tactical and operational environment.[64]

Field Regulations and Tactics

Despite the post-1905 ferment—or perhaps because of it— the Imperial Russian War Ministry was unable to produce a new field regulation, or *polevoi ustav* (PU), until 27 April 1912. When General N. V. Ruzskii's special commission finally published the regulation, it represented the culmination of an elaborate process involving extensive discussion both in the open military press and behind closed doors. In contrast with its 1904 counterpart, PU-1912 was clearer in essentials and more to the point, with fewer burdensome details. It consisted of five sections: direction, reconnaissance, bivouac, field movement, and conduct of all-arms battle. In accordance with much of the prevailing sentiment for unity of thought and action, the new regulation asserted that success in war resulted from both clear establishment of objectives and orchestration of the common effort to achieve them. In turn, successful military actions depended upon the mutual support of all types of troops. Like the past regulation, the 1912 regulation upheld the basic conviction that the offensive was the primary means of bringing about decision in war. In deference to the concept of individual initiative, the regulation provided only general guidance, leaving to commanders a high degree of independence in choice of means. Also, perhaps in deference to defenders of the cultural patrimony, the writers of doctrine studiously skirted foreign military terminology, substituting Russian words even for long-standing cognates which had gradually crept into Russian usage.[65]

The first three sections of PU-1912 corrected many of the deficiencies evident in the 1904 version. For example, the 1912 regulation devoted an extensive section to reconnaissance in which discussion ranged from fundamentals to more complex issues such as maintaining constant linkage between infantry security elements and mounted scouts. The section on bivouac repeated the basics of organizing quarters in the field not only to prepare for movement and imminent battle but now also to guard against new threats, such as the distinct possibility of air reconnaissance. The regulation of 1912 also called for organization of security in depth and breadth with firm contacts between various kinds of security posts and detachments. Probably because of the greater likelihood for meeting battles, the section on field movement reiterated the importance of rapidity of execution in changing from march formation to combat formation. Indeed, PU-1912 spelled out in some detail the kinds of formations appropriate to a battlefield approach march.[66]

The section "Actions in Battle of Formations (Detachments) of All Types of Troops" occupied the most prominent place within the new regulation. In keeping with convention, PU-1912 acknowledged only two forms of battle, offensive and defensive, although the regulation also devoted considerable attention to a variation of offensive battle, the meeting battle, which had scarcely received passing mention in the 1904 regulation. Primary stress naturally fell on offensive action: "The most effective means for destruction of the enemy is to attack him." The only way to shatter the enemy was to deal him a strong blow in a vital place. This meant that the attacker or defender had to be stronger than the enemy along the direction of the decisive blow and during its delivery. The preferred method of conducting the attack was to launch a combination of blows from the front and one or both flanks.[67]

Whether on the offense or the defense, PU-1912 admonished commanders to take into account the qualities of contemporary fire and, in so doing, to work out march-maneuver schemes while taking into account available means of fire support, including machine guns, howitzers, and heavy field artillery. What had changed since 1904 was the novel emphasis on the predominance of firepower: fire not only prepared the way but now also determined the very possibility of closing for the bayonet attack. The basic infantry attack now consisted of a combination of movement toward the enemy and fire from rifle positions. PU-1912 asserted that "the better concealed and more rapid the movement from one position to the next, the less infantry [would] suffer loss and achieve the best results from its fire, thanks to the unexpectedness of its coming from new positions." When advancing in various elements by various stages, infantrymen were to move in a manner that "did not present significant targets" and that called for them "to remain in the open for only short periods." If need be, at close range infantrymen were to advance on hands and knees.[68]

Offensive battle now consisted of four phases: the approach (*sblizhenie*), the attack (*nastuplenie*), the final assault (*ataka*), and the pursuit (*presledovanie*). Troops in the approach march assumed a kind of precombat formation in column with security detachments to the flanks and reconnaissance elements forward.[69] In healthy recognition of artillery's destructive power, PU-1912 stipulated that troops approaching within three to five kilometers of the enemy deploy into smaller columns, at which point higher commanders worked out their general plans for the attack, assigned objectives, and issued orders. When the distance closed within range of effective rifle and machine gun fire (1–1.5 kilometers), subordinate commanders deployed their units under cover of the advance guard into battle formation (*boevoi poriadok*). Prescribed frontages for advance into the attack ranged from about one-half kilometer for a battalion, one kilometer for a regiment, two kilometers for a brigade, three for a division, and up to five or six for a corps. Even during the approach, the regulation recommended that artillery be pressed as far forward as possible to assist in suppressing enemy defensive fires. At this stage, the chief mission of the artillery was to achieve superiority over enemy artillery.[70]

The attack phase opened as the infantry occupied its first firing positions, from which they would move forward assisted by their own rifle and supporting artillery fires. The most advantageous form of the attack was to move forward in skirmish formation with two to ten paces separating individual soldiers. Utilizing a combination of fire and movement, small units and individuals worked their way forward in leaps and bounds covering each other and facilitating the advance. Insofar as possible, artillery was to support the infantry by suppressing enemy fires throughout the attack. When infantrymen had reached the forward limits of their adversaries' close-in defenses (about 150 meters from the enemy), they were to take up their last firing position to await reinforcement from company and battalion reserves, all the while pouring rifle fire into the defense.

The final assault began at the moment when the defense appeared suppressed to the maximum degree possible by offensive rifle, machine-gun, and artillery fire. When infantry rose to assault, the artillery either shifted or withheld its fire to avoid hitting friendly troops—unless separate batteries and guns had been pushed sufficiently far forward to support the assault by direct fire. The regulation prescribed that the final assault had to be carried out as aggressively and energetically as possible. This meant that the attackers rushed forward from their final firing positions to engage the enemy with cold steel. Meanwhile, insofar as possible, cavalry moved against the enemy's flanks and rear. Following successful assault, the attackers went over to the pursuit, continuing physical or fire destruction of the enemy until the requirement for consolidation forced a halt to forward movement. Cavalry took over the primary role in pressing the pursuit.[71]

Despite emphasis on the attack, the regulation of 1912 also ascribed great significance to the defense, which might become the dominant form of battle in the event that "an objective could not be accomplished by the attack." In light of the experiences of 1904–5, the regulation required that defense assume "an active character" and that logically the defense be considered a prelude to offensive action: "while defending, it is necessary to strive with all forces and means to thwart the enemy's fire and, after undermining his moral forces, to go over to the attack and destroy him." PU-1912 surmised that defenders would construct entrenchments and strong points consisting of field fortifications covered by obstacles and interlocking crossfires.[72]

Although PU-1912 represented an advance in Russian military doctrine, it also retained a number of deficiencies. It failed to advocate hasty entrenchment in the attack against fortified positions. Probably for reasons of difficulty in fire control, it also failed to ascribe any direct suppressive role to artillery in the final assault. Ironically, this meant that the Russians, who had heretofore underestimated the role of rifle fire in their offensive doctrine, now overestimated its ability without substantial artillery assistance to suppress the defense during the most critical phase of the attack. In addition, because the writers of doctrine apparently considered some Manchurian circumstances unique, PU-1912 failed to note that future battles might find defenders forced either to attack or to devise for themselves extensive and interlocking networks of continuous entrenchments and field fortifications.

Despite these and other shortcomings, PU-1912 marked a notable advance in its attempt to integrate recent combat experience with current practice, process, procedure, and force structure. The regulation reflected a much improved—if still imperfect—understanding of the role which firepower played in modern battle. In its prescriptions for the approach phase to combat, the new doctrine provided an important point of departure for the development of further theory and practice on the meeting battle. The appearance of PU-1912 probably accounted for the reassertion of audacity and maneuver which the military historian A. M. Zaionchkovskii found prevalent among smaller prewar formations.[73] Thus, although the regulation left unsettled many issues of battlefield command and control, General Ruzskii's team of doctrine writers correctly emphasized the importance of individual initiative as a major part of the solution to modern challenges of time, distance, and decentralized execution.[74]

Infantry Tactics

To provide lower-level supplementation for PU-1912, the War Ministry on 27 February 1914 published a separate *Instruction for the Action of Infantry in Battle* (*Nastavlenie dlia deistvii pekhoty v boiu*). It acknowledged that "the chief role in battle belonged to the infantry; other types of troops must by all means cooperate with it in the accomplishment of its combat missions and selflessly assist it at the difficult moment." The new instruction elaborated on PU-1912 to describe the relationship between fire, maneuver, and shock: "the force of infantry in battle consists both in rifle and machine gun fire with decisive movement forward and in the bayonet blow."[75] The chief variables governing the infantry attack involved terrain and enemy resistance, in accordance with which the commander chose his formations not only to limit his losses to fire and account for terrain but also to facilitate control, flexibility, and delivery of fire. Within range of rifle fire, these requirements were best met by deployment in open-order skirmish formation. Behind the skirmish formation, the more densely configured reserve advanced in deployed formation; but, also in accordance with circumstances, the reserve possessed considerable latitude in adopting various forms of the open order to reduce losses and facilitate the advance.[76]

Primary difficulties were those of control and mass. Even a platoon deployed in skirmish formation presented a frontage of one hundred paces and more, a fact which both imposed added responsibilities on squad and section leaders and reemphasized the need for independence and initiative among the lower ranks. At platoon and company levels, a combination of voice commands, flags, and hand signals were the primary means of communication. And, although open formations reduced vulnerability and presented new opportunities for supporting the attack by fire, the same formations rendered ever more difficult the ability to achieve either the physical or fire mass necessary for delivery of a decisive bayonet assault. Here under altered circumstances was the classic dilemma which had confronted infantry since the introduction of improved weaponry in the mid–nineteenth century: how to achieve timely concentration in a situation under which survival required dispersion.

To resolve this dilemma, a few innovative Russian tacticians toyed with new ideas. The least imaginative called for increased attention to the volume of the attackers' rifle fire in hopes that the infantry alone could shoot its way through the defense to facilitate a successful advance. Others pinned their hopes on the novel use of extant weaponry. M. D. Bonch-Bruevich and A. A. Buniavskii, for instance, perceived the need for additional firepower and recommended that under certain circumstances commanders might move heavy machine guns forward to add their staccato fire to suppression of the defense.[77] However, since the infantry table of organization called for only eight machine guns per regiment, the likelihood of their availability for augmentation at company and platoon level was problematic at best. Moreover, the sheer weight and size of heavy machine guns—even as modified in 1910—rendered doubtful their ability to accompany foot soldiers into all phases of the attack. Another answer lay with the forward deployment of field artillery and the timely application of its fire weight to the offense, especially in the final assault, but more experience and coordinating efforts were required before artillerymen and their guns proved equal to the task.[78]

Meanwhile, conventional infantry wisdom apportioned traditional resources in ways which at least offered incremental advantage. Commanders allocated their infantry to the attack in combat formation by task, dividing them into two parts by function. Each was to employ maneuver to the furthest extent possible, but the mission of the first was primarily to support the attack by fire in the advance, while the mission of the second was to make good the losses of the first and then combine with remnants of the first to launch the decisive final assault with cold steel. As before 1905, the first was called the combat element (*boevaia chast'*), while the second was tabbed the reserve.[79] Thus the functional understanding of the reserve remained restrictive; rather than serve as a force suitable for discretionary commitment, the reserve's specific mission was usually to lend its force to the attainment of mass in the final assault.

This division of functions within the combat formation characterized all levels of infantry task organization for offensive action from company level up to division. In accordance with the instruction, the company combat formation consisted of a platoon skirmish line and a company reserve, at battalion level of a company skirmish line and a battalion reserve, and so on, through brigade and up to division, where the commander retained greater flexibility and more assets than his subordinates in designating and weighting the composition of forces that made up the overall combat formation.[80]

Interestingly, in view of the experiences of 1904–5, the infantry instruction of 1914 allocated scant space to issues of reconnaissance and maneuver. Although the narrative touched on the deployment of mounted and dismounted scouts, there was no special section to govern use of the regimental scout detachment, a curious oversight in light of its importance in Manchuria. The instruction did, however, devote a special section to maneuver, which it defined as "placing the infantry detachment in the most advantageous position for the accomplishment of a given mission." Maneuver was accomplished at the appropriate time and in accordance with weather conditions by proper utilization of movement, speed, security, and terrain. The instruction enjoined the commander to pursue the correct combination of frontal and enveloping movements that would put him at an advantageous position in regard to the enemy.[81]

In accordance with the precedent laid down by PU-1912, firepower received greater emphasis than in earlier instructions. "Destruction of the enemy is achieved by a combination of rifle and machine-gun fire," the instruction of 1914 intoned. In contrast with earlier prescriptions, individual fire was preferred over volley fire, and circumstances under which the latter received preference were limited. There followed detailed descriptions under which rifle fire could be most effective. For its part, "machine-gun fire produced a suppressing moral effect on the enemy as a result of the great losses inflicted upon him in a short period of time." Machine guns were most effective when employed on the flank and by surprise at ranges of a kilometer or less. They were to be employed with an eye to easy occupation of alternate positions and with due care exercised so their fire would not be masked by the presence of friendly troops. They might be retained under regimental control or allocated to battalions, and in some cases battalion commanders might incorporate them into a company battle area.[82] Ammunition expenditures from rifles and machine guns created problems of accountability, responsibility, and resupply, to which the instruction devoted several pages describing procedures and processes for conserving ammunition and replenishing stocks in the battle area.[83]

A major section of the instruction dealt with the bayonet assault, defining first of all the term *final assault* (*ataka*, in contrast with *nastuplenie*, the "attack" in English) as "movement to strike a blow with the bayonet." It was noted that the final assault must be started either when, according to the objective, the situation, or the results achieved, the time had come to surge forward to strike with the bayonet or when it was noticed that the moral force had shifted in favor of the attack. Indicators of the latter included confusion, disorder, and slackening of defensive fire. The instruction prescribed that the final assault be "fast, decisive, and violent as a hurricane." Assaulting infantry were to attack not only with the bayonet but also with rifle and machine-gun fire, accompanied by the throwing of grenades.[84] The latter injunction probably resulted more from wishful thinking than from reality, for the infantry did not receive routine issue of hand grenades until World War I. The instruction also called for the destruction of the enemy formation so that it could not reconstitute itself to renew battle. In contrast with prescriptions for the preceding attack phase, the instruction did not call for integration of artillery fire into the final assault fires.

Infantry instructions also dealt with subjects related to withdrawal in close proximity to the enemy, cavalry attacks, and night combat, to include use of searchlights. Withdrawal might occur either in the event an assault had been unsuccessful or in a situation in which it was considered unnecessary to maintain close-quarters contact with the enemy. Withdrawal occurred by stages, ideally with a clear understanding of the enemy's situation and friendly missions and dispositions. Against cavalry, the infantryman must hold a firm belief "in the power of his fire and bayonet" and be convinced that "while he remained facing the horseman and had not lost control of himself that the cavalryman was no danger to him." The idea was to not change formations precipitously, for "dislocated infantry was easy pickings for cavalry," but to remain in position and deliver maximum fire against attacking cavalry. Ideally, infantry formations dealt with cavalry assisted by the flanking fires of machine guns and neighboring units.[85]

In some instances, to reduce losses from enemy fire, night action might provide the best means of conducting the infantry attack. However, "the plan for night actions had to be simple and its execution carefully prepared." Deployment took place in closer proximity to the enemy and under conditions of "complete silence." Firing during a night attack was not permitted because, "on closing with the enemy, the affair must be concluded with the bayonet." Only in the defense was Russian infantry permitted to ward off a night attack with a combination of the bayonet and rifle fire. Searchlights proved useful in the attack and defense, but the instruction stressed that illumination often rendered ranges deceptive and targets confusing and fleeting. When attacking in the face of transitory or local illumination, friendly commanders were to employ shallow formations while stressing uninterrupted movement unless the attacking force found itself silhouetted directly in the rays of a searchlight. Under conditions of general illumination, the attack and assault approximated those conducted under daylight conditions. In the defense, searchlights were subordinated to the local commander, tied into his defensive plan, and linked to him by telephone.[86]

Cavalry Tactics

Although the field regulation of 1912 considered cavalry an integral part of all-arms battle, the evolution of the empty battlefield had rendered the role of the horseman more problematic than ever, especially below division and corps level, where mounted scouts had usurped many of cavalry's traditional reconnaissance and security functions. Conventional wisdom held that cavalry's primary purpose was to cooperate with the other arms in order to achieve the common objective. In practice, this meant that cavalry moved in the van and on the flanks of troops on the march both to screen and secure the movement of large formations and to prepare the ground ahead for infantry combat. Cavalry also fought cavalry to deprive the enemy of information and security. Mishchenko's raid and lesser ill-fated enterprises aimed at striking deeply to the enemy rear had apparently cooled but not entirely dampened Russian ardor for launching mass mounted incursions into enemy-held territory in the manner of the great cavalry raids of the American Civil War.

From these assumptions and from evolutionary changes favoring utility and simplification flowed the sometimes contradictory doctrinal requirements for cavalry organization and action in the field. Apparently at least in part because of the Mishchenko precedent, the Russians had temporarily discarded the idea of independent cavalry operations in formations any larger than division or regiment. Or, as the British noted, "the idea, formerly prevailing, of independent action has been set in the background and the employment of cavalry in large masses is exceptional." In the odd event that cavalry forces were committed to deep raiding, regulations stated that the conditions of success lay in secrecy, rapidity of movement, and the commander's ability. Raids were to be undertaken only in intervals between main operations or when the enemy's cavalry was so inferior that Russian cavalry could be spared without jeopardizing fulfillment of other missions.[87]

In keeping with the post-1905 emphasis on utility, the cavalry instruction of 1912 dropped earlier varied and complex drill formations and evolutions. The fighting formation of the squadron now consisted of two or three troops extended in line with one or more in reserve. Cavalry attacking cavalry began with the slow trot in extended order, speeded to a gallop at 400 meters' distance from the enemy, then closed within 100 meters at the charge. Cavalry attacking infantry assumed a single-rank formation in the first echelon, while the rear echelon assumed double-rank formation, keeping open order. Simplicity governed the nature of the attack formation itself, with the pre-1914 regular Russian cavalry adopting the traditional Cossack "lava," or swarm. In the squadron, the lava consisted of a wide extension of two or three troops, often in crescent form, with a reserve following 100 or 200 meters to the rear.[88]

The revised regulations of 1912 also provided for dismounted combat, with two-thirds of the cavalry force engaged as infantry while the remaining one-third served as horse-holders. In 1914, the British handbook on the Russian army noted that "while mounted action is held to be the principal means of fighting employed by cavalry, great importance is now attached to dismounted combat when the nature of the task involved, or the terrain, renders mounted action insufficient or unsuitable." Regular cavalrymen were issued bayonets for their carbines, and on going into dismounted action, troopers fixed bayonets, for, as in the infantry, "the ultimate aim of the attack [was] to get to close quarters with the bayonet." Cossacks, who were not equipped with bayonets, slung their rifles and advanced on foot with drawn sabers. Naturally, in view of complications associated with dual-purpose combat, a simplified version of infantry regulations governed battlefield deployment of dismounted cavalry formations.[89]

A final odd twist, considering the emphasis on ground combat and firepower, occurred around 1912, when the Russians systematically began equipping their cavalry with lances. This probably represented a kind of atavistic impulse recalling the glory days of the past, when the *armé blanche* threw fear into the hearts of infantrymen equipped with slow-firing shoulder weapons of limited range. Only the experience of World War I and the Russian Civil War would teach the Russians both the limits and possibilities of the mobile arm in modern warfare.

Artillery Tactics

Doctrinal problems of artillery employment were only slightly less perplexing in the period between 1905 and 1914. Along with PU-1912, the Russians published a new artillery instruction. Its chief features included emphasis on the necessity for careful reconnaissance before and during action, the importance ascribed to observation and communication, and the distinct limits prescribed for the authority and duties of various commanders. To facilitate reconnaissance, each eight-gun battery commander was provided with eleven mounted and eighteen dismounted scouts, observers, and signalers. When action was probable, the artillery *divizion* commander sent advanced artillery patrols forward. They consisted of officers and scouts from the constituent batteries of the *divizion*, and their task was to work with forward infantry and cavalry units and to find suitable observation points as soon as the action was imminent. Communication relied extensively on field telephones, but the artillery instruction insisted on the necessity of maintaining alternate means of communication.[90]

Although retention of the eight-gun battery reflected larger problems of organization and a mentality emphasizing physical concentration to mass fires, the Russians continued to stress both maneuver and flexibility. To facilitate the attack, *divizion* and battery commanders might send forward to accompany the troops half-batteries, sections, and even single guns. In the attack phase, artillery's function was to engage hostile artillery and machine guns, seal off approaches, breach obstacles by fire, and oppose enemy counterattacks.

The artillery commander, in accomplishing his mission, could dispose of his guns in three kinds of positions: open, covered, and semi-covered. The latter position was one in which the gun remained hidden until it opened fire. The preferred deployment was by covered position because it conferred maximum freedom of action. Doctrine prescribed that fire be delivered in short, rapid bursts, with field guns firing over the heads of advancing infantry. Unless observation posts were located in close proximity to maneuver force command posts, communication between supporting artillery and supported infantry and cavalry was by signal flags or other suitable expedient means.[91]

Constraints

Subsequent commentators, especially Soviet military historians, have noted that Russian military doctrine on the eve of World War I reflected substantial advances over both earlier Russian and comparable European doctrine. But as always, the question remained of how well the Russians could effectively put that doctrine into practice when confronted with the challenges of contemporary combat. Members of the Russian General Staff who were responsible for tracing corresponding German doctrinal developments evidently believed that the German army lagged behind the Russian in translating contemporary understanding of battle into effective tactical doctrine. However, in the next breath, the Russian specialists on the German threat also held that in the event of war the Germans would rapidly close the gap between doctrine and reality because of their overall high levels of discipline, morale, and education.[92] Whether the same would be true of the Russians remained in no small measure problematic: B. M. Shaposhnikov noted that his class at the Nicholas Academy graduated only forty-eight officers to the General Staff, where they would serve as interpreters of doctrine and coordinators of tactical planning and preparation for a peacetime army of 1,300,000. To be sure, these officers were not alone, but it was not without reason he lamented that they represented "a drop in the bucket."[93]

There were other incongruities. One was the role played by military perceptions of psychology, in which Dragomirov's legacy continued to figure in Russian calculations. Even after 1905, Russian tacticians viewed battle as a kind of ultimate contest in which the interplay of material and moral factors dominated the outcome, with the latter assuming the ascendancy in a triumph of will. To accept battle on defensive terms was to acknowledge the adversary's superiority and surrender the initiative. In this respect, perhaps the Russians had learned all too well the lessons of Japanese initiative in the Far East.[94]

The problem was to retain the initiative in the face of a changing technology which had opened yawning gaps in battlefield control and communications. No magical solutions presented themselves. Thus one of the commander's last hopes for controlling a battle lay in emphasizing the offensive, thereby retaining the initiative in a heroic effort to act decisively with sufficient mass to keep the enemy off balance and constantly responding to new developments. In effect, the energetic commander's best formula for success was to seize the initiative, thereby willing himself to retain control.[95]

Other tactical calculations emphasized maneuver to avoid costly frontal assaults against prepared positions. To their credit the Russian writers of what Westerners now call doctrine never lost sight of the importance of the meeting engagement, the flank attack, and the deep envelopment as means of avoiding (or postponing) by maneuver the long rush through rifle, machine-gun, and artillery fire toward a distant line of entrenchments. However, maneuver depended upon knowledge of the enemy and the terrain and the ability to communicate that knowledge along with offensive designs across greater distances with a greater degree of surety than was possible before 1914.[96]

The problematic nature of control and communications can be multiplied upward in the hierarchy from company and battalion through regiment and on to division, corps, army, front, and supreme headquarters. What was difficult in battle at the tactical level became still more difficult at successively higher levels and nearly impossible at the strategic level. Here, Russian military theory and regulation provided a structural framework for answers, but again, technology was not sufficiently developed to endow embryonic concepts with the palpable benefits of what is now called operational-level command and control. The absence of such linkages had already been aptly illustrated by the Russian war game of April 1914.

More ominous and less noticeable were genuine defects in force structure. Although the collective firepower of infantry units had become awesome, infantry alone could not carry a defensive position, even against a hasty defense, without the suppressive and destructive effects of coordinated artillery fire. At division level, forty-eight light field guns were pathetically inadequate to support the offensive maneuver of sixty-four infantry companies. Twelve howitzers at corps level to support 128 infantry companies appeared still more pathetic without even considering issues of ammunition allocation and resupply.[97]

Maneuvers and Exercises

Just as in the two previous prewar periods, various kinds of exercises and maneuvers between 1907 and 1914 provided important insight into how the Imperial Russian Army might perform during wartime. In addition, they illustrated just how well the Russians had absorbed and seemed willing to apply the principal lessons of their Far Eastern experience. Finally, exercises and maneuvers displayed the presence or absence of important linkages among theory, doctrine, force structure, planning, and practice.

Two observations should preface remarks about Russian exercises and maneuvers after the War of 1904–5. The first is that in the wake of post-1905 disturbances and demobilization the army was in no condition until after 1910 to conduct anything like exercises on the scale of the Kursk maneuvers of 1902. The second is that various exercises must be seen in the context not only of maneuvers held in the vicinity of the capital and Krasnoe selo but also within the various military districts. The legacy of Miliutin's calculated decentralization meant that district commanders still possessed a significant degree of latitude in the training of their own troops, and diverse styles were reflected both in varying approaches and rigor and, consequently, in varying degrees of readiness.[98]

The maneuvers which actually occurred between 1907 and 1910 remained limited in scale and commitment of troops. Those taking place in the vicinity of summer training camps and garrisons were "more or less perfunctory" and "rather of the nature of drill than of maneuver" in the words of one British observer.[99] It was the annual August maneuvers in St. Petersburg to which foreign attachés usually devoted the lion's share of attention. These maneuvers began in 1907 with modest division-size exercises, expanded to corps-size in 1908 and 1909, then contracted to brigade and division-scale in 1910.[100]

Although the field problems varied from year to year, the comments of British observers often focused on common perceptions. Reports invariably emphasized the powerful physiques and stamina of the Russian infantry. True, observers saw many guardsmen who were picked for their outstanding physical attributes, but the reader must draw favorable conclusions about the care and training of troops.[101] A second item of commentary emphasized the degree to which commanders and troops alike exercised due care to protect themselves from direct fire in various phases of the attack and final assault. Some observers apparently believed that the Russians reacted too strongly to rifle fire in the advance, slowing forward movement by extending into skirmish lines and moving forward rather slowly as individuals or in small groups by leaps and bounds.[102] Regardless of the merits ascribed to tactics, the Russians evidently adhered to the prescribed formations for the attack. Although firing by the individual soldier predominated, occasionally the Russians reverted to volley fire.[103]

Application of maneuver and fire support techniques drew less positive commentary. Thanks to Russian emphasis on massing artillery fire mechanically rather than directionally, there was a pronounced tendency to retain batteries and guns closer to the rear and under centralized control, whether in the movement to contact or in the actual attack. This emphasis, along with crude means of communications and signaling, meant that the guns were slow to respond to changing circumstances. Under conditions of actual field firing, Russian artillery in the first post-1905 maneuvers fired primarily from concealed positions, using indirect lay. By 1909, the Russians had become bold enough once again to employ a portion of the guns far enough forward to support the attack with direct fire. Still, the majority of the guns fired indirectly from concealed positions. Machine guns were pressed forward as far as possible to support advancing infantry, but there was difficulty attempting to duplicate the effects of machine-gun fire. Apparently, no means had been developed to fire blanks from machine guns, so Russians adopted an expedient in the form of noise from large rattles. Needless to say, the expedient did not do justice to the guns' tactical effects.[104]

The record on maneuver also revealed anomalies and inconsistencies. In 1907, one battalion-level maneuver against an entrenched position involved positioning troops for a simultaneous frontal assault and an enveloping flank attack. About 2,000 yards separated the maneuver forces, and neither the communications nor the attack were well coordinated.[105] In 1908, observers noted that the Russians devoted much attention to counterattacks and that the purpose of tactical maneuver was usually a flank attack (shallow envelopment) rather than a turning movement (deep envelopment).[106]

While infantry and artillery often experienced difficulties in coordinating movement and action, the Russian cavalry revealed even more disturbing tendencies. At first, in accordance with Far Eastern experience, cavalry concerned itself with dismounted action and fire tactics. However, by 1910 cavalry had reverted to more atavistic habits. By that time observers reported that "the preference for mounted as opposed to dismounted tactics was clearly seen." The overall impression created by Russian cavalry was "that its leaders neither take advantage of ground nor adapt their tactics to the requirements of the moment." Lances had begun to reappear, and cavalry commanders were now bold enough to attack opposing cavalry supported by infantry. When one such attack was judged a failure, its leader, General Bezobrazov, "considered that he would have been successful in war."[107]

Observers' reports on Russian exercises for 1912 and 1913 varied appreciably in geographical scope and substance from earlier years. British attachés now had the opportunity to view maneuvers in several of the military districts, thus affording additional perspective. In addition, both the St. Petersburg maneuvers and those within the various districts involved both larger formations and appreciable numbers of reservists. More information was also available on the amount of ammunition supply allocated for field artillery and machine guns. Although now uniformly equipped with lances, cavalrymen were once again frequently observed in dismounted combat roles.[108]

The Russian maneuvers of 1912 within the St. Petersburg district involved 112 battalions, 54 squadrons, 370 guns, eight aircraft, one dirigible, and two balloons. For the first time since 1904–5, British attachés recorded the presence of reservists in these maneuvers at the rate of sixty to eighty per company. This change was apparently a direct result of Sukhomlinov's reforms of 1910, and in the British view, "the efficiency of units was not affected by the presence of reservists in the ranks." Also in the British view, the Russian method of calculating the effects of direct and indirect fire was cumbersome, rendering it impractical for implementation in field problems. The result was that "umpires interfered very little, and casualties were not practiced at all."[109]

In the St. Petersburg maneuvers of 1912, observers noted that in preparation for the meeting engagement, maneuver commanders held significant portions of their forces in general reserve even though the situation was relatively clear. Once the opposing forces made contact, the smaller "Blue" force assumed a hasty defensive position, while the "Red" force conducted a textbook attack. One of six "Red" artillery batteries advanced to support the infantry attack with direct fire at 400 yards' range, but observers criticized the battery commander for carelessly exposing his men to defensive rifle fire. Three other batteries supported from open positions 1,700 yards to the left rear, when they could have occupied covered positions at closer range. As had been the case in previous maneuvers, the infantry attack and final assault unfolded too slowly, with insufficient care devoted to a buildup of the attackers' suppressive rifle fire. Individual soldiers labored under loads of more than seventy pounds, including about one hundred rounds of ammunition.[110]

The implications of the various maneuvers were clear. Although the Tsar insisted upon formal parades, gone for the most part was the pre-1904 preoccupation with formal drills and reviews, or the worst aspect of "paradomania" which had earlier afflicted the army. At the same time, officers seemed intent on alleviating the army's worst tactical shortcomings as they had been demonstrated in Manchuria. Due attention was devoted to the effects of rapid-fire artillery and small-arms fire. However, maneuvers also demonstrated that many tasks required greater attention. For example, troops were still not trained to use their entrenching tools in the attack. The volume of fire which could be brought to bear on the defense required more attention to integrate the destructive and suppressive effects of rifle, machine-gun, and artillery fire, especially in the final assault. Although the Russians generally acknowledged the importance of machine guns, their numbers and use fell far short of what the Manchurian experience seemed to mandate.[111]

British observers' reports on the 1913 maneuvers were far more sketchy than those of previous years. However, what the reports lacked on the maneuvers' substance they made up with commentary on troop numbers involved and new technologies. In addition to maneuvers involving 108 battalions, 71 squadrons, and 326 guns at Krasnoe selo, the Kiev Military District held maneuvers approximately half the size of those in the environs of the imperial capital. In regard to Kiev, the comment was that "the forces opposed must, therefore, have amounted to 27,000–28,000 each, a very much larger proportion of their war strength than is usually seen on manoeuvres."[112]

In 1913, the British attachés also began to note changes in various kinds of armament and equipment. They reported the reorganization of artillery at corps level, including the appearance of 4.8-inch howitzers in heavy artillery *diviziony*. Attachés also reported the appearance of Schneider pack guns within mountain artillery units. Although various types of balloons and aircraft had been observed in maneuvers since 1910, in 1913 the Russians utilized French Nieuports and Henri Farmans in maneuvers around both Kiev and St. Petersburg. The Russians also pressed nine automobiles into staff service at Krasnoe selo, but French officers reported that "the soldier chauffeurs are indifferent, and between roads and bad driving, the cars will probably have a short life."[113]

The application of modern technology to communications remained haphazard and inconsistent. While it was rumored in 1912 that the Russians had ordered 110 sets of cart-mounted wireless apparatuses, they were not much in evidence either in 1912 or in 1913. Units sometimes purchased experimental wireless sets and telephones privately. In the few instances when wireless was used, no arrangements existed to prevent interference on competing broadcast frequencies. In view of these and other difficulties, "telephones, mounted orderlies and motor cyclists were the means of communication chiefly employed."[114] Telephones were everywhere in evidence, but so were their chief drawbacks: limited range (regiment and division carried only ten to fifteen miles of wire), the vulnerability of wire nets, and the difficulty of displacement to account for movement.[115] When technology failed, the Russians reverted to mounted and dismounted messengers, a state of affairs which offered no appreciable advance in communications since 1904–5.

Lapses in communication emphasized a strong unifying thread of commentary which emerged from the exercise and maneuver experience: the increasing difficulty under contemporary circumstances of orchestrating sound command and control linkages. Although junior officers and noncommissioned officers demonstrated improved control of formations in the extended order attack and defense, the enlarged scale of the battlefield mandated by improved technology imposed tougher tasks on battalion and regimental commanders who were now out of sight and sound of their subordinate units. The same kind of difficulties afflicted higher levels of command, and the small number of telephones and primitive wireless apparatus could not fill the ever-growing communications and control gap. This state of affairs only worsened coordination of various kinds of support for the combat commander. At the same time, given the importance ascribed to offensive action, units on the move needed better reconnaissance than mounted scouts could give. Yet cavalry—suffering its own identity crisis—seemed both ill equipped and ill disposed to provide the continuous reconnaissance coverage required by commanders looking for meeting engagements. At a different level, artillery could scarcely be expected to provide continuous support under rapidly changing circumstances with the assistance of only limited telephone support and supplemental voice and flag signaling systems.

In writing about the eastern front in World War I, Norman Stone has noted that historians and other commentators have viewed Russia's overall economic weakness as a major cause for the lackluster performance of the Russian army. Stone also notes, however, that economic inadequacies were often poor indicators of actual combat performance, with the implication that the historian must look elsewhere for explanations of the army's failure in the field.[116]

The pre-1914 record strongly suggests that the historian must also look to the significance of linkages, intellectual and structural, physical and military-political. Failures in linkages, many of which were already evident in the gaming and maneuver experiences, help explain why prewar preparation would not fully correspond with wartime requirements. Thus, although Russian military art at the tactical level had incorporated many hard-learned lessons from the experience of 1904–5, gaps in planning and higher-level execution meant that every division's sixty-four companies would often expend their tactical energies in vain. Tactics properly linked to the conduct of modern operations would have enabled the Russian army in 1914 to put into practice what its theorists and writers of doctrine had been preaching. Once the army failed to link tactics with operations, discrepancies in logistical linkages would make themselves felt even more strongly, leading to what Bernard Pares called "a war of men against metal."[117] The irony was that although tactical thinkers had learned to transcend Dragomirov's legacy, shortages would once again force the Russians to rely on bayonets before bullets.

Looming over everything else was a fundamental misunderstanding of offensive-defensive correlations in modern war. For a variety of reasons—some not fully understood—the Russians largely ignored defensive calculations to emphasize the importance of offensive military action. To be sure, historical consciousness probably played an important role. In 1912, the Russians celebrated the centenary of Napoleon's failed invasion with great fanfare. One of the lessons of 1812 was that a defensive strategy was not only an admission of weakness but also a willing acceptance of the prospect of giving battle on one's own territory. No one seemed anxious to repeat the experience. Indeed, as I. I. Rostunov has noted, Russian war plans from the time of Obruchev and Miliutin to Alekseev and Danilov had always included an offensive component. Depending upon circumstances, the Russians might temporarily assume the defensive, then shift to the offensive. The significant variables in this calculation were considerations of timing and means. Changing perceptions of French and German strengths and vulnerabilities were important aspects of the planning calculus. New confidence in the Imperial Russian Army aside, what was missing—and noted by such commentators as Svechin—was a sense of how meaningfully to relate the defensive and the offensive to one another within the framework of changing relationships between capabilities and alliance priorities. Here, Neznamov's ringing declaration that "we must know what we want" assumed added importance. No matter how vast the resources appeared, hard decisions had to be made about priorities, for everything could not be done simultaneously. For Russia in the summer of 1914, the postulation of rapid and simultaneous advances against both the Austrians and the Germans meant that a mixture of wishful thinking and opportunism had triumphed over military common sense.

Conclusions

DURING THE half-century separating the reform programs of D. A. Miliutin and V. A. Sukhomlinov, the Imperial Russian Army and its military art had undergone a number of important transformations. Organization, theory, and field regulation had all responded—albeit imperfectly—to many of the military challenges inherent in the age of the industrial revolution. As the army sought accommodation with new military realities, it had also come to reflect the altered circumstances first of reform-era Russia, then of alternately conservative and revolutionary Russia. In the field after 1861, the army had waged two large-scale wars during a period in which the combat experience of other major European powers was limited to colonial wars. In contrast with the first century and a half of the Russian army's existence, during which change usually came episodically and incrementally, the period between 1861 and 1914 witnessed a dramatic increase in the pace and nature of transformation.[1]

Paradoxically, as prelude to any assessment of fundamental change, the historian cannot avoid concluding that the Russian experience from Alexander II to Nicholas II underscored the importance of a number of unchanging military verities. These included the necessity for sound leadership, realistic and rigorous training, high morale, and well-defined command relationships. On a grander scale, another verity inherent in the Russian experience between 1861 and 1914 was the requirement to anticipate, define, and coordinate the needs within projected theaters of military action for joint naval and ground force operations. As the scope and scale of modern, industrial-style war increased and as the Russians extended the periphery of their empire, mass land forces increasingly required naval support, if only to extend and secure the limits of continental theaters of military actions. In the political-military realm, still another verity, embracing both the military and naval spheres, was the need to bring into alignment the objectives of state policy and overall military capabilities.

The Russians responded better to changing military circumstances when they operated on a less grand scale in matters of organization, preparation, and equipment. In accordance with perceptions of change and need, the Russians proceeded to recruit, train, and arm a ground force which was among the most powerful in the world. They successfully implemented a program of universal military service necessary to fill the ranks of a growing army which increasingly came to resemble the cadre and reserve forces of the European powers. Although commentators might legitimately take issue with the specific content and quality of troop training, various programs did a reasonably good job of bridging the gap between civilian and military life, including—at least at first—the teaching of literacy when necessary. Despite nagging difficulties, the combination of training and indoctrination was sufficient to produce motivated soldiers who, after initial periods of shock and adjustment, responded well to the challenges of combat. Thanks to acquisition policies which stretched the state budget to the breaking point, the armament of Russian soldiers improved until the beginning of the twentieth century, when they were uniformly equipped with the latest magazine-fed rifles firing smokeless powder cartridges. The machine gun emerged haltingly to lend its staccato voice to the chorus of technological change. Analogous advances in artillery produced the quick-firing field gun, the advent of which left the Russians lagging only in the development of large-caliber howitzers and the associated techniques and technologies necessary to take full advantage of the new firepower.[2]

The Russians had also worked out the force structure to answer what they felt were the needs of contemporary and future warfare. In an age of infantry supremacy, that arm remained the centerpiece of field forces. To their credit, the Russians neglected neither cavalry nor artillery, although theorists and tacticians remained less sure of the manner of employing these arms either with each other or with infantry. Except for heavy artillery, which would receive limited attention after 1905, artillery organization compared favorably with that of other European powers both in armament and proficiency. In 1914, the Russians lagged behind the Germans only in the amount of heavy artillery allocated to corps-size formations.

Modernization and mobilization were two areas which revealed the weakness of the economic infrastructure. Throughout the period, the War Ministry labored under the handicap of limited revenues, and it was only through the careful husbanding of resources that the Imperial Russian Army was able to make the transition to smokeless powder weaponry. Nicholas II's well-known international initiatives to limit the spread of new artillery technology during the 1890s sprang from budgetary considerations. The same considerations linked to railroad construction produced even in the 1860s the beginning of a "mobilization gap" which would plague Russian staff officers far into the twentieth century. That is, the empire's vast distances and spare rail network combined in a way that meant Russian mobilization would lag behind Germany and Austria in the event of war with either or both. This realization figured prominently not only in the evolution of diplomacy and mobilization schedules but also in continuing debates over the proper relationship between field forces and border fortresses.

The same hardships imposed by technological necessity and scale also presented the gravest military challenges. At the level of war that observers in the late twentieth century would firmly identify as strategic, Russian military planners of the late imperial period confronted problems of concentrating larger forces across greater distances with greater speed than had been the case before the age of the railroad and telegraph. These factors greatly increased the complexity of mobilization for confrontation, a complexity rendered more challenging by the possibility that Imperial Russia faced simultaneous combat in the Far East and on the borders of European Russia. Indeed, one commentator, John Erickson, has long insisted that this possibility forced the Russians, in effect, to maintain two armies, a European and a Far Eastern.[3]

Still, this situation was not without promise. Completion of the Trans-Siberian Railroad in 1902 established a land link between European Russia and the Far East, the strategic implications of which were obscured by Russian military failures in 1904–5. Precarious as it was, this link would continually evolve, enabling the Russians to shift ground forces internally across the Eurasian continent faster than hostile naval powers could shift fleets between the Atlantic and the Pacific. In an ironic twist of history, the same industrial revolution which made steam navies masters of the ocean now endowed Russia with the advantage of interior lines in future conflict around the periphery of "the world island." Or, as the historian Paul Kennedy has so eloquently put it, the heirs of the geostrategist MacKinder quietly emerged to challenge those of the naval strategist Mahan.[4]

If issues of technology and scale profoundly affected strategy, they were felt even more deeply in the realms of operations and tactics. Mass armies required greatly enlarged areas of operation. At the same time, despite the requirement to achieve mass and concentration for decisive results, the greater lethality of weapons dictated a contradictory need for dispersion. Much of the history of modern warfare can be written as a function of attempts to reconcile these seemingly contradictory elements. Obviously, the Russians—just like everyone else—had their share of difficulty making the reconciliation.[5]

Perhaps nowhere were the consequences of changing technology and scale more apparent than in a comparison of the actions around Plevna and Mukden. During the third storm of Plevna, the Russian and Romanian positions extended in a great semicircle measuring about sixteen kilometers from flank to flank. Total engaged forces numbered about 100,000. Despite the growing lethality of breechloaders, the attentive observer might occasionally catch a glimpse of sunlight reflected on his adversaries' bayonets several kilometers distant. Although the battlefield might be obscured by fog and smoke, an observer slightly elevated and to the rear of engaged forces could have traced the course of action within at least one of the major sectors of the battlefield. Although artillery preparations might stretch for several days, troops were actually engaged in the attack or defense only a day or two at a time.

Almost none of these circumstances held true for Mukden. Before the Mukden operation drew to a close, Kuropatkin's positions stretched more than one hundred kilometers from flank to flank. Although the two sides often occupied trench lines and other field fortifications within close proximity to one another, neither was visible to the other. The range, power, and accuracy of smokeless powder weaponry had created a battlefield populated by invisible warriors.[6] Except for rare moments in the attack or defense, soldiers had to remain under cover, usually in field fortifications, which grew increasingly more elaborate. Although warriors were invisible, they were more numerous than ever. Before the battles around Mukden wound their course, more than half a million men would be engaged. And their encounters would sputter across the front in a seemingly disconnected series of battles, the aggregate result of which determined the outcome of the entire operation.

To confront the awesome changes wrought by technology, scale, and mass, the Russians brought to Mukden much the same training, education, communications, and staff procedures they had brought to Plevna. Dragomirov's tactical conceptions—even as altered after 1877–78—determined the agenda for training the combat arms. But despite the advent of repeating rifles, machine guns, and quick-firing artillery, these conceptions relied too much on shock action in the offensive. Or, to phrase the situation another way, Dragomirov had failed to realize that the shock of the bayonet had been largely supplanted by the shock of the .30-caliber bullet. Yet, in fairness to Dragomirov, it must be pointed out that his tactical system was not static. The basic problems were two: his thought changed only incrementally, and over time his disciples' slavish devotion to the master's system translated itself into ossification of thought and regulation.

Even more seriously in the eyes of some observers, General Leer's continuing domination of strategic studies at the Nicholas Academy failed to address the needs of planning and troop direction at higher levels in the field. Between 1878 and 1904, a gap opened between Dragomirov's tactical concepts at the lowest level and Leer's strategic principles at the highest level.[7] It was a gap that the Russians gradually began filling with their own understanding of what the Soviets later called "operational art." Actions within a theater were no longer limited to detachments and other formations improvised after the outbreak of hostilities but fell within the province first of corps and army groups, then of armies and fronts. Post-1905 speculation would begin the slow process of theoretical evolution which would culminate during the 1920s with recognition of operational art as a vital concept linking military strategy on one hand with battlefield tactics on the other.[8]

H. G. Wells wrote that history is "a race between education and catastrophe." In the military realm, his notion might be paraphrased to assert that adaptation is a race between education and catastrophe. In this respect the Russian system flirted with catastrophe by clinging throughout the last quarter of the nineteenth century to Leer's outmoded interpretations of warfare in the tradition of Napoleon. Unfortunately, the experience of the Russo-Turkish War—fought for the most part against a passive foe in theaters relatively untouched by technology (except what the armies themselves imported)—posed little challenge to traditional interpretations and conceptions. Worse, the Russians failed to profit systematically from the lessons of military history narrowly interpreted as combat experience. It was no coincidence that the officers of Kuropatkin's generation, who had last seen large-scale action a quarter century earlier in Bulgaria, brought with them in 1904 to the Far East habits learned against a less energetic foe at a different time and place. Thanks in part to Leer's domination of the curriculum at the Nicholas Academy, it was also no coincidence that younger officers failed substantially to challenge their superiors' concepts or to require intellectual revalidation of outmoded combat experience.

A. A. Neznamov understood the problem very well: neither remote experience nor Napoleonic principles as interpreted by Jomini and Leer provided Kuropatkin and his fellow officers with an appreciation of the stepped-up pace of modern battle. Traditional conceptions of battle envisioned the orderly unfolding of the approach march, calculated movement to contact, concentration of forces, the pause for last-minute reconnaissance and issuance of orders, and main engagement. Since the 1850s, however, there had developed an increasingly pronounced tendency to speed forces from the march directly into main battle without pause for concentration and deliberate development of the battle. In place of the set-piece confrontation, commanders, especially German commanders (and their Japanese students), used momentum and expansion of the meeting engagement to seek a confusing and seemingly disparate set of battles which allowed maximum latitude for the development of initiative and controlled staff improvisation. More than ever, whirl came to characterize battles and evolving operations, and the premium went to those officers prepared intellectually and organizationally to accept and profit from discordance in the field. Little in Russian staff-level education and combat experience (chiefly 1877–78) prepared Kuropatkin and his officers to deal with the escalating pace of battle as adversaries sought to turn Russian strengths, including mass and inflexibility, to Russian disadvantage. Once Neznamov and his contemporaries understood the new calculus, they found themselves after 1905 hopelessly caught up in their own army's race between adaptation and catastrophe.

The chaotic nature of modern battle, together with problems of scale, resource mobilization, and technological application, underscored the importance not only of speed but also of decentralized command, control, and communications arrangements. The lethality of smokeless powder weaponry called for dispersion on the battlefield and open-order tactics, but few contemporary observers and later critics of seemingly outmoded formations and techniques have suggested alternate forms which might have incorporated satisfactory and workable methods of control.[9] Between 1877 and 1914, the Russians sought to replace the battalion with the company as the primary tactical formation. To a great degree, however, tactical devolution depended upon the caliber of junior officers and noncommissioned officers, and for various reasons, the Russians incurred shortages of both types, especially under conditions of prolonged combat. Successful devolution also depended upon the development of new technology, including the widespread use of telephones and the later introduction of wireless communications. In the event, the Russians tended to be slow in availing themselves of existing communications technology, and their dawdling often prevented them from responding in a timely manner to rapidly changing situations on the battlefield.

Meanwhile, neither experience nor education prepared the Russians (and, indeed, their various adversaries) to deal effectively with the embarrassing reappearance under modern guise of traditional siege warfare. For all the theoretical emphasis on the offensive, maneuver, and the rapid development of decisive battle, the attacks at Plevna and Port Arthur suggested that slowly unfolding and costly assaults on prepared and field fortifications might consume anywhere from one-third to one-half of the resources of a modern nation's field armies. Plevna and Port Arthur also laid bare fundamental difficulties of coordinating and massing the effects of modern combat arms, especially field artillery. Resolving these difficulties would require not only additional experience but also the development and mass application of new techniques and new technologies.

At the same time, the Russian dream of using mobility and mass in the form of large mounted formations to overcome operational and tactical obstacles remained largely unrealized. Although successes such as that of General Gurko's forward detachment in 1877 showed great promise, failures such as General Mishchenko's raid of 1904-5 exercised a strong countervailing influence.[10] In retrospect, what remains notable was the degree of tenacity with which the Russians clung to their vision of the deep strike in the face of recurring difficulties with conception, organization, and application.

At the end of the twentieth century, a well-worn cliché among military officers is the assertion that "amateurs talk tactics while professionals talk logistics." In light of the Russian experience between 1861 and 1914, this cliché might be modified to assert that "amateurs talk numbers while professionals talk doctrine, linkages, and systems." Although physical mass and material assets are important military attributes, the consistent generation of combat power relies just as much on unity of concept and will, connectivity of parts, and integration of organization, plan, and technology. These are the very qualities which the Imperial Russian Army consistently displayed at its lower levels. Or, to paraphrase A. M. Zaionchkovskii and Neznamov, every good Russian division was made up of sixty-four fine companies. The explicit criticism was that failure occurred more often than not at higher reaches of the command and organizational hierarchy.

By 1914, persistent failure at higher levels had prompted the Russians to evolve a reasonably efficient military administration, perhaps the most efficient since the era of Miliutin. Nevertheless, the political context robbed it of effectiveness, thanks in part to Sukhomlinov's inability to generate confidence inside and outside the military and also to shortcomings inherent in the tsarist political system. Although historians differ among themselves about the inevitability of the Russian Revolutions of 1917, few would question the assertion that upheaval stemmed in no small part from a need for the high government to relate better with the populace. As the military experience indicated, there was also a need for the high government to relate better with its army. One of the most fundamental tasks of the post-1917 regime was to rebuild the military structure in a way that would guarantee firm linkages between the political leadership and its military forces.

Among the legacies that the new Soviet military establishment would inherit from its predecessor was a penchant for rationalizing success and failure, a penchant to which the Russians and Soviets owe the origins of their modern military science. The search for system and explanation in complex military phenomena also explained the origins of institutionally sanctioned military-historical research. While discussing the burning issues of their time, Imperial Russian theorists worked out both an intellectual framework and a precise military terminology, many elements of which retain currency even in the late twentieth century. Contemporary Soviet concepts of military art, embracing strategy, operational art, and tactics, owe their origins to prerevolutionary Russian military theory. This was no historical accident: a number of officers who had won their spurs as military theorists for the imperial regime, including Svechin, Neznamov, Mikhnevich, and Zaionchkovskii, would serve the early Soviet regime as *voenspetsy*, or military specialists.[11] More than mere theorists and transcribers of the past, they represented continuous living links between the old and the new. In large part because of their legacy, Russian and Soviet concepts of operational art traced a clear line of development between the nineteenth and twentieth centuries.

A final observation relates to the role that military-historical studies played in the organizational and tactical development of the Imperial Russian Army between 1861 and 1914. For several reasons, the entire era witnessed a flowering of military history, but not always to the benefit of the Imperial Russian Army. This was because studies failed to be timely, failed to relate to the army's current concerns in the field, and failed to maintain a vital link with those who developed doctrine. These are lessons which reach beyond the Imperial Russian experience.

Notes

To avoid repetition, Moscow, St. Petersburg, and Leningrad are rendered as "M," "SPB," and "L."

Introduction

1. Foreword by Peter Paret in Harold R. Winton, *To Change an Army* (Lawrence, Kans., 1988), vii.
2. S. Kozlov, "Voennaia nauka i voennye doktriny v pervoi mirovoi voine," *Voenno-istoricheskii zhurnal*, no. 11 (November 1964), 31.
3. The exception is A. A. Kersnovskii's dated and highly idiosyncratic *Istoriia russkoi armii*, 4 pts. (Belgrade, 1933–38).
4. See, for example, V. A. Zolotarev, *Rossiia i Turtsiia. Voina 1877–78 gg.* (M, 1983), K. F. Shatsillo, *Rossiia pered pervoi mirovoi voinoi (Vooruzhennye sily tsarizma v 1905–1914 gg.* (M, 1974), I. I. Rostunov (ed.), *Istoriia russko-iaponskoi voiny 1904–1905 gg.* (M, 1977), P. A. Zaionchkovskii, *Samoderzhavie i russkaia armiia na rubezhe XIX-XX stoletii* (M, 1973), L. G. Beskrovnyi, *Armiia i flot Rossii v nachale XX v.* (M, 1986), John Bushnell, *Mutiny amid Repression: Russian Soldiers in the Revolution of 1905* (Bloomington, Ind., 1985), William C. Fuller, Jr., *Civil-Military Conflict in Imperial Russia 1881–1914* (Princeton, N.J., 1985), and Joseph Bradley, *Guns for the Tsar: American Technology and the Small Arms Industry in Nineteenth-Century Russia* (DeKalb, Ill., 1990).
5. John L. H. Keep, *Soldiers of the Tsar: Army and Society in Russia 1462–1874* (Oxford, 1985), John Shelton Curtiss, *The Russian Army under Nicholas I, 1825–1855* (Durham, N.C., 1965), and John Shelton Curtiss, *Russia's Crimean War* (Durham, 1979).
6. Norman Stone, *The Eastern Front 1914–1917* (New York, 1975), and I. I. Rostunov, *Russkii front pervoi mirovoi voiny* (M, 1976).
7. A broad sampling of relevant scholarship includes Robert Fred Baumann, "The Debates over Universal Military Service in Russia 1870–1874," Ph.D. diss., Yale University, 1982, L. G. Beskrovnyi, *Russkaia armiia i flot v XIX veke* (M, 1973), L. G. Beskrovnyi, *Russkoe voennoe iskusstvo XIX v.* (M, 1974), E. Willis Brooks, "D. A. Miliutin: Life and Activity to 1856," Ph.D. diss., Stanford University, 1970, John Bushnell, "The Tsarist Officer Corps 1881–1914: Customs, Duties, Inefficiency," *American Historical Review*, 86, no. 4 (October 1981), 753–780, John Erickson, "The Russian Imperial/Soviet General Staff," Center for Strategic Technology, Texas A&M University, College Station Papers 3 (College Station, Tex., 1981), David R. Jones, "Russia's Armed Forces at War, 1914–1918: An Analysis of Military Effectiveness," Ph.D. diss., Dalhousie University, 1986, A. G. Kavtaradze, "Iz istorii russkogo general'nogo shtaba," *Voenno-istoricheskii zhurnal*, no. 12 (December 1971), 75–80, no. 7 (July 1972), 87–92, G. P. Meshcheriakov, *Russkaia voennaia mysl' v XIX v.* (M, 1973), Forrestt A. Miller, *Dmitrii Miliutin and the Reform Era in Russia* (Nashville, Tenn., 1963), Walter M. Pintner, "Russian Military Thought and the Shadow of Suvorov," in Peter Paret (ed.), *Makers of Modern Strategy from Machiavelli to the Nuclear Age* (Princeton, N.J., 1986), 354–375, Carl Van Dyke, "Culture of Soldiers: The Role of the Nicholas Academy of the General Staff in the Development of Russian Imperial Military Science, 1832–1912," M.Phil. thesis, University of Edinburgh, 1989, Peter Von Wahlde, "Military Thought in Imperial Russia," Ph.D. diss., Indiana University, 1966, and P. A. Zaionchkovskii, *Voennye reformy 1860–1870 godov v Rossii* (M, 1952).

8. For a survey of the more important sources, see L. G. Beskrovnyi, *Ocherki po istochnikovedeniiu voennoi istorii Rossii* (M, 1957), 348–450, V. A. Zolotarev, *Russko-turetskaia voina 1877–78 gg. v otechestvennoi istoriografii* (M, 1978), P. A. Zaionchkovskii (ed.), *Spravochniki po istorii dorevoliutsionnoi Rossii*, 2d ed. (M, 1978), 236–309, and L. G. Beskrovnyi, *Ocherki voennoi istoriografii Rossii* (M, 1962), 152–315.

1. The Army of D. A. Miliutin and M. I. Dragomirov

1. *Opisanie russko-turetskoi voiny 1877–78 gg. na Balkanskom poluostrove*, 9 vols. in 14 books (SPB, 1901–13), I, 151.

2. According to the terms of the Congress, Russia lost southern Bessarabia, was precluded from interfering in the affairs of Wallachia and Moldavia, and was forced to accept demilitarization and neutralization of the Black Sea.

3. Alfred J. Rieber has argued strongly that the impulse for social reform proceeded from military imperative; see his edited work, *The Politics of Autocracy: Letters of Alexander II to Prince A. I. Bariatinskii, 1857–1864* (Paris and The Hague, 1966), 15–31; see also Keep, *Soldiers of the Tsar*, 351–53.

4. On the Suvorov tradition, see Bruce W. Menning, "Train Hard, Fight Easy: The Legacy of A. V. Suvorov and His 'Art of Victory,' " *Air University Review*, 38, no. 1 (November–December 1986), 77–88.

5. On the more negative aspects of tsarist militarism, see John Keep, "The Military Style of the Romanov Rulers," *War & Society*, 1, no. 2 (September 1983), 60–72, and John Shelton Curtiss, "The Army of Nicholas I: Its Role and Character," *American Historical Review*, 63, no. 4 (July 1958), 884–86.

6. John L. H. Keep, "The Russian Army's Response to the French Revolution," *Jahrbücher für Geschichte Osteuropas*, H. 4, Bd. 28 (1980), 510, 514–16, 520.

7. Brooks, "D. A. Miliutin: Life and Activity to 1856," 67.

8. On the link between reform and the military frontier, see Zaionchkovskii, *Voennye reformy*, 36, and Bruce W. Menning, "The Army and Frontier in Russia," in Carl W. Reddel (ed.), *Transformation in Russian and Soviet Military History* (Washington, D.C., 1990), 25–38. Miliutin himself felt the frontier required unconventional solutions to vexing military problems; see his memoirs, D. A. Miliutin, *Vospominaniia*, repr. ed. (Newtonville, Mass., 1979), 305–6.

9. L. G. Beskrovnyi (ed.), *Russkaia voenno-teoreticheskaia mysl' XIX i nachala XX vekov* (M, 1960), 20–21.

10. Zaionchkovskii, *Voennye reformy*, 37.

11. P. A. Zaionchkovskii (ed.), *Dnevnik D. A. Miliutina*, 4 vols. (M, 1947–50), I, 26–27; for a perspective on the same phenomenon among the rank and file, see Elise Kimerling Wirtschafter, *From Serf to Russian Soldier* (Princeton, N.J., 1990), esp. chap. 3.

12. *Opisanie russko-turetskoi voiny*, I, 139–40; E. A. Prokof'ev, *Voennye vzgliady dekabristov* (M, 1953), 99–101.

13. See, for example, Zaionchkovskii, *Voennye reformy*, 50, and Zaionchkovskii (ed.), *Dnevnik D. A. Miliutina*, I, 20–21. Miliutin has been the subject of numerous studies, many of which have been reviewed by Peter Von Wahlde in "Dmitrii Miliutin: Appraisals," *Canadian Slavic Studies*, 3, no. 2 (Summer 1969), 400–414.

14. E. Willis Brooks, "Reform in the Russian Army, 1856–1861," *Slavic Review*, 43, no. 1 (Spring 1984), 74–77.

15. A. Bil'derling, "Graf Dmitrii Alekseevich Miliutin," *Voennyi sbornik*, no. 2 (February 1912), 5–6.

16. M. Osipov, "Obzor voenno-nauchnykh trudov D. A. Miliutina," *Voenno-istoricheskii zhurnal*, no. 9 (September 1972), 103–5.

17. P. A. Zaionchkovskii, "Vydaiushchiisia uchenyi i reformator russkoi armii," *Voenno-istoricheskii zhurnal*, no. 12 (December 1965), 32–34; see also V. N. Kopylov, "Miliutin kak voennyi deiatel'," *Voennaia mysl'*, no. 3 (March 1945), 60–61.

18. E. Willis Brooks, "The Improbable Connection: D. A. Miljutin and N. G. Cernyshevskij, 1848–1862," *Jahrbücher für Geschichte Osteuropas*, H. 1, Bd. 37 (1989), 30–31, 34–35, and John Keep, "Chernyshevskii and the 'Military Miscellany,' " in Inge Auerbach, Andreas Hilgruber, and Gottfried Schramm (eds.), *Felder und Vorfelder russischer Geschichte. Studien zu Ehren von Peter Scheibert* (Freiburg, 1985), 115–17.

19. Miller, *Dmitrii Miliutin*, 30–32.

20. Miliutin's report is reprinted in its entirety in D. A. Skalon (ed.), *Stoletie Voennogo Ministerstva*, 13 vols. in 51 books (SPB, 1901–13), supplement II to I, 70–183; see esp. the commentary on 70–71, and 73–75.

21. Ibid., supplement II to I, 76.

22. For general background, see William C. Fuller, Jr., "Strategy and Power in Russia, 1600–1914," manuscript, 1990, chap. 7.

23. Barbara Jelavich, *St. Petersburg and Moscow: Tsarist and Soviet Foreign Policy, 1814–1974* (Bloomington, Ind., 1974), 133–36.

24. Ibid., 163–66.

25. Richard A. Pierce, *Russian Central Asia 1867–1914* (Berkeley, Calif., 1960), 17–45.

26. See, for example, David MacKenzie, *The Lion of Tashkent: The Career of M. G. Cherniaev* (Athens, Ga., 1974), 99–104.

27. According to M. I. Dragomirov, Central Asia was a "vast factory for upstarts," advancing the careers of "Tashkentsy," that is, officers whose "technique [had] been spoiled by the experience of small expeditions . . . where the adversary is not a formidable warrior [and] where tactics takes on a secondary priority to the organization of [supply] trains, the sanitary service, and the hygiene of troops. . . ." See Carl Van Dyke, "Russia's Military Professionals: The Development of a Unified Military Doctrine at the Nicholas Academy of the General Staff, 1832–1914," manuscript, 1990, 140.

28. Kersnovskii, *Istoriia russkoi armii*, pt. 2, 411–14.

29. Jelavich, *St. Petersburg and Moscow*, 159.

30. Miller, *Dmitrii Miliutin*, 184–85.

31. Erickson, "The Russian Imperial/Soviet General Staff," 6

32. [D. A. Miliutin,] "Voennye reformy Imperatora Aleksandra II," *Vestnik evropy*, 17, no. 1 (January 1882), 21–32.

33. Erickson, "The Russian Imperial/Soviet General Staff," 7; see also A. Rediger, *Uchebnye zapiski po voennoi administratsii* (SPB, 1888), 438, and Kersnovskii, *Istoriia russkoi armii*, pt. 2, 398.

34. Skalon (ed.), *Stoletie Voennogo Ministerstva*, I, 483–85.

35. [Miliutin,] "Voennye reformy Imperatora Aleksandra II," 18; Peter Von Wahlde, "Russian Military Reform: 1862–1874," *Military Review*, 39, no. 10 (January 1960), 63–64.

36. On the developing Bariatinskii-Miliutin feud, see Zaionchkovskii, *Voennye reformy*, 126–27.

37. A. Kavtaradze, "Iz istorii russkogo general'nogo shtaba," *Voenno-istoricheskii zhurnal*, no. 12 (December 1971), 75–76.

38. Dallas D. Irvine, "The Origin of Capital Staffs," *Journal of Modern History*, 10, no. 2 (June 1938), 162, 165.

39. Ibid., 177.

40. David R. Jones, "Administrative System and Policy-Making Process, Central Military (before 1917)," in *The Military-Naval Encyclopedia of Russia and the Soviet Union*, II, 112.
41. [Miliutin,] "Voennye reformy Imperatora Aleksandra II," 13–14; Zaionchkovskii, *Voennye reformy*, 100.
42. Skalon (ed.), *Stoletie Voennogo Ministerstva*, I, 484, 486; see also Great Britain War Office, *The Armed Strength of Russia* (London, 1873), 86–87.
43. See, for example, M. G., "Ustroistvo i sluzhba general'nogo shtaba," *Voennyi sbornik*, 92, no. 7 (July 1873), 87.
44. A. Barbasov, "Russkii voennyi deiatel' N. N. Obruchev," *Voenno-istoricheskii zhurnal*, no. 8 (August 1973), 101–2; Keep, *Soldiers of the Tsar*, 358, 362.
45. A. V. Fedorov, *Russkaia armiia v 50—70-kh godakh XIX veka* (L, 1959), 130; Skalon (ed.), *Stoletie Voennogo Ministerstva*, I, 550–53.
46. Quoted in Baumann, "The Debates over Universal Military Service," 10.
47. Kersnovskii, *Istoriia russkoi armii*, pt. 2, 402.
48. Skalon (ed.), *Stoletie Voennogo Ministerstva*, I, 561–62.
49. Zaionchkovskii, *Voennye reformy*, 281.
50. Quoted in ibid., 258.
51. See Fuller, "Strategy and Power in Russia, 1600–1914," chap. 7; Obruchev's analysis is summarized in Zaionchkovskii, *Voennye reformy*, 280–88.
52. Zaionchkovskii, *Voennye reformy*, 282–83.
53. Ibid., 285–87.
54. For background, see A. M. Zaionchkovskii, *Podgotovka Rossii k imperialisticheskoi voine* (M, 1926), 29–31.
55. Miller, *Dmitrii Miliutin*, 196–200.
56. The financial argument is summed up in Dietrich Beyrau, *Militär und Gesellschaft im vorrevolutionären Russland* (Cologne, 1984), 303–4.
57. Baumann, "The Debates over Universal Military Service," 211.
58. Quoted in Fedorov, *Russkaia armiia*, 213n.
59. Miller, *Dmitrii Miliutin*, 218.
60. Zaionchkovskii, *Voennye reformy*, 304.
61. The most comprehensive treatment of the legislative process is Baumann, "The Debates over Universal Military Service," chaps. 2 and 3.
62. The legislation is reprinted as "Ustav o voinskoi povinnosti," in *Voennyi sbornik*, 95, no. 2 (February 1874), 107–50.
63. Keep, *Soldiers of the Tsar*, 376–77; see also Nicholas N. Golovine, *The Russian Army in the World War* (New Haven, Conn., 1931), 18–19.
64. On the problem of exceptions, see Robert F. Baumann, "Universal Service Reform and Russia's Imperial Dilemma," *War & Society*, 4, no. 2 (September 1986), 31–48.
65. Keep, *Soldiers of the Tsar*, 370–71.
66. Beskrovnyi, *Russkaia armiia i flot v XIX veke*, 45–46.
67. K. I. Druzhinin, "Voennoe delo v Rossii pri vstuplenii na prestol Imperatora Aleksandra II i pered voinoi 1877–78 gg.," in A. S. Grishinskii, V. P. Nikol'skii, and N. L. Klado (eds.), *Istoriia russkoi armii i flota*, 16 vols. (M, 1911–13), XI, 17.
68. *Opisanie russko-turetskoi voiny*, I, 85.
69. Ibid., 104.
70. Great Britain War Office, *The Armed Strength of Russia* [Official Copy] (London, 1882), 58.
71. Kersnovskii, *Istoriia russkoi armii*, pt. 2, 404.
72. Beskrovnyi, *Russkaia armiia i flot v XIX veke*, 54.
73. *Opisanie russko-turetskoi voiny*, I, 90–91.

74. Ibid., 55.
75. Beskrovnyi, *Russkaia armiia i flot v XIX veke*, 42–52.
76. *Opisanie russko-turetskoi voiny*, I, 88.
77. Ibid., 110.
78. Bruce W. Menning, "A. I. Chernyshev: A Russian Lycurgus," *Canadian Slavonic Papers*, 30, no. 2 (June 1988), 196–200.
79. Robert H. McNeal, "The Reform of Cossack Military Service in the Reign of Alexander II," in Béla K. Király and Gunther E. Rothenberg (eds.), *War and Society in East Central Europe*, Brooklyn College Studies on Society in Change, no. 10 (New York, 1979), 416–20.
80. *Opisanie russko-turetskoi voiny*, I, 118–19.
81. Ibid., 96.
82. Beskrovnyi, *Russkaia armiia i flot v XIX veke*, 442–43.
83. *Opisanie russko-turetskoi voiny*, I, 102–3.
84. Ibid, 120.
85. M. I. Bogdanovich (comp.), "Istoricheskii ocherk deiatel'nosti voennogo upravleniia v pervoe dvadtsati-piatiletie tsarstvovaniia Gosudaria Imperatora Aleksandra Nikolaevicha (1855–1880 gg.)," *Voennyi sbornik*, 132, no. 3 (March 1880), 22, 28, 33, 35; Beskrovnyi, *Russkaia armiia i flot v XIX veke*, 40–41.
86. V. V. Mavrodin and Va. V. Mavrodin, *Iz istorii otechestvennogo oruzhiia. Russkaia vintovka* (L, 1981), 32–34.
87. V. G. Fedorov, *Vooruzhenie russkoi armii za XIX stoletie* (SPB, 1911), 195–209.
88. Ibid., 224–26; for an excellent discussion, see Bradley, *Guns for the Tsar*, 107–11.
89. N. I. Beliaev, *Russko-turetskaia voina 1877–1878 gg.* (M, 1956), 26.
90. Bradley, *Guns for the Tsar*, 114–16.
91. The difficulties confronting artillerymen are ably summarized by Dennis E. Showalter, *Railroads and Rifles: Soldiers, Technology, and the Unification of Germany* (Hamden, Conn., 1975), chap. 9; see also *Opisanie russko-turetskoi voiny*, I, 188–90.
92. Beskrovnyi, *Russkaia armiia i flot v XIX veke*, 44–47.
93. Beliaev, *Russko-turetskaia voina*, 32–33.
94. Quoted in P. A. Zaionchkovskii, "Perevooruzhenie russkoi armii v 60—70-kh godakh XIX v.," *Istoricheskie zapiski*, 36 (1951), 86.
95. Beskrovnyi, *Russkaia armiia i flot v XIX veke*, 347.
96. Zaionchkovskii, "Perevooruzhenie russkoi armii v 60—70-kh godakh XIX v.," 85.
97. On the recruitment and training of officers before 1856, see J. E. O. Screen, *The Helsinki Yunker School 1846–1879* (Helsinki, 1986), 11–20.
98. Miller, *Dmitrii Miliutin*, 96–101.
99. Screen, *The Helsinki Yunker School 1846–1879*, 23–27.
100. Oliver Allen Ray, "The Imperial Russian Army Officer," *Political Science Quarterly*, 76, no. 4 (December 1961), 579–81.
101. Quoted in Beskrovnyi, *Russkaia armiia i flot v XIX veke*, 160.
102. Erickson, "The Russian Imperial/Soviet General Staff," 12.
103. Great Britain War Office, *The Armed Strength of Russia* (1873 ed.), 107–10.
104. Erickson, "The Russian Imperial/Soviet General Staff," 12–13; see also A. Gololobov, "Nasha akademiia general'nogo shtaba," *Voennyi sbornik*, 79, no. 5 (May 1871), 90–107.
105. G. Leer, "Sovremennoe sostoianie strategii," *Voennyi sbornik*, 57, no. 10 (October 1867), 230.
106. G. Leer, "Strategicheskii i takticheskii obzor franko-prusskoi voiny (do Sedana)," *Voennyi sbornik*, 81, no. 10 (October 1871), 165, 168.
107. The merits of the two legacies are discussed at length in chap. 4; see also F. A.

Maksheev, "Neskol'ko slov o Dragomirovskom i Leerovskom periodakh nachal'stvovaniia akademiei," *Voennyi sbornik*, no. 12 (December 1907), 234–38.

108. Erickson, "The Russian Imperial/Soviet General Staff," 15–16.

109. Van Dyke, "Russia's Military Professionals," 68–69.

110. Erickson, "The Russian Imperial/Soviet General Staff," 14.

111. *Sovetskaia voennaia entsiklopediia*, II, s.v. "Dragomirov Mikhail Ivanovich."

112. V. N. Lobov, "Russkii voennyi myslitel' i pedagog," *Voennaia mysl'*, no. 2 (February 1990), 44–45.

113. Gary Vincent Malanchuk, "The Training of an Army: M. I. Dragomirov and the Imperial Russian Army, 1860–1905," M.A. thesis, Miami University, 1978, 15–17.

114. Ibid., 19–21; M. I. Dragomirov (comp.), *Kurs taktiki dlia gg. ofitserov uchebnogo pekhotnogo bataliona* (SPB, 1867), 69; see also A. M. Komissarov, "M. I. Dragomirov o voinskoi distsipline," *Voenno-istoricheskii zhurnal*, no. 6 (June 1989), 89–90.

115. P. M. Gudim-Levkovich (comp.), *Kurs elementarnoi taktiki*, 3 vols. (SPB, 1887–90), I, 22–23; Meshcheriakov, *Russkaia voennaia mysl'*, 143.

116. Beskrovnyi, *Russkaia voenno-teoreticheskaia mysl'*, 20–21.

117. Meshcheriakov, *Russkaia voennaia mysl'*, 202–5.

118. M. I. Dragomirov, *Sbornik original'nykh i perevodnykh statei 1858–1880*, 2 vols. (SPB, 1881), I, 192.

119. Beskrovnyi, *Russkaia armiia i flot v XIX veke*, 143–44.

120. Ibid.

121. Von Wahlde, "Military Thought in Imperial Russia," 120–22; M. A. Gareev, *Obshchevoiskovye ucheniia* (M, 1983), 53–54.

122. Dragomirov (comp.), *Kurs taktiki*, 67–69.

123. Ibid., 290–95; F. V. Greene, *Report on the Russian Army and Its Campaigns in Turkey in 1877–1878* (New York, 1879), 441–43.

124. *Opisanie russko-turetskoi voiny*, I, 164.

125. Quoted in I. I. Rostunov (ed.), *Russko-turetskaia voina 1877–1878* (M, 1977), 45–46.

126. Ibid., 46.

127. Fligel'-ad"iutant baron Zeddeler, "Pekhota, artilleriia i kavaleriia v boiu i vne boia v Germano-frantsuzskoi voine 1870–1871 godov," *Voennyi sbornik*, 86, no. 7 (July 1872), 62–68.

128. *Opisanie russko-turetskoi voiny*, I, 165–67.

129. Meshcheriakov, *Russkaia voennaia mysl'*, 212–13; see also Dragomirov, *Kurs taktiki*, 184–85.

130. Meshcheriakov, *Russkaia voennaia mysl'*, 212–13.

131. For a reaction to Sukhotin's views at a slightly later date, see V. A. Sukhomlinov, *Vospominaniia* (Berlin, 1924), 53, 62–63.

132. Beskrovnyi, *Russkaia armiia i flot v XIX veke*, 154; see also Zaionchkovskii, *Voennye reformy*, 207; and cf. Jay Luvaas, *The Military Legacy of the Civil War* (Chicago, 1959), 112–13.

133. Zaionchkovskii, *Voennye reformy*, 210.

134. Ibid.

135. Beskrovnyi, *Russkaia armiia i flot v XIX veke*, 343–44; see also Thomas Vernon Moseley, "Evolution of the American Civil War Infantry Tactics," Ph.D. diss., University of North Carolina, 1967, 71–72.

136. Beskrovnyi, *Russkaia armiia i flot v XIX veke*, 155–57; see also *Opisanie russko-turetskoi voiny*, I, 188–90.

137. Meshcheriakov, *Russkaia voennaia mysl'*, 210–11.

138. *Opisanie russko-turetskoi voiny*, I, 190.

139. Ibid., 196.
140. Beskrovnyi, *Russkaia armiia i flot v XIX veke*, 159–61; *Opisanie russko-turetskoi voiny*, I, 202–3.
141. *Opisanie russko-turetskoi voiny*, I, 203–04; Gareev, *Obshchevoiskovye ucheniia*, 52.
142. V. N. Kopylov, "Taktika russkoi armii v russko-turetskoi voine 1877–1878 gg.," in D. V. Pankov (comp.), *Razvitie taktiki russkoi armii* (M, 1957), 194–95, and P. Kondrat'ev, "Takticheskoe obrazovanie ofitserov," *Voennye besedy ispol'nenyia v shtabe Gvardii i Peterburgskogo okruga v 1896–97 g.*, 18 (1897), 69–70.
143. *Sbornik materialov po russko-turetskoi voine*, 97 vols. (SPB, 1898–1911), XXI, pt. 2, 217–19, 225, 227, 238–39, 248–49, 354–55, and 416.
144. V. Potto, "Sovremennoe obrazovanie i vospitanie voisk," *Voennyi sbornik*, 105, no. 12 (December 1875), 464–68; Kopylov, "Taktiki russkoi armii," 195.

2. Russo-Turkish War, 1877–1878

1. Greene, *Report on the Russian Army*, 455.
2. For an overview, see Edward C. Thaden, *Russia Since 1801* (New York, 1971), 264–68.
3. Jelavich, *St. Petersburg and Moscow*, 176–78.
4. Beskrovnyi, *Russkoe voennoe iskusstvo XIX v.*, 303–7.
5. *Osoboe pribavlenie k opisaniiu russko-turetskoi voiny 1877–78 gg. na Balkanskom poluostrove*, 4 vols. (SPB, 1899–1901), IV, 81–83; Beliaev, *Russko-turetskaia voina*, 67–68; and Fuller, "Strategy and Power in Russia, 1600–1914," chap. 7.
6. *Sbornik materialov po russko-turetskoi voine*, XI, 121–22.
7. I. I. Rostunov, "Boevye deistviia russkoi armii na Balkanakh v 1877–1878 gg.," in G. L. Arsh, V. N. Vinogradov, et al. (eds.), *Russko-turetskaia voina 1877–1878 gg. i Balkany* (M, 1978), 10–25; Beskrovnyi, *Russkoe voennoe iskusstvo XIX v.*, 315–18.
8. Beliaev, *Russko-turetskaia voina*, 57–65.
9. Rostunov (ed.), *Russko-turetskaia voina 1877–1878*, 92–96.
10. G. Zander (comp.), *Takticheskie primery dlia voennykh i iunkerskikh uchilishch* (Odessa, 1897), 60–61.
11. *Opisanie russko-turetskoi voiny*, II, 135.
12. Greene, *Report on the Russian Army*, 444; see also Beliaev, *Russko-turetskaia voina*, 119, 121.
13. *Opisanie russko-turetskoi voiny*, II, 154–55.
14. Kopylov, "Taktika russkoi armii," 202.
15. *Opisanie russko-turetskoi voiny*, II, 184–85; Zaionchkovskii (ed.), *Dnevnik D. A. Miliutina*, II, 189.
16. Beliaev, *Russko-turetskaia voina*, 87–88, 117n.
17. Colonel Epauchin [N. Epanchin], *Operations of General Gurko's Advance Guard in 1877*, trans. H. Havelock (London, 1900), 3–7, chap. 3.
18. Rostunov (ed.), *Russko-turetskaia voina 1877–1878*, 104–5.
19. Kopylov, "Taktika russkoi armii," 203–4; Beliaev, *Russko-turetskaia voina*, 159–60; William McElwee, *The Art of War: Waterloo to Mons* (Bloomington, Ind., 1974), 196–99; and Chris Bellamy, "Antecedents of the Modern Soviet Operational Manoeuvre Group (OMG)," *RUSI*, 129, no. 3 (September 1984), 51–54.
20. Beliaev, *Russko-turetskaia voina 1877–1878*, 104–5.
21. Kopylov, "Taktika russkoi armii," 205.
22. P. Voronov, "Nachalo Plevny," *Voennyi sbornik*, 252, no. 4 (April 1900), 263–65.
23. *Opisanie russkoi-turetskoi voiny*, II, 306–17.

24. Beliaev, *Russko-turetskaia voina*, 146.
25. Ibid., 162.
26. Rostunov (ed.), *Russko-turetskaia voina 1877–1878*, 108–9.
27. *Sbornik materialov po russko-turetskoi voine*, XVI, 354–55; Beliaev, *Russko-turetskaia voina*, 165–66.
28. Beliaev, *Russko-turetskaia voina*, 167–68; *Sbornik materialov po russko-turetskoi voine*, XXVI, 110.
29. Nicholas Nikolaevich to Alexander II, 22 July 1877, in *Osoboe pribavlenie*, III, 54.
30. Quoted in Beliaev, *Russko-turetskaia voina*, 170.
31. Ibid., 172.
32. *Opisanie russko-turetskoi voiny*, IV, pt. 1, 132–37; F. V. Greene, *Sketches of Army Life in Russia* (New York, 1880), 41–42.
33. Beskrovnyi, *Russkoe voennoe iskusstvo XIX v.*, 328–30.
34. Greene, *Report on the Russian Army*, 210.
35. *Opisanie russko-turetskoi voiny*, IV, pt. 1, 160–61.
36. Greene, *Report on the Russian Army*, 214; Dragomirov's wounding is recounted in Greene, *Sketches of Army Life in Russia*, 50.
37. Beliaev, *Russko-turetskaia voina*, 232; "Zapiski P. M. Gudim-Levkovicha o voine 1877–1878 goda," *Russkaia starina*, 124, no. 12 (December 1905), 527–28.
38. Zander (comp.), *Takticheskie primery*, 62–63.
39. *Opisanie russko-turetskoi voiny*, V, 25–27.
40. Ibid., 33–35.
41. Ibid., 41–44; A. M. Zaionchkovskii, *Nastupatel'nyi boi po opytu deistvii generala Skobeleva v srazheniiakh Lovchei, Plevnoi (27 i 30 avgusta) i Sheinovo* (SPB, 1893), 18–23.
42. Beliaev, *Russko-turetskaia voina*, 236–37.
43. Beskrovnyi, *Russkoe voennoe iskusstvo XIX v.*, 332–34.
44. "Zapiski P. M. Gudim-Levkovicha," 537.
45. *Opisanie russko-turetskoi voiny*, V, 181–87.
46. Ibid., 195–206.
47. Ibid., 224–32.
48. D. I. Shor, "Inzhenernoe obespechenie osady Plevny v russko-turetskoi voine 1877–1878 gg.," in V. I. Zheleznykh (ed.), *Voenno-inzhenernye voiska russkoi armii* (M, 1958), 140–45.
49. Beskrovnyi, *Russkoe voennoe iskusstvo XIX v.*, 335–36.
50. Zeddeler survived to assume a postwar position with the Ministry of War; see *Voennaia entsiklopediia*, X, s. v. "Zeddeler, bar. Login Logginovich."
51. A. K. Puzyrevskii, *Desiat' let nazad* (SPB, 1887), 136–47.
52. Ibid., 147–53; E. V. Bogdanovich, *Gvardiia russkogo Tsaria na sofiiskoi doroge* (SPB, 1879), 83–92.
53. *Sbornik materialov po russko-turetskoi voine*, LVIII, 84.
54. Bogdanovich, *Gvardiia russkogo Tsaria*, 109–13.
55. Greene, *Sketches of Army Life in Russia*, 145; *Dnevnik Vysochaishogo prebyvaniia Imperatorskogo Aleksandra II-go za Dunaem v 1877 godu* (SPB, 1878), 227.
56. Zaionchkovskii (ed.), *Dnevnik D. A. Miliutina*, II, 231–32; Beliaev, *Russko-turetskaia voina*, 237.
57. M. Gazenkampf, *Moi dnevnik 1877–78 gg.* (SPB, 1908), 152–54.
58. Beskrovnyi, *Russkoe voennoe iskusstvo XIX v.*, 336–37.
59. Greene, *Report on the Russian Army*, 324.
60. P. Kartsov, "Traianov pereval," *Russkii vestnik*, 194, no. 1 (January 1888), 79.

61. Ibid., 77.
62. Rostunov, "Boevye deistviia russkoi armii," 22–23.
63. Zaionchkovskii, *Nastupatel'nyi boi*, 44–48.
64. Rostunov (ed.), *Russko-turetskaia voina 1877–1878*, 166–67.
65. Zaionchkovskii, *Nastupatel'nyi boi*, 54–9.
66. Greene, *Report on the Russian Army*, 356.
67. Ibid.
68. Rostunov (ed.), *Russko-turetskaia voina 1877–1878*, 200.
69. Rostunov, "Boevye deistviia russkoi armii," 24.
70. Ibid., 337–39.
71. Beliaev, *Russko-turetskaia voina*, 299–301.
72. Zaionchkovskii (ed.), *Dnevnik D. A. Miliutina*, II, 237.
73. Beliaev, *Russko-turetskaia voina*, 382–86.
74. Greene, *Report on the Russian Army*, 399–401.
75. V. I. Vinogradov, *Russko-turetskaia voina 1877–1878 gg. i osvobozhdenie Bolgarii* (M, 1978), 209–10.
76. Rostunov (ed.), *Russko-turetskaia voina 1877–1878*, 232–33.
77. Beskrovnyi, *Russkoe voennoe iskusstvo XIX v.*, 351n.
78. See, for example, Curtiss, *The Russian Army under Nicholas I, 1825–1855*, 250–51, 360.
79. Beliaev, *Russko-turetskaia voina*, 196, 304, 410–11.
80. Vsevolod Garshin, *From the Reminiscences of Private Ivanov*, trans. Peter Henry, Liv Tudge, et al. (London, 1988), 25–35; although Garshin is relatively unknown to Western readers, his depictions of Russian army existence, especially "From the Reminiscences of Private Ivanov," appear less stylized than those of the better-known Alexander Kuprin.
81. Beliaev, *Russko-turetskaia voina*, 196–97, 411–12.
82. Ibid., 304–5, 412–13.
83. Ibid., 411.
84. Beskrovnyi, *Russkoe voennoe iskusstvo XIX v.*, 315–17.
85. Beyrau, *Militär und Gesellschaft im vorrevolutionären Russland*, 29–30.
86. *Sbornik materialov po russko-turetskoi voine*, XCVI, 395–400.
87. Quoted in Beliaev, *Russko-turetskaia voina*, 405–6.
88. Ibid., 302.
89. *Opisanie russko-turetskoi voiny*, VIII, pt. 2, 518; *Sbornik materialov po russko-turetskoi voine*, XCVI, 165–71.
90. See Miliutin's commentary in Zaionchkovskii (ed.), *Dnevnik D. A. Miliutina*, III, 127; the gap between aspirations and capability is ably described by Jacob W. Kipp in "Tsarist Politics and the Naval Ministry, 1876–1881: Balanced Fleet or Cruiser Navy?" *Canadian-American Slavic Studies*, 17, no. 2 (Summer 1983), 166–68.
91. Greene, *Report on the Russian Army*, 435–36.

3. The Army of P. S. Vannovskii and A. N. Kuropatkin

1. Zaionchkovskii, *Samoderzhavie i russkaia armiia*, 39.
2. Commentators as diverse as P. A. Zaionchkovskii and A. A. Kersnovskii speak with a single voice when they label the period between 1881 and 1904, especially the reign of Alexander III, as one of "stagnation" for the army. See their respective works, *Samoderzhavie i russkaia armiia*, 62, 347–50, and *Istoriia russkoi armii*, pt. 3, 493.
3. Quoted in *Opisanie russko-turetskoi voiny*, IX, pt. 2, v-vi; see also Zolotarev, *Rossiia i Turtsiia*, 7–8, and Carl Van Dyke, "Culture of Soldiers: The Role of the Nicholas Academy

of the General Staff in the Development of Russian Imperial Military Science, 1832–1912," 113.

4. Zaionchkovskii (ed.), *Dnevnik D. A. Miliutina*, IV, 67–68.

5. Zaionchkovskii, *Samoderzhavie i russkaia armiia*, 50–59; and P. A. Zaionchkovskii, *Rossiiskoe samoderzhavie v kontse XIX stoletiia* (M, 1970), 35–46.

6. Fuller, *Civil-Military Conflict in Imperial Russia*, 40; Zaionchkovskii, *Rossiiskoe samoderzhavie*, 46–48.

7. In 1903, Kuropatkin patiently had to explain to the Tsar why the War Minister had more power than the commander of the Kiev Military District! See "Dnevnik A. N. Kuropatkina," *Krasnyi Arkhiv*, II (1922), 49.

8. On militarism, see John Keep, "The Origins of Russian Militarism," *Cahiers du Monde Russe et Sovietique*, 26, no. 1 (January–March 1985), esp. 7–10; on the military and the two Tsars, see Zaionchkovskii, *Samoderzhavie i russkaia armiia*, 36–50.

9. The change in style is aptly illustrated in P. A. Zaionchkovskii, *The Russian Autocracy in Crisis, 1878–1882*, ed. and trans. G. M. Hamburg (Gulf Breeze, Fla., 1979), 204–15; cf. Jacob W. Kipp and W. Bruce Lincoln, "Autocracy and Reform: Bureaucratic Absolutism and Political Modernization in Nineteenth-Century Russia," *Russian History*, 6, pt. 1 (1979), 13, 18.

10. See Vannovskii's biographical sketch in Skalon (ed.), *Stoletie Voennogo Ministerstva*, I, 573–75.

11. See General M. A. Domantovich's remarks as quoted in Zaionchkovskii, *Samoderzhavie i russkaia armiia*, 60.

12. Quoted in Kersnovskii, *Istoriia russkoi armii*, pt. 3, 500n.

13. Zaionchkovskii, *Samoderzhavie i russkaia armiia*, 60–64.

14. Zaionchkovskii (ed.), *Dnevnik D. A. Miliutina*, IV, 106–7.

15. Quoted in *Ocherk deiatel'nosti voennogo ministerstva za istekshee desiatiletie blagopolochnogo Tsarstvovaniia Gosudaria Imperatora Aleksandra Alekseevicha 1881–1891* (SPB, 1892), ii.

16. Fuller, *Civil-Military Conflict in Imperial Russia*, 47–49. Army expenditures for 1881 were 255,627,000 rubles, or about 30 percent of the state budget. Not until 1893 did annual expenditures exceed the 1881 figure, while the army's relative share of the state budget actually decreased between 1895 and 1903—with the exception of 1900—to less than 20 percent annually.

17. Thaden, *Russia since 1801*, 263–64.

18. Fuller, "Strategy and Power in Russia, 1600–1914," chap. 8.

19. Jelavich, *St. Petersburg and Moscow*, 204–13.

20. Ibid., 213–21.

21. A. M. Zaionchkovskii, "Russia's Preparation for the World War: International Relations," unpublished trans. by Mary Ada Underhill and Lewis K. Underhill, 65; see also Fuller, "Strategy and Power in Russia, 1600–1914," chap. 8.

22. Zaionchkovskii, *Samoderzhavie i russkaia armiia*, 65.

23. Ibid., 67–8; see Kuropatkin's biographical sketch in Skalon (ed.), *Stoletie Voennogo Ministerstva*, I, 640–45.

24. S. Iu. Vitte, *Vospominaniia*, 3 vols. (M, 1960), II, 156.

25. See Anton I. Denikin, *Career of a Tsarist Officer*, trans. M. Patoski (Minneapolis, 1975), 56–57.

26. Zolotarev, *Russko-turetskaia voina 1877/78 gg.*, 40.

27. P. A. Zhilin (ed.), *Russkaia voennaia mysl' konets XIX-nachalo XX v.* (M, 1982), 38; Zolotarev, *Rossiia i Turtsiia*, 78–81.

28. Until the Military Historical Commission began publication in 1898, the most

comprehensive survey of combat experience (*boevoi opyt*) remained a collection of eight articles, "Obzor mnenii vyskazannykh v nashei voennoi literature po raznym voennym voprosam," published in *Voennyi sbornik*, commencing with 132, no. 3 (March 1880), 173–210, and concluding with 136, no. 11 (November 1880), 32–53.

29. Zhilin (ed.), *Russkaia voennaia mysl'*, 35–36.

30. Zolotarev, *Russko-turetskaia voina 1877/78 gg.*, 32–34.

31. M. V. Grulev, *Zapiski generala-evreia* (Paris, 1930), 132–36; see also Hans-Peter Stein, "Der Offizier des russischen Heeres im Zeitabschnitt zwischen Reform und Revolution (1861–1905)," *Forschungen zur osteuropäischen Geschichte*, 13 (1967), 370–71.

32. Quoted in Zaionchkovskii, *Samoderzhavie i russkaia armiia*, 175.

33. Notes (149 pages) compiled in 1887 on strategy lectures by officers at the Nicholas Academy included only one page on the Russo-Turkish War of 1877–78! See *Konspekt strategii sostavlen ofitserami starshago klassa Nikolaevskoi Akademii General'nogo shtaba* (SPB, 1887), 72, 74.

34. Quoted in V. A. Avdeev, "Voenno-istoricheskie issledovaniia v Akademii General'nogo shtaba russkoi armii," *Voenno-istoricheskii zhurnal*, no. 12 (December 1987), 79.

35. *Opisanie russko-turetskoi voiny*, IV, pt. 1, v.

36. Quoted in Zaionchkovskii, *Samoderzhavie i russkaia armiia*, 93.

37. Skalon (ed.), *Stoletie Voennogo Ministerstva*, I, 579–85.

38. Ibid., 585.

39. A half-century later, A. A. Svechin offered a poignant critique of the system in *Strategiia*, 2d ed. (M, 1927), 208.

40. For the opinion of an especially influential field commander, see [General-ad"iutant Skobelev,] *Mnenie Komandira 4-go Armeiskogo korpusa general-ad"iutanta Skobeleva ob organizatsii mestnogo voennogo upravleniia i o korpusakh*, n.p., n.d, 1–9.7

41. Quoted in Zaionchkovskii (ed.), *Dnevnik D. A. Miliutina*, IV, 109.

42. Skalon (ed.), *Stoletie Voennogo Ministerstva*, I, 647–49; cf. Irvine, "The Origin of Capital Staffs," 177–79.

43. Quoted in Zaionchkovskii, *Samoderzhavie i russkaia armiia*, 109.

44. Jones, "Administrative System and Policy-Making Process, Central Military (Before 1917)," 120.

45. Kavtaradze, "Iz istorii russkogo general'nogo shtaba" (December 1971), 78; see also Great Britain War Office, General Staff, *Reorganisation of the Russian Head-Quarters Staff* [Confidential] (London, 1904), 3–12.

46. Fuller, "Strategy and Power in Russia, 1600–1914," chap. 8.

47. Skalon (ed.), *Stoletie Voennogo Ministerstva*, I, 620–21.

48. Kersnovskii, *Istoriia russkoi armii*, pt. 3, 501.

49. Kavtaradze, "Iz istorii russkogo general'nogo shtaba" (December 1971), 79.

50. Zaionchkovskii, *Samoderzhavie i russkaia armiia*, 123.

51. Ibid., 126.

52. Ibid., 152.

53. Quoted in ibid., 173.

54. Georgii Shavel'skii, *Vospominaniia poslednogo protopresvitera russkoi armii i flota*, 2 vols. (New York, 1954), I, 98.

55. Great Britain War Office, Intelligence Division, *System of Training of Staff Officers in Foreign Armies* [Confidential] (London, 1901), 89–95; see also I. V. Derevianko, "Mozg armii," *Voenno-istoricheskii zhurnal*, no. 10 (November 1989), 79–80.

56. A. A. Ignatyev, *A Subaltern in Old Russia*, trans. I. Montagu (London, 1944), 99–100.

57. Grulev, *Zapiski general-evreia*, 146, 167–68.

58. Zaionchkovskii, *Samoderzhavie i russkaia armiia*, 179, 181.
59. Matitiahu Mayzel, *Generals and Revolutionaries* (Osnabruck, 1979), 17–18, 37–39.
60. Fuller, *Civil-Military Conflict in Imperial Russia*, 22.
61. See John Bushnell's classic account, "The Tsarist Officer Corps, 1881–1914: Customs, Duties, Inefficiency," esp. 753–58 and 763–66.
62. Stein, "Der Offizier des russischen Heeres," 377.
63. Peter Kenez, "A Profile of the Prerevolutionary Officer Corps," *California Slavic Studies*, 7 (1973), 127–28.
64. Zaionchkovskii, *Samoderzhavie i russkaia armiia*, 176.
65. Beskrovnyi, *Russkaia armiia i flot v XIX veke*, 185–86.
66. Fuller, *Civil-Military Conflict in Imperial Russia*, 23–26.
67. N. N. Ianushkevich, "Sostoianie vooruzhennykh sil Rossii so vremeni vvedeniia obshchobiazatel'noi voinskoi povinnosti. Vazhneishie reformy v voennom dele v tsarstvovanie Imperatora Aleksandra III," in Grishinskii et al. (eds.), *Istoriia russkoi armii i flota*, XIII, 26.
68. A. B. Zhuk, *Vintovki i avtomaty* (M, 1987), 27–32.
69. Mavrodin and Mavrodin, *Iz istorii otechevestvennogo oruzhiia*, 91–94; Zhuk, *Vintovki i avtomaty*, 47–49.
70. Beskrovnyi, *Russkaia armiia i flot v XIX veke*, 319, 321; A. B. Zhuk, *Revol'very i pistolety* (M, 1983), 35.
71. Beskrovnyi, *Russkaia armiia i flot v XIX veke*, 354–55.
72. *Obzor deiatel'nosti voennogo ministerstva v tsarstvovanie Imperatora Aleksandra III-ogo 1881–1894* (SPB, 1903), 151–53; see also Christopher Bellamy, *Red God of War* (London, 1986), 24.
73. Bellamy, *Red God of War*, 22–24.
74. The financial implications of rearmament with quick-firing guns were an important justification behind the Tsar's initiative to convoke a special international disarmament conference at the Hague in 1899. See Dan L. Morrill, "Nicholas II and the Call for the First Hague Conference," *Journal of Modern History*, 46, no. 2 (June 1974), 296–98.
75. Quoted in Zaionchkovskii, *Samoderzhavie i russkaia armiia*, 162–63.
76. G. E. Peredel'skii (ed.), *Otechestvennaia artilleriia 600 let* (M, 1986), 65; see also V. P. Nikol'skii, "Voennoe delo v Rossii pri vstuplenii na prestol Imperatora Nikolaia II i pered voinoi 1904–05 gg." in Grishinskii et al. (eds.), *Istoriia russkoi armii i flota*, XIII, 39–40.
77. Beskrovnyi, *Russkaia armiia i flot v XIX veke*, 321.
78. Ibid.
79. Ibid., 323.
80. Ibid., 323–24.
81. Quoted in ibid., 429.
82. Zaionchkovskii, *Samoderzhavie i russkaia armiia*, 123–25; Great Britain War Office, Intelligence Division, *Distribution of the Russian Military Forces in Asia* [Confidential] (London, 1903), 23.
83. Beskrovnyi, *Russkaia armiia i flot v XIX veke*, 49; Great Britain War Office, Intelligence Division, *Twenty Years of Russian Army Reform and the Present Distribution of Russian Land Forces* [Confidential] (London, 1893), 10.
84. Zaionchkovskii, *Samoderzhavie i russkaia armiia*, 129–30; cf. Great Britain War Office, *The Armed Strength of Russia* [Official Copy] (London, 1886), 108.
85. Zaionchkovskii, *Samoderzhavie i russkaia armiia*, 131.
86. Ibid., 132n.
87. Beskrovnyi, *Russkaia armiia i flot v XIX veke*, 48–49; cf. N. G. Korsun and P. Kh.

Kharkevich, "Taktika russkoi armii v russko-iaponskoi voine 1904–1905 gg.," in Pankov (comp.), *Razvitie taktiki russkoi armii*, 256.

88. Zaionchkovskii, *Samoderzhavie i russkaia armiia*, 132–34, 143; see also F. Chenevix-Trench, *Report on the Russian Army* [Confidential], 16th October 1886 (n.p.), 1–2.

89. Beskrovnyi, *Russkaia armiia i flot v XIX veke*, 49–50; Zaionchkovskii, *Samoderzhavie i russkaia armiia*, 133, 139.

90. Beskrovnyi, *Russkaia armiia i flot v XIX veke*, 50.

91. Great Britain War Office, Intelligence Division, *Reports on Changes in Various Foreign Armies during the Year 1899, Together with Their Peace and War Strengths* [Confidential] (London, 1900), 45.

92. Zaionchkovskii, *Samoderzhavie i russkaia armiia*, 133–34.

93. K. Druzhinin, "Izsledovanie strategicheskoi deiatel'nosti germanskoi kavalerii v kompaniiu 1870 goda, do srazheniia Mars-la-Ture vkliuchitel'no," *Voennye besedy ispolnennyia v shtabe Gvardii i Peterburgskogo voennogo okruga v 1896–97 gg.*, 18–19; see also Great Britain War Office, Intelligence Division, *Twenty Years of Russian Army Reform*, 6, and Ianushkevich, "Sostoianie vooruzhennykh sil Rossii," 16.

94. Beskrovnyi, *Russkaia armiia i flot v XIX veke*, 53.

95. Zaionchkovskii, *Samoderzhavie i russkaia armiia*, 134–36.

96. Ibid., 136.

97. Beskrovnyi, *Russkaia armiia i flot v XIX veke*, 53–56.

98. Great Britain War Office, Intelligence Division, *Handbook of the Russian Army* [For Official Use Only], 3rd ed. (London, 1902), 172–73.

99. Zaionchkovskii, *Samoderzhavie i russkaia armiia*, 139.

100. Ibid., 137–40; Beskrovnyi, *Armiia i flot Rossii v nachale XX v.*, 20–21.

101. Beskrovnyi, *Russkaia armiia i flot v XIX veke*, 57–60.

102. Zaionchkovskii, *Samoderzhavie i russkaia armiia*, 142–43.

103. Beskrovnyi, *Russkaia armiia i flot v XIX veke*, 64–66; see also Robert H. McNeal, *Tsar and Cossack, 1855–1914* (London, 1987), 66–74, and A. Rediger (comp.), *Komplektovanie i ustroistvo vooruzhennoi sily* 4th ed., 2 vols. (SPB, 1913), I, 88.

104. Chenevix-Trench, *Report on the Russian Army*, 7.

105. Great Britain War Office, *The Armed Strength of the Russian Empire* [Confidential], 2 vols. (London, 1894), II, 226.

106. For a refined version of this plan, see *Voenno-morskaia strategicheskaia igra 1902 g. na temu "Zaniatie russkami silami verkhnogo Bosfora"* [Ves'ma sekretno] (SPB, 1902), 199–205; for a survey of Russian war planning during the period, see Zaionchkovskii, *Podgotovka Rossii k imperialisticheskoi voine*, 31–46, and Fuller, "Strategy and Power in Russia, 1600–1914," chap. 8.

107. In this regard, see esp. Great Britain War Office, Intelligence Division, *Russia's Power to Concentrate Troops in Central Asia* [Secret] (London, 1888), 14–15, and Great Britain War Office, Intelligence Branch, *Analysis of General Kuropatkin's Scheme for the Invasion of India* [Secret] (London, 1886), 1–3.

108. E. Willis Brooks, "The Military and Industrialization in Reform Russia: The Railroad Connection," unpublished paper, 7–9; this conclusion is supported by Fuller, *Civil-Military Conflict in Imperial Russia*, 62–65.

109. Beskrovnyi, *Russkaia armiia i flot v XIX veke*, 430–31.

110. For an excellent description of the Trans-Siberian Railroad before its completion, see Great Britain War Office, Intelligence Division, *Notes on the Siberian Railway between Cheliabinsk and Lake Baikal* [Confidential] (London, 1897), 1–12.

111. One verst equals 1.09 kilometer.

112. Beskrovnyi, *Armiia i flot Rossii v nachale XX v.*, 114–15.
113. Quoted in ibid., 115.
114. Quoted in Zaionchkovskii, *Samoderzhavie i russkaia armiia*, 127.
115. Fuller, "Strategy and Power in Russia, 1600–1914," chap. 8; on foreign reaction and assessment, see Great Britain War Office, Intelligence Division, *Twenty Years of Russian Army Reform and the Present Distribution of the Russian Land Forces*, 6–9, 11–13.
116. Quoted in Beskrovnyi, *Russkaia armiia i flot v XIX veke*, 445–46.
117. *Obzor deiatel'nosti voennogo ministerstva*, 187.
118. Zaionchkovskii, "Russia's Preparation for the World War: International Relations," 85.
119. Zaionchkovskii, *Podgotovka Rossii k imperialisticheskoi voine*, 46–48.
120. Zaionchkovskii, "Russia's Preparation for the World War: International Relations," 113, 115.
121. Zaionchkovskii, *Samoderzhavie i russkaia armiia*, 147–48.
122. Ibid., 148.
123. Zhilin (ed.), *Russkaia voennaia mysl'*, 32.
124. Quoted in Zaionchkovskii, *Samoderzhavie i russkaia armiia*, 148. In part, the lag in placing Russian railroads on a full mobilization footing stemmed from shortages of locomotives and rolling stock, which themselves had to be concentrated for mobilization. See A. L. Sidorov, "Zheleznodorozhnyi transport Rossii v pervoi mirovoi voine i obostrenie ekonomicheskogo krizisa v strane," *Istoricheskie zapiski*, 26 (1948), 12–13.
125. Zaionchkovskii, *Samoderzhavie i russkaia armiia*, 149.
126. Ibid., 153.
127. Ibid. Perhaps this unfinished work is explained by E. A. Nikol'skii, who, while working in the Main Staff shortly after the turn of the century, remembered shortened working hours and an "unhurried, unpressured existence." See his "Sluzhba v Glavnom shtabe i v Glavnom upravlenii General'nogo shtaba (1903–1908 gg.)." manuscript, Hoover Institution Archives, 8–9.
128. Quoted in ibid.
129. "Dnevnik A. N. Kuropatkina," 35.
130. Sukhomlinov, *Vospominaniia*, 99–101.
131. Zaionchkovskii, *Samoderzhavie i russkaia armiia*, 152; cf. Great Britain War Office, Intelligence Division, *Distribution of the Russian Military Forces in Asia*, 5, 23.
132. Beskrovnyi, *Russkaia armiia i flot v XIX veke*, 415; cf. Zhilin (ed.), *Russkaia voennaia mysl'*, 31–32, which places the figure at three pairs, an estimate close to that of British intelligence. In the latter regard, see Great Britain War Office, Intelligence Division, *Notes on the Siberian Railway between Cheliabinsk and Lake Baikal*, 5.

4. The Legacy of G. A. Leer and M. I. Dragomirov

1. A. A. Svechin (ed.), *Strategiia v trudakh voennykh klassikov*, 2 vols. (M, 1926), II, 273.
2. Quoted in E. I. Martynov, *Iz pechal'nogo opyta Russko-Iaponskoi voiny*, 2d ed. (SPB, 1907), 86.
3. Gunther E. Rothenberg, "Moltke, Schlieffen, and the Doctrine of Strategic Envelopment," in Paret (ed.), *Makers of Modern Strategy*, 296.
4. N. Pavlenko, "Iz istorii razvitii teorii strategii, *Voenno-istoricheskii zhurnal*, no. 10 (October 1964), 111–13; see also F. F. Gaivoronskii (ed.), *Evoliutsiia voennogo iskusstva: Etapy, tendentsii, printsipy* (M, 1987), 94–96.
5. On Sheinovo, see chap. 2; on Sedan, see the classic account of the Franco-Prussian War in English, Michael Howard, *The Franco-Prussian War: The German Invasion of France*,

1870–1871 (London, 1961), 203–23; on the legacy of Moltke, see McElwee, *The Art of War: Waterloo to Mons*, 107–9, 162, 184–87.

6. Svechin (ed.), *Strategiia v trudakh voennykh klassikov*, II, 102–4.

7. Voide's translation began with "Voina," *Voennyi sbornik*, 245, no. 1 (January 1899), 5–37, and concluded twenty-four issues later in 246, no. 12 (December 1900), 387–404. See V. Borisov, "Znachenie Klauzevitsa v sovremennoi voennoi nauke," *Voennyi sbornik*, 244, no. 12 (December 1898), 241–47.

8. See, for example, N. P. Glinoetskii, *Istoricheskii ocherk Nikolaevskoi Akademii general'nogo shtaba* (SPB, 1882), 2–3.

9. Svechin (ed.), *Strategiia v trudakh voennykh klassikov*, II, 103.

10. Ibid., 101.

11. Beskrovnyi (ed.), *Russkaia voenno-teoreticheskaia mysl'*, 21–22.

12. *Voennaia entsiklopediia*, XIV, s.v. "Leer, G. A."; see also *Sovetskaia voennaia entsiklopediia*, IV, s.v. "Leer Genrikh Antonovich," and Peter Von Wahlde, "A Pioneer of Russian Strategic Thought: G. A. Leer, 1829–1904," *Military Affairs*, 35, no. 4 (December 1971), 1–2.

13. Meshcheriakov, *Russkaia voennaia mysl'*, 234–36.

14. Beskrovnyi, *Ocherki po istochnikovedeniiu voennoi istorii Rossii*, 379; R. Savushkin, "K voprosu o vozniknovenii i razvitii operatsii," *Voenno-istoricheskii zhurnal*, no. 5 (May 1979), 79–80, and Von Wahlde, "A Pioneer of Russian Strategic Thought," 3–4.

15. Svechin (ed.), *Strategiia v trudakh voennykh klassikov*, II, 272, 276.

16. G. A. Leer (comp.), *Zapiski strategii*, 3rd ed. (SPB, 1877), 2–6.

17. Quoted in Savushkin, "K voprosu o vozniknovenii i razvitii operatsii," 80.

18. Leer (comp.), *Zapiski strategii*, 1.

19. Beskrovnyi, *Ocherki po istochnikovedeniiu voennoi istorii Rossii*, 378; cf. Von Wahlde, "A Pioneer of Russian Strategic Thought," 4–5.

20. Svechin (ed.), *Strategiia v trudakh voennykh klassikov*, II, 273.

21. Beskrovnyi (ed.), *Russkaia voenno-teoreticheskaia mysl'*, 273–74.

22. A. V. Gerua, *Posle voiny: O nashei armii* (SPB, 1907), 6–7.

23. E. I. Martynov, *Strategiia v epokhu Napoleona i v nashe vremia* (SPB, 1894), 272.

24. "Mysli o tekhnike voin budushchago," *Voennyi sbornik*, 211, no. 5 (May 1893), 38–39, 212, no. 8 (August 1893), 222–23, passim.

25. Zhilin (ed.), *Russkaia voennaia mysl'*, 99–100.

26. Ibid., 101–2; A. A. Gulevich, "Voina i narodnoe khoziaistvo," *Voennyi sbornik*, 241, no. 1 (January 1898), 60.

27. A. Gulevich, "Voina i narodnoe khoziaistvo," *Voennyi sbornik*, 241, no. 6 (June 1898), 300–301, 308–9.

28. On the origins of the two schools, see Zolotarev, *Russko-turetskaia voina 1877/78 gg.*, 21–29, and Beskrovnyi, *Ocherki voennoi istoriografii Rossii*, 185–88, 203–69.

29. Quoted in Zhilin (ed.), *Russkaia voennaia mysl'*, 148.

30. Quoted in ibid., 149.

31. Von Wahlde, "Military Thought in Imperial Russia," 100–112.

32. Von Wahlde, "A Pioneer of Russian Strategic Thought," 5.

33. Von Wahlde, "Military Thought in Imperial Russia," 112–13.

34. A. Ageev, "Voenno-teoreticheskie vzgliady N. P. Mikhnevicha," *Voenno-istoricheskii zhurnal*, no. 1 (January 1975), 91.

35. Quoted in ibid., 91–92.

36. N. P. Mikhnevich, *Vliianie noveishikh tekhnologicheskikh izobretenii na taktiku voisk* (SPB, 1893), 120–22, 146.

37. Ibid., 92; Zhilin (ed.), *Russkaia voennaia mysl'*, 90–92.

38. Beskrovnyi, *Russkaia voenno-teoreticheskaia mysl'*, 35–37; Pintner, "Russian Military Thought: The Western Model and the Shadow of Suvorov," 372–73.
39. *Sovetskaiā voennaia entsiklopediia*, V, s.v. "Mikhnevich Nikolai Petrovich."
40. G., "Popytka k vyiasneniiu veriatneishago znacheniia malodymnogo porokha v polevom boiu," *Voennyi sbornik*, 188, no. 4 (April 1891), 215, 223, 225.
41. Meshcheriakov, *Russkaia voennaia mysl'*, 268.
42. Quoted in ibid., 269.
43. See "lessons learned" respectively in Zaionchkovskii, *Nastupatel'nyi boi*, 60, 66–67, 75, 81, 87, and Puzyrevskii, *Desiat' let nazad*, 225–28.
44. K. N. Durop (comp.), *Taktika. Podrobnyi konspekt dlia voennykh i iunkerskikh uchilishch*, 2 pts. (SPB, 1886).
45. Meshcheriakov, *Russkaia voennaia mysl'*, 269; for a concise survey of Dragomirov's later tactical thought, see M. Dragomirov, "Konspekt lektsii po taktike," *Voennyi sbornik*, no. 9 (September 1912), 1–28.
46. I. Maslov, "Issledovanie o sredstvakh k umen'sheniiu poter' v voiskakh ot pekhotnogo ognia," *Voennyi sbornik*, 183, no. 10 (October 1888), 325.
47. Meshcheriakov, *Russkaia voennaia mysl'*, 270.
48. I. Maslov, "O svoistvakh pekhotnogo ognia i o sredstvakh k razvitiiu ego sily," *Voennyi sbornik*, 174, no. 3 (March 1887), 87, 117–18.
49. Although Maslov relied heavily on empirical data, he unequivocally asserted, "If with the aid of these two powerful sources of investigation [experience and mathematical analysis], we are convinced in the correctness of a conclusion made from the history of military art, then such a conclusion acquires the value almost of absolute truth and we attain the possibility of knowing the very rationale of a phenomenon." See his first article, "Issledovanie o sredstvakh k umen'sheniiu poter' v voiskakh ot pekhotnogo ognia," *Voennyi sbornik*, 183, no. 9 (September 1888), 61; in his subsequent article he relied heavily on Puzyrevskii's historical analysis.
50. Great Britain War Office, *The Armed Strength of Russia* (1882 ed.), 257.
51. Mikhnevich, *Vliianie noveishikh tekhnicheskikh izobretenii na taktiku voisk*, 69.
52. Beskrovnyi, *Ocherki po istochnikovedeniiu voennoi istorii Rossii*, 355–56.
53. Beskrovnyi, *Russkaia armiia i flot v XIX veke*, 150.
54. The frontage stipulated would have resulted, on the average, in placing one man every 1.16 paces! See Great Britain War Office, *The Armed Strength of Russia* (1882 ed.), 258n.
55. Ibid., 260.
56. Beskrovnyi, *Russkaia armiia i flot v XIX veke*, 151.
57. Ibid.
58. Great Britain War Office, *The Armed Strength of Russia* (1882 ed.), 262.
59. Cf. John A. English, *A Perspective on Infantry* (New York, 1981), 4–8.
60. Maslov, "Issledovanie o sredstvakh k umen'sheniiu poter'," 60.
61. "Obzor mnenii, vyskazannykh v nashei voennoi literature po razhnym voennym voprosam za 1877–1879 gody," *Voennyi sbornik*, 132, no. 4 (April 1880), 420–21, and N. Iunakov, "Nochnoi boi: Takticheskoe issledovanie," *Voennye besedy ispolnennye v shtabe Gvardii i Peterburgskogo voennogo okruga v 1887–1888 gg.*, V (1888), 67–68.
62. Great Britain War Office, Intelligence Division, *Handbook of the Military Forces of Russia* [For Official Use Only], 2d ed. (London, 1898), 146–47; cf. Adasovskii, "Vliianie malokalibernogo oruzhiia i bezdymnogo porokha na deistviia artillerii v boiu," *Voennye besedy ispolnennyia v shtabe voisk Gvardii i Peterburgskogo voennogo okruga v 1892–1893 gg.* XIII (1893), 127–28.
63. N. V. D—v, "Nashi boevye instruktsii i nastavleniia," *Voennyi sbornik*, 241, no. 5 (May 1898), 98.

64. Beskrovnyi, *Ocherki po istochnikovedeniiu voennoi istorii Rossii*, 358.
65. Korsun and Kharkevich, "Taktika russkoi armii," 236.
66. Great Britain War Office, Intelligence Division, *Handbook of the Russian Army*, 3rd ed., 162.
67. *Voinskie ustavy dlia pekhoty* (Warsaw, 1902), 147, 151.
68. Korsun and Kharkevich, "Taktika russkoi armii," 236–37.
69. Ibid., 237.
70. Ibid., 239.
71. *Voinskie ustavy dlia pekhoty*, 152.
72. Korsun and Kharkevich, "Taktika russkoi armii," 239–40.
73. N. K., "Po povodu proekta ustava polevoi sluzhby i nastavleniia dlai deistviia v boiu otriadov iz vsekh rodov oruzhiia," *Voennyi sbornik*, no. 6 (June 1903), 79.
74. A. Baumgarten, "Sovremennoe polozhenie voprosa ob artileriiskikh massakh," *Voennyi sbornik*, 177, no. 10 (October 1887), 177.
75. A. Baumgarten, "Artileriia v polevom boiu," *Voennyi sbornik*, 184, no. 8 (August 1890), 344.
76. The complexities of indirect fire are described in Bellamy, *Red God of War*, 28–30.
77. Meshcheriakov, *Russkaia voennaia mysl'*, 277–79.
78. Great Britain War Office, Intelligence Division, *Handbook of the Russian Troops in Asia* (London, 1890), 44–46.
79. Great Britain War Office, Intelligence Division, *Handbook of the Military Forces of Russia*, 2d ed., 165–66, 168–69.
80. Pototskii, "Posledniia novovvedeniia i predlozheniia v polevoi artillerii," *Voennye besedy ispolnennyia v shtabe voisk Gvardii i Peterburgskogo voennogo okruga za 1892–1893 gg.*, XIV (1893), 67–68.
81. Bellamy, *Red God of War*, 30.
82. N. Sukhotin, "Amerikanskaia konnitsa," *Voennyi sbornik*, 175, no. 5 (May 1887), 97–99; cf. Luvaas, *The Military Legacy of the Civil War*, 112–13, and McElwee, *The Art of War: Waterloo to Mons*, 153–59.
83. N. P. Mikhnevich, "Partizanskie deistviia kavalerii v 1812 i 1813 godakh," *Voennye besedy ispolnennye v shtabe voisk Gvardii i Peterburgskogo voennogo okruga v 1887–1888 g.g.*, IV (1888), 265–66.
84. Dragun, "Zametki o konnitse," *Voennyi sbornik*, 179, no. 2 (February 1888), 281; Sukhomlinov, *Vospominaniia*, 13, 29, 53, 62.
85. Kersnovskii, *Istoriia russkoi armii*, pt. 3, 504–5.
86. *Voinskii ustav peshogo stroia kavalerii* (SPB, 1884), 55–65.
87. Druzhinin, "Issledovanie strategicheskoi deiatel'nosti germanskoi kavalerii v kampaniiu 1870 goda, do srazheniia Mara-la-Ture vkliuchitel'no," 19.
88. Great Britain War Office, *The Armed Strength of Russia* (1882 ed.), 264.
89. Great Britain War Office, Intelligence Division, *Handbook of the Russian Troops in Asia*, 44.
90. Bruce W. Menning, "The Deep Strike in Russian and Soviet Military History," *Journal of Soviet Military Studies*, 1, no. 1 (April 1988), 20.
91. *Extracts from the New Drill Regulations of the Russian Cavalry*, trans. L. C. Scherer (Washington, D.C., 1898), 3–4.
92. F. M. Nirod, "Prozhitoe," typescript, n.d., Hoover Institution Archives, 61–63.
93. *Extracts from the New Drill Regulations of the Russian Cavalry*, 54–55, and Great Britain War Office, Intelligence Division, *Handbook of the Military Forces of Russia*, 2d ed., 158–59.
94. Meshcheriakov, *Russkaia voennaia mysl'*, 277.

95. Ianushkevich, "Sostoianie vooruzhennykh sil Rossii," 26. On the general problem of scouts, see Thomas G. Fergusson, *British Military Intelligence, 1870–1914* (Frederick, Maryland, 1984), 156.
96. "Mysli o tekhnike voin budushchago," *Voennyi sbornik*, 211, no. 6 (June 1893), 271.
97. A. Pevnev, *Voiskovaia konnitsa i ee boevoe ispol'zovanie* (M, 1926), 5–6.
98. See, for example, M. A. Sul'kevich, "Takticheskaia podgotvka polevoi artilerii," *Voennyi sbornik*, 244, no. 11 (November 1898), 154–57.
99. Meshcheriakov, *Russkaia voennaia mysl'*, 280–81, Gareev, *Obshchevoiskovye ucheniia*, 57–58.
100. Durop (comp.) *Taktika*, pt. 2, 17–20.
101. *Ocherk deiatel'nosti Voennogo Ministerstva*, 81–83.
102. Ianushkevich, "Sostoianie vooruzhennykh sil Rossii," 25.
103. Gareev, *Obshchevoiskovye ucheniia*, 57.
104. *Ocherk deiatel'nosti Voennogo Ministerstva*, 87–88; Nikol'skii, "Voennoe delo v Rossii pri vstuplenii na prestol Imperatora Nikolaia II i pered voinoi 1904–05 gg.," 40.
105. Zaionchkovskii, *Samoderzhavie i russkaia armiia*, 40, 254–56, 265–68.
106. Great Britain War Office, Intelligence Division, *Foreign Manoeuvres, 1893. Extracts from the Reports of Various Officers on the Manoeuvres in Austria, France, Germany, Italy, Russia and Servia* [Confidential] (London, 1894), 53–54; Great Britain War Office, Intelligence Division, *Foreign Manoeuvres, 1894. Extracts from the Reports of Various Officers on the Manoeuvres in Austro-Hungary, Belgium, Bulgaria, France, Germany, Italy, Roumania, Russia, Servia and Switzerland* [Confidential] (London, 1895), 152–53.
107. Great Britain War Office, Intelligence Division, *Foreign Manoeuvres, 1900. Extracts from the Reports of Various Officers on the Manoeuvres in Austria-Hungary, Belgium, France, Germany, Holland, Italy, Russia and Switzerland* [Confidential] (London, 1901), 91–92.
108. Great Britain War Office, Intelligence Division, *Foreign Manoeuvres, 1895. Extracts from the Reports of Various Officers on the Manoeuvres in Austria, Belgium, France, Germany, Italy, Roumania, Russia and Servia* [Confidential] (London, 1896), 101–2.
109. Great Britain War Office, Intelligence Division, *Handbook of the Russian Troops in Asia*, 42.
110. Great Britain War Office, Intelligence Division, *Foreign Manoeuvres, 1899. Extracts from the Reports of Various Officers on the Manoeuvres in Austria-Hungary, Denmark, Germany, Holland, Italy, Russia, and Switzerland* [Confidential] (London, 1900), 48.
111. Zaionchkovskii, *Samoderzhavie i russkaia armiia*, 283–90.
112. Gareev, *Obshchevoiskovye ucheniia*, 62.
113. Nikol'skii, "Voennoe delo v Rossii pri vstuplenii na prestol Imperatora Nikolaia II i pered voinoi 1904–05 gg.," 40.
114. Gareev, *Obshchevoiskovye ucheniia*, 59–60.
115. Ibid., 60.
116. Quoted in Sukhomlinov, *Vospominaniia*, 101.

5. Russo-Japanese War, 1904–1905

1. Quoted in A. I. Sorokin, *Oborona Port-Artura* (M, 1954), 3.
2. *Russko-iaponskaia voina 1904–1905 gg.*, 9 vols. in 16 books (SPB, 1910), I, 430–31.
3. Nikol'skii, "Sluzhba v Glavnom shtabe i v Glavnom upravlenii General'nogo shtaba (1903–1908 gg.)," 10; see also I. V. Derevianko, "Russkaia agenturnaia razvedka v 1902–1905 gg.," *Voenno-istoricheskii zhurnal*, no. 5 (May 1989), 76.
4. Rostunov (ed.), *Istoriia*, 93–95.

5. The number of machine guns would grow quickly, thanks to internal reallocations and rush purchase orders abroad.

6. S. M. Belitskii, *Strategicheskie rezervy* (M-L, 1930), 42–43; *Russko-iaponskaia voina*, I, 356–59.

7. Rostunov (ed.), *Istoriia*, 95–97.

8. Quoted in *Russko-iaponskaia voina*, I, 290; M. I. Dragomirov put it another way, when he asked, "Who will be his Skobelev?" See Sukhomlinov, *Vospominaniia*, 101.

9. See, for example, A. N. Kuropatkin, *The Russian Army and the Japanese War*, trans. A. B. Lindsay, 2 vols. (New York, 1909), II, 207–9.

10. N. A. Levitskii, *Russko-iaponskaia voina 1904–1905 gg.* (M, 1938), 24–25; India Division of the Chief of Staff, *Lessons of the Russo-Japanese War* [For Official Use Only] (Simla, 1906), 6–7.

11. Ibid., 18–32; see also A. Noskov, "Komplektovanie armii vo vremia voiny," *Voennaia mysl'*, no. 9 (September 1938), 137.

12. Shatsillo, *Rossiia pered pervoi mirovoi voinoi*, 6.

13. Ian Hamilton, *A Staff Officer's Scrap Book* (London, 1905), 88.

14. *Russko-iaponskaia voina*, II, 197; A. A. Svechin, *Russko-iaponskaia voina 1904–1905 gg.* (Oranienbaum, 1910), 54, 56.

15. Svechin, *Russko-iaponskaia voina*, 56.

16. Ibid., 67; Hamilton, *A Staff Officer's Scrap Book*, 61, 87, 105–7.

17. Rostunov (ed.), *Istoriia*, 150–53; see also Freiherr von Tettau, *Achtzehn Monate mit Russlands Heeren in der Mandschurei*, 2 vols. (Berlin, 1907), I, 90–97; a good summary in English is A. H. Burne, *The Liao-yang Campaign* (London, 1936), 36–41.

18. Oliver Ellsworth Wood, *The Yalu to Port Arthur* (Kansas City, Mo., 1905), 29–32.

19. *Russko-iaponskaia voina*, VIII, pt. 1, 250–52, 257.

20. Ibid., 255–56.

21. N. A. Tretyakov, *My Experiences at Nan Shan and Port Arthur*, trans. A. C. Alford (London, 1911), 56–61.

22. Rostunov (ed.), *Istoriia*, 170–71.

23. Levitskii, *Russko-iaponskaia voina*, 36.

24. *Russko-iaponskaia voina*, VIII, pt. 1, 192, 196–97; see also V. Frolov, "Geroicheskaia oborona Port-Artura," *Voenno-istoricheskii zhurnal*, no. 9 (September 1984), 69.

25. Quoted in Sorokin, *Oborona Port-Artura*, 77; General P. I. Kondratenko commanded Port Arthur's ground defenses.

26. E. K. Nojine [Nozhin], *The Truth about Port Arthur*, trans. A. B. Lindsay (London, 1908), x-xi; for a detailed overview, see, *Voennaia entsiklopediia*, XVIII, s.v. "Port-Artur"; on engineering aspects of the Port Arthur siege, see V. A. Zakharov, "Sostoianie i razvitie russkogo voenno-inzhenernykh voisk s nachala XX v. do Velikoi Oktiabr'skoi sotsialisticheskoi revoliutsii," in Zheleznykh et al. (eds.), *Voenno-inzhenernoe iskusstvo i inzhenernye voiska russkoi armii*, 152–56.

27. Tadayoshi Sakurai, *Human Bullets*, trans. M. Honda and A. Bacon (Tokyo, 1907), 84; *Russko-iaponskaia voina*, VIII, pt. 1, 522–23.

28. Rostunov (ed.), *Istoriia*, 180–83; *Russko-iaponskaia voina*, VIII, pt. 1, 534.

29. *Russko-iaponskaia voina*, VIII, pt. 2, 132.

30. Sorokin, *Oborona Port-Artura*, 103.

31. *Russko-iaponskaia voina*, VIII, pt. 2, 154.

32. Ibid., 142–43.

33. Ibid., 144–45.

34. Quoted in Sorokin, *Oborona Port-Artura*, 109.

35. Ibid., 111–13.
36. V. Frolov, "Russko-iaponskaia voina 1904–1905 gg.," *Voenno-istoricheskii zhurnal*, no. 2 (February 1974), 86.
37. Sorokin, *Oborona Port-Artura*, 117; *Russko-iaponskaia voina*, VIII, pt. 2, 284, 297–98.
38. *Russko-iaponskaia voina*, VIII, pt. 2, 328–29.
39. Ibid., 337–39.
40. Frolov, "Geroicheskaia oborona Port-Artura," 73.
41. B. W. Norregaard, *The Great Siege: The Investment and Fall of Port Arthur* (London, 1906), 165–66; Sorokin, *Oborona Port-Artura*, 123.
42. *Russko-iaponskaia voina*, VIII, pt. 2, 349–50.
43. Ellis Ashmead-Bartlett, *Port Arthur: The Siege and Capitulation*, 2d ed. (Edinburgh, 1906), 233.
44. Ibid., 163–66.
45. Norregaard, *The Great Siege*, 192.
46. Ashmead-Bartlett, *Port Arthur*, 268–69.
47. Sorokin, *Oborona Port-Artura*, 140–41.
48. Rostunov, *Istoriia*, 237–38.
49. Nojine, *The Truth About Port Arthur*, 256.
50. Ibid., 257–58.
51. Ashmead-Bartlett, *Port Arthur*, 330.
52. Rostunov (ed.), *Istoriia*, 248.
53. Kuropatkin, *The Russian Army and the Japanese War*, II, 213–14.
54. Burne, *The Liao-yang Campaign*, 51–56.
55. Levitskii, *Russko-iaponskaia voina*, 42–43.
56. *Russko-iaponskaia voina*, II, pt. 2, 39–42.
57. Ibid., 50–56.
58. Hamilton, *A Staff Officer's Scrapbook*, 210–11, 221–22.
59. Ibid., 257–58, 268–70.
60. Rostunov (ed.), *Istoriia*, 262–65.
61. Burne, *The Liao-yang Campaign*, 93–94; for an overview, see *Voennaia entsiklopediia*, xv, s.v. "Liaoian."
62. Kuropatkin, *The Russian Army and the Japanese War*, II, 226–27.
63. *Russko-iaponskaia voina*, III, pt. 3, 5–9.
64. An excellent overview is C. D. Bellamy, "The Russian Artillery and the Origins of Indirect Fire," *Army Quarterly and Defence Journal*, 112, no. 3 (July 1982), 331–33; see also Great Britain War Office, General Staff, *Notes upon Russian Artillery Tactics in the War of 1904–1905* [For Official Use Only] (London, 1907), 42–47.
65. Rostunov (ed.), *Istoriia*, 273.
66. *Russko-iaponskaia voina*, III, pt. 3, 343–45.
67. M. Kazakov, "Ispol'zovanie rezervov v russko-iaponskoi voine 1904–1905 gg.," *Voenno-istoricheskii zhurnal*, no. 4 (April 1971), 46.
68. Rostunov (ed.), *Istoriia*, 276; Kuropatkin, *The Russian Army and the Japanese War*, II, 230.
69. Burne, *The Liao-yang Campaign*, 104–5.
70. Kuropatkin, *The Russian Army and the Japanese War*, II, 230–31.
71. Freiherr von Tettau, *Achtzehn Monate*, I, 331–44.
72. Ibid., 234.
73. *Russko-iaponskaia voina*, III, pt. 3, 245–47.

74. Rostunov (ed.), *Istoriia*, 282–83.
75. Kuropatkin, *The Russian Army and the Japanese War*, II, 241–42.
76. Levitskii, *Russko-iaponskaia voina*, 54–55.
77. *Sovetskaia voennaia entsiklopediia*, VIII, s.v. "Shakhe."
78. *Russko-iaponskaia voina*, IV, pt. 1, 445–48.
79. *The Battle on the Scha Ho*, trans. Karl von Donat (London, 1906), 40.
80. Svechin, *Russko-iaponskaia voina*, 251–53.
81. A. A. Svechin, *Evoliutsiia voennogo iskusstva s drevneishikh vremen do nashikh dnei*, 2 vols. (M-L, 1927–28), II, 504.
82. *The Battle on the Scha Ho*, 82–83.
83. *Russko-iaponskaia voina*, IV, pt. 1, 344–45.
84. Rostunov (ed.), *Istoriia*, 293.
85. V. A. Cheremisov, *Russko-iaponskaia voina 1904–1905 goda* (Kiev, 1907), 208–10.
86. Svechin, *Russko-iaponskaia voina*, 323–24.
87. Ibid., 325; for a more favorable assessment, see N. A. Ukhach-Ogorovich, "Otsenka nabega na Inkou," *Voennyi sbornik*, no. 5 (May 1908), 61–70. See also Rediger (comp.), *Komplektovanie i ustroistvo vooruzhennoi sily*, II, 144; Bellamy, "Antecedents of the Modern Soviet Operational Manoeuvre Group (OMG)," 54–57; and Menning, "The Deep Strike in Russian and Soviet Military History," 21.
88. Kuropatkin, *The Russian Army and the Japanese War*, II, 255–56; *Russko-iaponskaia voina*, IV, 129–32.
89. Kuropatkin, *The Russian Army and the Japanese War*, II, 266–67; Frieherr von Tettau, *Achtzehn Monate*, II, 248–58.
90. Svechin, *Evoliutsiia voennogo iskusstva*, II, 508; Rostunov (ed.), *Istoriia*, 302.
91. Rostunov (ed.), *Istoriia*, 304–6.
92. Ibid., 306–8; for a detailed overview, see *Voennaia entsiklopediia*, XVI, s.v. "Mukden."
93. Svechin, *Russko-iaponskaia voina*, 341–44.
94. Rostunov (ed.), *Istoriia*, 309–10.
95. Svechin, *Russko-iaponskaia voina*, 350–52.
96. Ibid., 354–55.
97. Ibid., 363.
98. Ibid., 364.
99. *Russko-iaponskaia voina*, V, pt. 2, 237–39.
100. Ibid., 246.
101. Rostunov (ed.), *Istoriia*, 318–19.
102. *Russko-iaponskaia voina*, VI, 153–54.
103. K. P. Ushakov, *Podgotovka voennykh soobshchenii Rossii k mirovoi voine* (M-L, 1928), 19; see also Beskrovnyi, *Armiia i flot Rossii v nachale XX v.*, 150; Rostunov (ed.), *Istoriia*, 78–79.
104. Svechin, *Evoliutsiia voennogo iskusstva*, II, 461–62.
105. Rostunov, *Istoriia*, 80.
106. Svechin, *Evoliutsiia voennogo iskusstva*, II, 476–77.
107. Ibid., 476.
108. Ibid. See also Ignatyev, *A Subaltern in Old Russia*, 173, 189.
109. See, for example, Beskrovnyi, *Armiia i flot Rossii v nachale XX v.*, 151, and A. A. Neznamov, *Iz opyta russko-iaponskoi voiny (Zametki ofitsera General'nogo shtaba*, 2d ed. (SPB, 1906), 23.
110. Svechin, *Evoliutsiia voennogo iskusstva*, II, 409–10.

111. Even the obstreperous General Stessel was not blind to contemporary tactical change; see his "Glavneishie prakticheskie vyvody iz boevogo opyta na Kvantumskom Poluostrove otnositel'no deistvii voisk protiv Iapontsev," 20 August 1904, 1–15, in P. A. Bazarov Papers, Hoover Institution Archives.

112. For an excellent survey from the British perspective, see Great Britain War Office, General Staff, *Some Tactical Notes on the Russo-Japanese War* [Confidential] (London, 1906), 14–27.

113. Svechin, *Evoliutsiia voennogo iskusstva*, II, 576.

6. Theory and Structure, 1905–1914

1. A. A. Neznamov, *Sovremennaia voina*, 2d ed. (SPB, 1912), viii.

2. Sukhomlinov, *Vospominaniia*, 268.

3. Kersnovskii, *Istoriia russkoi armii*, pt. 3, 8–11; Stone, *The Eastern Front 1914–1917*, 21–24, 27–31.

4. Neznamov, *Iz opyta russko-iaponskoi voiny*, 129–30.

5. V. A. Avdeev, "Voenno-istoricheskie issledovaniia v russkoi armii," *Voenno-istoricheskii zhurnal*, no. 3 (March 1986), 82–83.

6. Nikol'skii, "Sluzhba General'nogo shtaba i v Glavnom upravlenii General'nogo shtaba (1903–1908 gg.)," 32–33; see also Zhilin (ed.), *Russkaia voennaia mysl'*, 40–41, 180.

7. A. Svechin, *Russko-iaponskaia voina 1904–1905 gg.* (Oranienbaum, 1910).

8. Pintner, "Russian Military Thought: The Western Model and the Shadow of Suvorov," 372; Von Wahlde, "Military Thought in Imperial Russia," 196–99.

9. A. Ageev, "Ofitsery russkogo general'nogo shtaba ob opyte russko-iaponskoi voiny 1904–1905 gg.," *Voenno-istoricheskii zhurnal*, no. 8 (August 1975), 99–104.

10. This is a theme which runs through much of the commentary, Russian and foreign, on the Russo-Japanese War; see Neznamov, *Iz opyta russko-iaponskoi voiny*, 26; Ignatyev, *A Subaltern in Old Russia*, 265; Cheremisov, *Russko-Iaponskaia voina 1904–1905 goda*, 290; and *Skazaniia inostrantsev o Russkoi Armii v voinu 1904–1905 g.g.* (SPB, 1912), 227–38.

11. Neznamov, *Iz opyta russko-iaponskoi voiny*, 2–3.

12. Ignatyev, *A Subaltern in Old Russia*, 247.

13. Neznamov, *Iz opyta russko-iaponskoi voiny*, 24–25.

14. Ibid., 36.

15. P. A. Bazarov, "Mneniia inostrannykh voennykh agentov po voprosam, kasaiushchimsia nastoiashchei kampanii," 2 December 1904, Pavel A. Bazarov Papers, Hoover Institution Archives, folders 1–3, pp. 3–4.

16. Ignatyev, *A Subaltern in Old Russia*, 119–20.

17. Ageev, "Ofitsery russkogo general'nogo shtaba," 102–3.

18. Dragomirov returned to Konotop to die in October 1905, but apparently not before he revised his text on tactics to account for changing battlefield conditions. See Zhilin (ed.), *Russkaia voennaia mysl'*, 184–85.

19. See I. Rostunov's retrospective analysis, "Uroki russko-iaponskoi voiny 1904–1905 gg.," *Voenno-istoricheskii zhurnal*, no. 2 (February 1984), esp. 77–79; cf. an anonymous contemporary perspective, E. U., "Uroki iaponskoi voiny," *Voennyi sbornik*, no. 8 (August 1908), esp. 97–98, in which the writings of two *genshtabisty*, M. V. Grulev and E. I. Martynov, were severely criticized.

20. Gerua, *Posle voiny: O nashei armii*, 17–18; see also Van Dyke, "Culture of Soldiers," 187–89.

21. A. A. Kersnovskii, *Filosofiia voiny* (Belgrade, 1939), 31; with that particularly Russian

penchant for precise military definition, Kersnovskii noted: "Strategy is the conduct of war. *Operatika* is the conduct of engagement [*srazhenie*]. Tactics—the conduct of battle [*boi*]." *Cf.* John Keegan, *Six Armies in Normandy* (New York, 1982), 243.

22. There is some speculation that Svechin, who is often credited with coining the term *operational art*, borrowed the concept from A. V. Gerua; see Svechin, *Strategiia*, 18–19.

23. Svechin, *Strategiia v trudakh voennykh klassikov*, II, 274.

24. Ibid., 175.

25. Von Wahlde, "Military Thought in Imperial Russia," 186–88.

26. Zhilin (ed.), *Russkaia voennaia mysl'*, 48.

27. Beskrovnyi, *Armiia i flot Rossii v nachale XX v.*, 38–39; see also Ageev, "Ofitsery russkogo general'nogo shtaba," 103, and Van Dyke, "Russia's Military Professionals," 139–41.

28. B. M. Shaposhnikov, *Vospominaniia. Voenno-nauchnye trudy* (M, 1974), 127.

29. Ibid., 143; for von Schlichting's influence, see "Sovremennaia strategiia," trans. by G. L. of General von Schlichting's *Osnovy sovremennoi taktiki i strategii*, in *Voennyi sbornik*, no. 6 (June 1908), 241–62.

30. Von Wahlde, "Military Thought in Imperial Russia," 182–83.

31. Ibid., 204–6.

32. N. P. Mikhnevich (comp.), *Strategiia*, 3rd ed. (SPB, 1911), 9, 43; see also Ageev, "Voenno-teoreticheskie vzgliady N. P. Mikhnevicha," 92–93.

33. Mikhnevich, *Strategiia*, 47.

34. Ibid., 98–106.

35. Ibid., 96, 144–45; see also Pavlenko, "Iz istorii razvitiia teorii strategii," 114–15.

36. Quoted in Beskrovnyi (ed.), *Russkaia voenno-teoreticheskaia mysl'*, 459.

37. Ibid., 446–51.

38. Quoted in ibid., 497.

39. Quoted in Von Wahlde, "Military Thought in Imperial Russia," 192.

40. Ibid., 193–94.

41. Neznamov, *Sovremennaia voina*, vii; see also Pintner, "Russian Military Thought: The Western Model and the Shadow of Suvorov," 368.

42. Quoted in Von Wahlde, "Military Thought in Imperial Russia," 223.

43. A. A. Neznamov, *Voennaia entsiklopediia*, XVIII, s.v. "Plan voiny."

44. Neznamov, *Sovremennaia voina*, 132, 145, 162, and passim; see also Beskrovnyi, *Russkaia voenno-teoreticheskaia mysl'*, 640, 642, 672.

45. Von Wahlde, "Military Thought in Imperial Russia," 225–26.

46. A. A. Neznamov, *Trebovaniia, kotorye pred'iavliaet sovremennyi boi k podgotovke (obucheniiu) nachal'nikov i mass* (SPB, 1909), 4.

47. Neznamov, *Sovremennaia voina*, 228–36.

48. A. A. Neznamov, *Voennaia entsiklopediia*, XVII, s.v. "Operatsiia (voennaia)."

49. L. G. Beskrovnyi has noted that this understanding is similar to contemporary Soviet operational art: "Neznamov provided a basis for understanding operational art which is rather close to our definition." See Beskrovnyi, *Russkaia voenno-teoreticheskaia mysl'*, 37; and Savushkin, "K voprosu vozniknovenii i razvitii operatsii," 80.

50. Neznamov, "Operatsiia (voennaia)," 132–33, and Neznamov, *Sovremennaia voina*, 10–13; see also A. Ageev, "Voenno-teoreticheskoe nasledie A. A. Neznamova," *Voenno-istoricheskii zhurnal*, no. 11 (November 1983), 86.

51. Ageev, "Voenno-teoreticheskoe nasledie A. A. Neznamov," 86.

52. Neznamov, *Sovremennaia voina*, 21–23.

53. Neznamov, *Trebovaniia*, 6.

54. See, for example, A. A. Neznamov, "Boevaia podgotovka armii," *Russkii invalid*, no. 38 (18 February 1912), 4–5; for an overview, see Van Dyke, "Culture of Soldiers," 213–16.
55. Zhilin (ed.), *Russkaia voennaia mysl'*, 144–45.
56. A. Adaridi, "Liubiteliam doktriny," *Russkii invalid*, no. 14 (17 January 1912), 3, and N. Chetyrkin, "Doktrina v ustave polevoi sluzhby," *Russkii invalid*, no. 41 (22 February 1912), 2–3.
57. Neznamov, "Boevaia podgotovka armii," 4–5.
58. P. Zhilin, "Diskussii o edinoi voennoi doktriny," *Voenno-istoricheskii zhurnal*, no. 5 (May 1961), 62–64.
59. A. M. Zaionchkovskii, "Edinstvo voennoi doktriny (Opasnosti i ubelichenii)," *Russkii invalid*, no. 280 (31 December 1911), 2–3; cf. V. N. Domanevskii, "Samodeiatel'nost' nachal'nikov," *Russkii invalid*, no. 48 (1 March 1912), 5.
60. Quoted in Zhilin (ed.), *Russkaia voennaia mysl'*, 146.
61. Denikin, *Career of a Tsarist Officer*, 182.
62. Jones, "Imperial Russia's Armed Forces at War," 20.
63. A. A. Svechin, "Bol'shaia voennaia programma," *Russkaia mysl'*, bk. VIII (August 1913), 19–29.
64. John Bushnell, "The Revolution of 1905–1906 in the Army: The Incidence and Impact of Mutiny," *Russian History*, 12, no. 1 (Spring 1985), 72, notes that in 1905 the army experienced more than 234 mutinies, and in 1906 at least 211.
65. Shatsillo, *Rossiia pered pervoi mirovoi voinoi*, 14.
66. Walter Thomas Wilfong, "Rebuilding the Russian Army, 1904–1914: The Question of a Comprehensive Plan for National Defense," Ph.D. diss., Indiana University, 1977, 42–43; see also Beskrovnyi, *Armiia i flot Rossii v nachale XX v.*, 64–66.
67. A. Kavtaradze, "Iz istorii russkogo general'nogo shtaba," *Voenno-istoricheskii zhurnal*, no. 7 (July 1972), 87–88.
68. Beskrovnyi, *Armiia i flot Rossii v nachale XX v.*, 48–49.
69. Iu. N. Danilov, *Rossiia v mirovoi voine* (Berlin, 1924), 33.
70. See esp. K. F. Shatsillo, *Russkii imperializm i razvitie flota* (M, 1968), 164–72, and D. C. B. Lieven, *Russia and the Origins of the First World War* (London, 1983), 50–61.
71. William C. Fuller, Jr., "The Russian Empire," in Ernest R. May (ed.), *Knowing One's Enemies: Intelligence Assessment before the Two World Wars* (Princeton, N.J., 1984), 98–102; see also Jones, "Imperial Russia's Armed Forces at War, 1914–1917," 14–17.
72. Shatsillo, *Russkii imperializm i razvitie flota*, 60–61.
73. The most extensive overview is Bushnell, *Mutiny amid Repression*; see also Fuller, *Civil-Military Conflict in Imperial Russia*, 133–41, 144–68.
74. Shatsillo, *Rossiia pered pervoi mirovoi voinoi*, 15.
75. K. Shatsillo, "Podgotvka Tsarizmom vooruzhennykh sil k pervoi mirovoi voine," *Voenno-istoricheskii zhurnal*, no. 9 (September 1974), 91–92.
76. Shatsillo, *Russkii imperializm i razvitie flota*, 57–63; see also Beskrovnyi, *Armiia i flot v nachale XX v.*, 193–94.
77. Rostunov, *Russkii front*, 42–43.
78. Ibid., 93.
79. Ibid., 34.
80. A. Kavtaradze, "Iz istorii russkogo general'nogo shtaba (1909-iiul' 1914 gg.)," *Voenno-istoricheskii zhurnal*, no. 12 (December 1974), 80.
81. Ibid., 80–81; see also Danilov, *Rossiia v mirovoi voine*, 33–34.
82. V. I. Gurko, *Features and Figures of the Past: Government and Opinion in the Reign of Nicholas II*, trans. Laura Matveev, Hoover Library on War, Revolution, and Peace, Publication No. 14 (Stanford, 1939), 551–53; Sukhomlinov, *Vospominaniia*, 265–66.

83. See Fuller's studied analysis, *Civil-Military Conflict in Imperial Russia*, 237–44, cf. Wilfong, "Rebuilding the Russian Army," 114–16, and Shatsillo, *Rossiia pered pervoi mirovoi voinoi*, 47–48.
84. Shatsillo, *Rossiia pered pervoi mirovoi voinoi*, 49–52.
85. Rostunov, *Russkii front*, 46–47; cf. Zaionchkovskii, *Samoderzhavie i russkaia armiia*, 119–20, and P. L. Lobko, *Zapiski voennoi administratsii*, 6th ed., 3 pts. (SPB, 1906), pt. 1, 57.
86. Rostunov, *Russkii front*, 47.
87. Ibid.
88. Rediger (comp.), *Komplektovanie i ustroistvo vooruzhennoi sily*, II, 96, 161, 163–64; see also Zaionchkovskii, *Podgotovka Rossii k imperialisticheskoi voine*, 107. These figures are corroborated by an unclassified German version of a Russian threat book, *Die Russische Armee* (Berlin, 1912), 27.
89. Great Britain War Office, General Staff, *Handbook of the Russian Army* [For Official Use Only], 6th ed. (London, 1914), 11–12.
90. C. D. Bellamy, "Sukhomlinov's Army Reforms 1908–1915," M.A. essay, King's College, University of London, 1978, 25–26.
91. Rostunov, *Russkii front*, 48–49.
92. Ibid., 50–51; Shatsillo, *Rossia pered pervoi mirovoi voinoi*, 43–44.
93. Rostunov, *Russkii front*, 52.
94. Rediger (comp.), *Komplektovanie i ustroistvo vooruzhennoi sily*, II, 163–64.
95. Wilfong, "Rebuilding the Russian Army," 152–53, and Shatsillo, *Rossiia pered pervoi mirovoi voinoi*, 43–45.
96. Rostunov, *Russkii front*, 55; for a slightly different set of figures, see Great Britain War Office, General Staff, *Handbook of the Russian Army*, 6th ed., 21.
97. Great Britain War Office, General Staff, *Handbook of the Russian Army*, 6th ed., 31; *Die Russische Armee*, 32.
98. Great Britain War Office, General Staff, *Handbook of the Russian Army*, 6th ed., 31.
99. N. G. Korsun and P. Kh. Kharkevich, "Taktika russkoi armii v perviui mirovuiu voinu 1914–1918 gg.," in Pankov (comp.), *Razvitie taktiki russkoi armii*, 281–82; see also Beskrovnyi, *Armiia i flot Rossii v nachale XX v.*, 18, and N. A. Danilov, *Lektsii po voennoi administratsii* (L, 1926), 101.
100. Danilov, *Rossiia v mirovoi voine*, 39–41; Bellamy, "Sukhomlinov's Army Reforms 1908–1915," 28–30.
101. Beskrovnyi, *Armiia i flot Rossii v nachale XX v.*, 20; Great Britain War Office, General Staff, *Handbook of the Russian Army*, 6th ed., 53.
102. Great Britain War Office, General Staff, *Handbook of the Russian Army*, 6th ed., 31.
103. Pevnev, *Voiskovaia konnitsa i ee boevoe ispol'zovanie*, 8.
104. Evgenii Barsukov, *Russkaia artilleriia v mirovuiu voinu*, 2 vols. (M, 1938), 18–19; Beskrovnyi, *Armiia i flot Rossii v nachale XX v.*, 21.
105. Glavnoe upravlenie General'nogo shtaba, *Vooruzhennye sily Germanii* [Sekret'no], 2 vols. (SPB, 1912–1914), II, 15–16.
106. Great Britain War Office, General Staff, *Report on Changes in Foreign Armies during 1910* [Confidential], n.p., n.d., 91; see also Great Britain War Office, General Staff, *Handbook of the Russian Army*, 6th ed., 68–69.
107. Barsukov, *Russkaia artilleriia v mirovuiu voinu*, 19–21.
108. Great Britain War Office, General Staff, *Handbook of the Russian Army*, 6th ed., 131–32.

109. A. A. Manikovskii (comp.), *Boevoe snabzhenie russkoi armii v voinu 1914–1918 gg.*, 3 pts. (Moscow, 1920–23), pt. 1, 89.

110. A. V. Vasil'ev, "Strelkovoe oruzhie russkoi armii v pervoi mirovoi voine," *Voenno-istoricheskii zhurnal*, no. 13 (November 1980), 82–83; Beskrovnyi, *Armiia i flot Rossii v nachale XX v.*, 80–81.

111. A. Kolenkovskii, *Manevrennyi period pervoi mirovoi imperialisticheskoi voiny 1914 g.* (M, 1940), 48–49; see also Stone, *The Eastern Front 1914–1917*, 31–32; Danilov, *Rossiia v mirovoi voine*, 41–43.

112. Beskrovnyi, *Armiia i flot Rossii v nachale XX v.*, 129–31.

113. Ibid., 135–36; David R. Jones, "The Beginnings of Russian Air Power, 1907–1922," in Robin Higham and Jacob W. Kipp (eds.), *Soviet Aviation and Air Power: A Historical View* (Boulder, Colo., 1977), 16.

114. Jones, "The Beginnings of Russian Air Power, 1907–1922," 17–18.

115. A. Zhilin, "Bol'shaia programma po usileniiu russkoi armii," *Voenno-istoricheskii zhurnal*, no. 7 (July 1979), 92, 96.

116. Shatsillo, *Rossiia pered mirovoi voinoi*, 97–98.

117. Zhilin, "Bol'shaia programma', 93–95.

118. Ibid., 95.

119. V. A. Melikov, *Problema strategicheskogo razvertyvaniia po opytu mirovoi i grazhdanskoi voin* (M, 1935), 257–58; Ushakov, *Podgotovka voennykh soobshchenii Rossii k mirovoi voine*, 15.

120. Jones, "Imperial Russia's Armed Forces at War, 1914–1917," 101–3.

121. Beskrovnyi, *Armiia i flot Rossii v nachale XX v.*, 59.

122. Ibid., 60–61.

123. N. Shvarts, *Ustroistvo voennogo upravleniia*, 2d ed. (M-L, 1927), 171–73; see also Jones, "Imperial Russia's Armed Forces at War, 1914–1917," 103–4.

124. Rostunov, *Russkii front*, 39–40.

125. A. Kavtaradze, "Voiskovoe upravlenie general'nogo shtaba russkoi armii," *Voenno-istoricheskii zhurnal*, no. 6 (June 1978), 78–79, 81.

126. Quoted in A. M. Zaionchkovskii, *Mirovaia voina 1914–1918*, 2d ed. (M, 1931), 23.

127. Ibid., 22; cf. Kolenkovskii, *Manevrennyi period pervoi mirovoi imperialisticheskoi voiny*, 41.

128. Danilov, *Rossiia v mirovoi voine*, 52.

7. Dilemmas of Design and Application, 1905–1914

1. Quoted in Von Wahlde, "Military Thought in Imperial Russia," 225–26.
2. Wilfong, "Rebuilding the Russian Army," 139–41.
3. Quoted in Rostunov, *Russkii front*, 89.
4. Ibid., 90.
5. Ibid., 91.
6. See Alekseev's notations of 17 December 1908 in Zaionchkovskii, *Podgotovka Rossii k imperialisticheskoi voine*, 348–54.
7. Quoted in Rostunov, *Russkii front*, 91–92.
8. Jack Snyder, *The Ideology of the Offensive* (Ithaca, N. Y., 1984), 166.
9. Rostunov, *Russkii front*, 92.
10. Snyder, *The Ideology of the Offensive*, 166.
11. Rostunov, *Russkii front*, 76; A. Svechin, "Evoliutsiia operativnogo razvertyvaniia," *Voina i revoliutsiia*, no. 5 (May 1926), 15.
12. Melikov, *Problema strategicheskogo razvertyvaniia*, 245.

13. Zaionchkovskii, *Podgotovka Rossii k imperialisticheskoi voine*, 320.
14. Ibid.
15. Snyder, *The Ideology of the Offensive*, 182; cf. Stone, *The Eastern Front 1914–1917*, 35.
16. Melikov, *Problema strategicheskogo razvertyvaniia*, 223–24.
17. Rostunov, *Russkii front*, 92–93.
18. Ibid., 94.
19. Ibid., 95.
20. Zaionchkovskii, *Podgotovka Rossii k imperialisticheskoi voine*, 320.
21. Danilov, *Rossiia v mirovoi voine*, 93–94.
22. Ibid., 94–95.
23. L. C. F. Turner, "The Russian Mobilization of 1914," in Paul M. Kennedy (ed.), *The War Plans of the Great Powers, 1880–1914* (London, 1979), 257.
24. Belitskii, *Strategicheskie rezervy*, 65–66.
25. Ibid., 67–68.
26. Stone, *The Eastern Front 1914–1917*, 54–55; Zaionchkovskii, *Mirovaia voina 1914–1918*, 63–64, 75–76.
27. Danilov, *Rossiia v mirovoi voine*, 97.
28. Jones, "Imperial Russia's Armed Forces at War, 1914–1917," 96.
29. A. M. Zaionchkovskii, *Lektsii po strategii, chitannye na voenno-akademicheskikh kursakh vysshego komsostava v Voennoi akademii RKKA v 1922–1923 gg.*, 2 pts. (M, 1923), pt. 1, 131.
30. Fuller, "The Russian Empire," 115.
31. A. A. Brusilov, *Moi vospominaniia* (M, 1983), 47n.
32. Zaionchkovskii, *Mirovaia voina 1914–1918*, 66.
33. Snyder, *The Ideology of the Offensive*, 176, 189–90.
34. Svechin, "Evoliutsiia operativnogo razvertyvaniia," 7–8, 21–22.
35. See esp. Danilov, *Rossiia v mirovoi voine*, 91–92.
36. Svechin, "Evoliutsiia operativnogo razvertyvaniia," 24.
37. For an excellent critique, see A. A. Neznamov, *Osnovy sovremennoi strategii* (M, 1919), 129–30.
38. Danilov, *Rossiia v mirovoi voine*, 99–100.
39. See the discussion in Jones, "Imperial Russia's Armed Forces at War," 122–24, 129–33.
40. Ibid., 103–4.
41. Stone, *The Eastern Front 1914–1917*, 51–52.
42. Ibid., 51, 58.
43. B. V. Gerua, *Vospominaniia o moei zhizni*, 2 vols. (Paris, 1969–70), I, 223–24.
44. Ibid., 225–27.
45. Sukhomlinov, *Vospominaniia*, 236.
46. Ibid., 236–38, and Danilov, *Rossiia v mirovoi voine*, 101.
47. A. N. Suvorov, "Voennaia igra starshikh voiskovykh nachal'nikov v aprele 1914 goda," *Voenno-istoricheskii sbornik*, vyp. 1 (1919), 9–11.
48. Ibid., 12.
49. Danilov, *Rossiia v mirovoi voine*, 101; Suvorov, "Voennaia igra," 14–15.
50. Quoted in Suvorov, "Voennaia igra," 15; cf. Brusilov, *Moi vospominaniia*, 47n.
51. Suvorov, "Voennaia igra," 16.
52. Melikov, *Problema strategicheskogo razvertyvaniia*, 275–76.
53. Ibid., 277.
54. Suvorov, "Voennaia igra," 18–19.
55. Ibid., 20–21.
56. Ibid., 22.

57. Ibid., 21.
58. Quoted in Rostunov, *Russkii front*, 103.
59. Manikovskii, *Boevoe snabzhenie russkoi armii*, pt. 3, 5–12.
60. Ibid, 105–7.
61. Jones, "Imperial Russia's Armed Forces at War," 40–41; see also 35–39 and 42–44.
62. See, for example, Great Britain War Office, General Staff, *Handbook of the Russian Army*, 6th ed., 36, and *Die Russische Armee*, 62–63.
63. Jones, "Imperial Russia's Armed Forces at War," 114–18; for a general overview, see B. E. Barskii, *Organizatsiia i upravlennie tylom* (M-L, 1926), chap. 5.
64. See, for example, the pioneering studies by N. Volotskii, "Teoriia veroiatnostei i boevoe snabzhenie artileri," *Voennyi sbornik*, no. 2 (February 1904), 139–52; no. 3 (March 1904), 135–44; and "Teoriia veroiatnostei i boevoe snabzhenie patronami," *Voennyi sbornik*, no. 11 (November 1904), 81–100.
65. Beskrovnyi, *Ocherki po istochnikovedeniiu voennoi istorii Rossii*, 360, 362.
66. Ibid., 361.
67. *Ustav polevoi sluzhby. Vysochaishche utverzhden 27 Aprelia 1912 g.* (M, 1916), 93, 104–5; see also the commentary in A. A. Strokov, *Vooruzhennye sily i voennoe iskusstvo v pervoi mirovoi voine* (M, 1974), 156–57.
68. *Ustav polevoi sluzhby* (1912 version), 108–9; Korsun and Kharkevich, "Taktika russkoi armii v pervuiu mirovuiu voinu 1914–1918 gg.," 272.
69. *Ustav polevoi sluzhby* (1912 version), 106–10; for additional details, see Christopher Bellamy, "Seventy Years On: Similarities between the Modern Soviet Army and Its Tsarist Predecessor," *RUSI*, 124, no. 3 (September 1979), 29–30, and David R. Jones, "Advanced Guard (Avangard)," *The Military-Naval Encyclopedia of Russia and the Soviet Union*, IV, 123–30.
70. *Ustav polevoi sluzhby* (1912 version), 109; Korsun and Kharkevich, "Taktika russkoi armii v pervuiu mirovuiu voinu 1914–1918 gg.," 270.
71. *Ustav polevoi sluzhby* (1912 version), 110; Zhilin (ed.), *Russkaia voennaia mysl'*, 121–22.
72. *Ustav polevoi sluzhby* (1912 version), 113–15; Korsun and Kharkevich, "Taktika russkoi armii v pervuiu mirovuiu voinu 1914–1918 gg.," 270–71.
73. Zaionchkovskii, *Mirovaia voina 1914–1918*, 22.
74. Strokov, *Vooruzhennye sily i voennoe iskusstvo v pervoi mirovoi voine*, 163.
75. *Nastavlenie dlia deistvii pekhoty v boiu* (SPB, 1914), 5.
76. Ibid., 11–13.
77. M. Bonch-Bruevich, *Boi* (Kiev, 1909), 76–79.
78. Great Britain War Office, General Staff, *Handbook of the Russian Army*, 6th ed., 137.
79. *Nastavlenie dlia deistvii pekhoty v boiu*, 5–6, 40–43.
80. Strokov, *Vooruzhennye sily i voennoe iskusstvo v pervoi mirovoi voine*, 160–61.
81. *Nastavlenie dlia deistvii pekhoty v boiu*, 22–24.
82. Ibid., 24–29.
83. Ibid., 33–35; the chief mechanism was by two- and four-wheel carts from the artillery park to units located forward.
84. Ibid., 36–37.
85. Ibid., 65–66.
86. Ibid., 67–74.
87. Great Britain War Office, General Staff, *Handbook of the Russian Army*, 6th ed., 135.
88. On the "lava," see P. Krasnov, "Lava," *Voennyi sbornik*, no. 3 (March 1911), 42–49, and Christopher Bellamy, "Heirs of Genghis Khan: The Influence of the Tartar-Mongols on the Imperial Russian and Soviet Armies," *RUSI*, 128, no. 1 (March 1983), 57–59.

89. Great Britain War Office, General Staff, *Handbook of the Russian Army*, 6th ed., 134–35.
90. Ibid., 136.
91. Ibid., 137.
92. Glavnoe upravlenie General'nogo shtaba, *Vooruzhennye sily Germanii*, II, 51; cf. Fuller, "The Russian Empire," 118.
93. Shaposhnikov, *Vospominaniia. Voenno-nauchnye trudy*, 167.
94. See, for example, N. A. Orlov, *Voennaia entsiklopediia*, XVIII, s.v. "Nravstvennyi element v voennom dele."
95. This is a theme emphasized very strongly, for instance, by A. A. Neznamov, *Voennaia entsiklopediia*, XVII, s.v. "Operatsiia (voennaia)."
96. See, for example, the critical commentary in Zhilin (ed.), *Russkaia voennaia mysl'*, 123–27.
97. See Brusilov's *Vospominaniia*, 56–60.
98. Gareev, *Obshchevoiskovye ucheniia*, 69.
99. Great Britain War Office, General Staff, *Report on Foreign Manoeuvres* [For Official Use Only] (n.p., 1907), 141.
100. Great Britain War Office, General Staff, *Report on Foreign Manoeuvres* [For Official Use Only] (n.p., 1910), 198.
101. Ibid., 200.
102. Great Britain War Office, General Staff, *Report on Foreign Manoevres* (1907), 142–44.
103. Great Britain War Office, General Staff, *Report on Foreign Manoeuvres* [For Official Use Only] (n.p., 1909), 190.
104. Ibid.
105. Great Britain War Office, General Staff, *Report on Foreign Manoeuvres* (1907), 143.
106. Great Britain War Office, General Staff, *Report on Foreign Manoeuvres* [For Official Use Only] (n.p., 1908), 153.
107. Great Britain War Office, General Staff, *Report on Foreign Manoeuvres* (1910), 204.
108. Great Britain War Office, General Staff, *Report on Foreign Manoeuvres in 1912* [Confidential] (n.p., 1912), 98, 106.
109. Ibid., 98, 100.
110. Ibid., 101.
111. Ibid., 105–6.
112. Great Britain War Office, General Staff, *Report on Foreign Manoeuvres in 1913* [Confidential] (n.p., 1913), 78.
113. Ibid., 80–82.
114. Ibid., 82.
115. Ibid.; Great Britain War Office, General Staff, *Report on Foreign Manoeuvres in 1912*, 110–11.
116. Norman Stone, "Organising an Economy for War: The Russian Shell Shortage, 1914–1917," in Geoffrey Best and Andrew Wheatcroft (eds.), *War, Economy and the Military Mind* (London, 1976), 109.
117. Quoted in ibid., 108.

Conclusions

1. See e:p. the comparison inherent in Keep, *Soldiers of the Tsar*, chaps. 14 and 15.
2. Cf. Shelford Bidwell and Dominick Graham, *Fire-Power: British Army Weapons and Theories of War 1904–1945* (London, 1982), 10–12.

3. "A Conference on the Soviet Military and the Future," Draft Proceedings, Texas A&M University, College Station, 1983, III-41.

4. Paul Kennedy, *Strategy and Diplomacy 1870–1945* (London, 1984), 51–52.

5. Jonathan M. House, "Towards Combined Arms Warfare," Research Survey No. 2, Combat Studies Institute, U.S. Army Command and General Staff College, 14–17.

6. Great Britain War Office, General Staff, *Notes upon Russian Artillery Tactics in the War of 1904–1905 with Conclusions Drawn Therefrom*, 7; on problems of mass, dispersion, and battlefield lethality, see James J. Schneider, "The Theory of the Empty Battlefield," *RUSI*, 132, no. 3 (September 1987), 37–42.

7. Gerua, *Posle voiny: o nashei armii*, 28–29, 35.

8. A definition of operational art appears in Svechin, *Strategiia*, 18.

9. On this issue see Paddy Griffith, *Forward into Battle* (Chichester, Sussex, 1981), 69–72.

10. Menning, "The Deep Strike in Russian and Soviet Military History," 19–21.

11. A. G. Kavtaradze, *Voennye spetsialisty na sluzhbe Respubliki Sovetov 1917–1920 gg.* (M, 1988), 224, 247.

Bibliography

Unpublished Documents

Bazarov, P. A. "Mneniia inostrannykh voennykh agentov po voprosam, kasaiushchimsia nastoiashchei kampanii." 2 December 1904. Pavel A. Bazarov Papers, folder 1–3.
Bazarov, Pavel Aleksandrovich. Papers (in Russian), 1904–5. Hoover Institution Archives.
Domanenko, General. "Sluzhba general'nogo shtaba v divisii i korpusie." N.d. Hoover Institution Archives.
Golovin, N. N. Papers (mostly in Russian), 1912–43. Boxes 11–13, Hoover Institution Archives.
———. "Sociology of War." Typescript (in English), 1938. N. N. Golovin Papers, Box 11.
———. "Strategiia v trudakh voennykh klassikov." N.d. N. N. Golovin Papers, Boxes 11–12.
"Japanese Army, Information on, Gathered by the Russian Military Forces" (in Russian). 1904. Pavel A. Bazarov Papers, folder 1–8.
Mylov, S. "Iz opyta tekushchei voiny." N.d. Pavel A. Bazarov Papers, folder 1–13.
Nikol'skii, E. A. "Sluzhba v Glavnom shtabe i v Glavnom upravlenii General'nogo shtaba (1903–1908 gg.)." 1934. E. A. Nikol'skii Memoirs.
Nikol'skii, Evgenii Aleksandrovich. Memoirs (in Russian), 1934. Hoover Institution Archives.
Stessel, General-Ad"iutant. "Glavneishie prakticheskie vyvody iz deistvii voisk protiv Iapontsev." 20 August 1904. Pavel A. Bazarov Papers, folder 1–13.
Zaionchkovskii, A. M. "Russia's Preparation for the World War: International Relations." Translation by Mary Ada Underhill and Lewis K. Underhill. U.S. Army Command and General Staff College Library.

Published Documents

Beskrovnyi, L. G., ed. *Russkaia voenno-teoreticheskaia mysl' XIX i nachala XX vekov.* Moscow, 1960.
Center for Strategic Studies, Texas A&M University. "A Conference on the Soviet Military and the Future." Draft Transcript of Conference Proceedings. College Station, Tex., 1983.
Chenevix-Trench, [Colonel] F. *Report on the Russian Army.* [Confidential]. 16 October 1886. N.p.
"Dnevnik A. N. Kuropatkina," *Krasnyi arkhiv.* II (1922), 9–112.
Dnevnik Vysochaishogo prebyvaniia Imperatora Aleksandra II-go za Dunaem v 1877 godu. St. Petersburg, 1878.
Erickson, John. "The Russian Imperial/Soviet General Staff."Center for Stategic Technology, Texas A&M University, College Station Papers 3. College Station, Tex., 1981.
Great Britain, General Staff, War Office. *Handbook of the Russian Army.* [For Official Use Only]. 3rd, 4th, 5th, and 6th eds. London, 1905, 1908, and 1914.
———. *Notes upon Company and Battalion Tactics and the Employment of Artillery in Battle. Based upon the Russo-Japanese War of 1904–5.* [For Official Use Only]. London, 1907.

310 / Bibliography

———. *Notes upon Russian Artillery Tactics in the War of 1904–5 with Conclusions Drawn Therefrom*. [For Official Use Only]. London, 1907.
———. *Reorganisation of the Russian Head-Quarters Staff*. [Confidential]. N.p., 1904.
———. *Report on Changes in Foreign Armies during 1908, 1909, and 1910*. [Confidential]. N.p., n.d.
———. *Report on Foreign Manoeuvres*. [For Official Use Only, Confidential]. N.p., 1907, 1908, 1909, 1910, 1911, 1912, 1913.
———. *Some Tactical Notes on the Russo-Japanese War*. [Confidential]. London, 1906.
———. *Twenty Years of Russian Army Reform and the Present Distribution of the Russian Land Forces*. [Confidential]. London, 1893.
Great Britain, India Division of the Chief of Staff, *Lessons of the Russo-Japanese War*. [Confidential]. Simla, 1906.
Great Britain, Intelligence Branch (Division), War Office. *Affairs in Turkestan*. [Secret]. London, 1886.
———. *Analysis of General Kuropatkin's Scheme for the Invasion of India*. [Secret]. London, 1886.
———. *Distribution of the Russian Military Forces in Asia*. [Confidential]. London, 1903.
———. *Foreign Manoeuvres, 1893, 1894, 1895, 1899, 1900, 1901*. [Confidential]. London, 1894, 1895, 1896, 1900, 1901, 1902.
———. *Handbook of the Military Forces of Russia*. 1st ed. London, 1894. 2d ed. London, 1898.
———. *Handbook of the Russian Army*. [For Official Use Only]. 3rd ed. London, 1902.
———. *Handbook of the Russian Troops in Asia*. London, 1890.
———. *Notes en Route from Srinagar to St. Petersburg (Overland) by Captain H. Picot*. London, 1892.
———. *Notes on the Siberian Railway between Cheliabinsk and Lake Baikal Made between the 28th September and the 15th October 1896 by Lieut.-Colonel G. F. Browne*. [Confidential]. London, 1897.
———. *Reports on Changes in Various Foreign Armies during the Year 1899, Together with Their Peace and War Strengths*. [Confidential]. London, 1900.
———. *Russia's Power to Concentrate Troops in Central Asia*. [Secret]. London, 1888.
———. *System of Training of Staff Officers in Foreign Armies. Russia*. [Confidential]. London, 1901.
———. *Twenty Years of Russian Army Reform and the Present Distribution of the Russian Land Forces*. [Confidential]. London, 1893.
———. *War Office Systems of Foreign Countries and India*. London, 1901.
Great Britain, War Office. *The Armed Strength of Russia*. Trans. of *Die Wehrkraft Russlands*, published in Vienna by the Austrian War Department. London, 1873.
———. *The Armed Strength of Russia*. [Official Copy]. London, 1882, 1886.
———. *The Armed Strength of the Russian Empire*. [Confidential]. 2 vols. London, 1894.
House, Jonathan M. "Towards Combined Arms Warfare." Research Survey No. 2, Combat Studies Institute, U.S. Army Command and General Staff College. Fort Leavenworth, Kans., n.d.
Konspekt strategii sostavlen ofitserami starshogo klassa Nikolaevskoi Akademii General'nogo Shtaba. St. Petersburg, 1887.
Lindsay, Captain A. B. *Russian Army List (Containing Composition and Stations of the Russian Army)*. 3rd ed. Simla, 1904.
Nastavlenie dlia deistvii pekhoty v boiu. vysochaishe utverzhdeno 27 fevralia 1914 g. St. Petersburg, 1914.

Nastavlenie dlia vedeniia boia pekhotoiu (proekt). St. Petersburg, 1910.
Osoboe pribavlenie k opisaniiu russko-turetskoi voiny 1877–78 g.g. na Balkanskom poluostrove. 4 vols. St. Petersburg, 1909–11.
Reglement für die taktische Ausbildung der russischen Fusstruppen und Instruction für das Verhalten der Campanie und des Battalions in Gesechte. Teschen, 1883.
Rieber, Alfred J., ed. *The Politics of Autocracy: Letters of Alexander II to Prince A. I. Bariatinskii, 1857–1864.* Paris and the Hague, 1966.
Russische Felddienst-Vorschrift I. Teil. Dienst der Truppen Erschienen Juni 1899. Vienna, 1899.
Sbornik materialov po russko-turetskoi voine. 97 vols. St. Petersburg, 1898–1911.
[Skobelev, General-Ad'iutant M. D.]. *Mnenine Komandira 4-go Armeiskogo korpusa general-ad"iutanta Skobeleva ob organizatsii mestnogo voennogo upravleniia i o korpusakh.* N.p., n.d.
"Ustav o voinskoi povinnosti." *Voennyi sbornik,* 95, no. 2 (February 1874), 107–50.
Ustav polevoi sluzhby. St. Petersburg, 1909.
Ustav polevoi sluzhby. Vysochaishe utverzhden 27 aprelia 1912 g. Moscow, 1916.
Verordnungen über den Dienst im Felde für die Russiche Armee vom Jahre 1881. Hannover, 1884.
Voinskie ustavy dlia pekhoty. Warsaw, 1902.
Voinskii ustav o sluzhbe v uchebnykh lagerakh i v pokhodakh v mirnoe vremia pekhoty, artillerii i kavalerii. St. Petersburg, 1871.
Voinskii ustav o stroevoi pekhotnoi sluzhbe. Pt. 2. St. Petersburg, 1865.
Voinskii ustav peshogo stroia kavalerii. St. Petersburg, 1884.
U.S. War Department, Adjutant General's Office. *Extracts from the New Drill Regulations of the Russian Cavalry (Edition of 1896).* Trans. First Lieut. L. C. Scherer. Washington, D.C., 1898.
Zaionchkovskii, P. A., ed. *Dnevnik D. A. Miliutina.* 4 vols. Moscow, 1947–50.

Theses, Dissertations, and Manuscripts

Baumann, Robert Fred. "The Debates over Universal Military Service in Russia 1870–1874." Ph.D. diss., Yale University, 1982.
Bellamy, C. D. "Sukhomlinov's Army Reforms 1908–1915." M.A. essay, King's College, University of London, 1978.
Brooks, Edwin Willis. "D. A. Miliutin: Life and Activity to 1856." Ph.D. diss., Stanford University, 1970.
———. "The Military and Industrialization in Reform Russia: The Railroad Connection." Manuscript, 1987.
Fuller, William C., Jr. "Strategy and Power in Russia, 1600–1914." Manuscript, 1990.
Jones, David R. "Russia's Armed Forces at War, 1914–1918: An Analysis of Military Effectiveness." Ph.D. diss., Dalhousie University, 1986.
Malanchuk, Gary Vincent. "The Training of an Army: M. I. Dragomirov and the Imperial Russian Army, 1860–1905." M.A. thesis, Miami University, 1978.
Menning, Bruce W. "Bayonets before Bullets: The Organization and Tactics of the Imperial Russian Army, 1861–1905." M.M.A.S. thesis, U.S. Army Command and General Staff College, 1984.
Moseley, Thomas Vernon. "Evolution of the American Civil War Infantry Tactics." Ph.D. diss., University of North Carolina, 1967.
Pintner, Walter M. "Russia as a Great Power, 1709–1856: Reflections on the Problem of

Relative Backwardness, with Special Reference to the Russian Army and Russian Society." Occasional Paper No. 33, Kennan Institute for Advanced Russian Studies, Wilson Center for International Scholars, Smithsonian Institution. Washington, D. C., 1978.

Van Dyke, Carl. "Culture of Soldiers: The Role of the Nicholas Academy of the General Staff in the Development of Russian Imperial Military Science, 1832–1912." M.Phil. thesis, University of Edinburgh, 1989.

———. "Russia's Military Professionals: The Development of a Unified Military Doctrine at the Nicholas Academy of the General Staff, 1837–1914." Manuscript, 1990.

Von Wahlde, Peter. "Military Thought in Imperial Russia." Ph.D. diss., Indiana University, 1966.

Wilfong, Walter Thomas. "Rebuilding the Russian Army, 1904–1914: The Question of a Comprehensive Plan for National Defense." Ph.D. diss., Indiana University, 1977.

Books

Andolenko, [General C. R.]. *Histoire de l'armée russe*. Paris, 1967.

Arsh, G. L., V. N. Vinogradov, et al., eds. *Russko-turetskaia voina 1877–1878 gg. i Balkany*. Moscow, 1978.

Ashmead-Bartlett, Ellis. *Port Arthur: The Siege and Capitulation*. 2d ed. Edinburgh, 1906.

Baikov, L. L. *Svoistva boevykh elementov i podgotovka voisk k voine i boiu*. Odessa, 1910.

Baiov, A., ed. *Russko-Iaponskaia voina v soobshcheniiakh v Nikolaevskoi Akademii General'nogo shtaba*. 2 vols. St. Petersburg, 1906.

Barskii, B. E. *Organizatsiia i upravlenie tylom*. Moscow-Leningrad, 1926.

Barsukov, Evgenii. *Russkaia artilleriia v mirovuiu voinu*. 2 vols. Moscow, 1938.

The Battle on the Scha Ho. Trans. Karl von Donat. London, 1906.

Bazhenov, Petr. *Deistviia russkoi kavalerii vo vremia Russko-turetskoi voiny 1877–78 gg. na Balkanskom poluostrove*. St. Petersburg, 1904.

Beliaev, N. I. *Russko-turetskaia voina 1877–1878 gg*. Moscow, 1956.

Belitskii, S. M. *Strategicheskie rezervy*. Moscow-Leningrad, 1930.

Bellamy, Christopher. *Red God of War*. London, 1986.

Beskrovnyi, L. G. *Armiia i flot Rossii v nachale XX v.: Ocherki voenno-ekonomicheskogo potentsiala*. Moscow, 1986.

———. *Ocherki po istochnikovedeniiu voennoi istorii Rossii*. Moscow, 1957.

———. *Ocherki voennoi istoriografii Rossii*. Moscow, 1962.

———. *Russkaia armiia i flot v XIX veke*. Moscow, 1973.

———. *Russkoe voennoe iskusstvo XIX v*. Moscow, 1974.

Best, Geoffrey, and Andrew Wheatcroft, eds. *War, Economy and the Military Mind*. London, 1976.

Beyrau, Dietrich. *Militär und Gesellschaft im vorrevolutionären Russland*. Cologne and Graz, 1984.

Bidwell, Shelford, and Dominick Graham. *Fire-power: British Army Weapons and Theories of War 1904–1945*. London, 1982.

Bogdanovich, E. V. *Gvardiia russkogo Tsaria na Sofiiskoi doroge 12 Oktiabria 1877 g*. St. Petersburg, 1879.

Bol'shie manevry v Vysochaishem prisutstvii pod gorodom Pskovskom 5–10 avgusta 1903 g. St. Petersburg, 1904.

Bonch-Bruevich, M. *Boi*. Kiev, 1909.

Bradley, Joseph. *Guns for the Tsar*. DeKalb, Ill., 1990.

Brusilov, A. A. *Moi vospominaniia*. Moscow, 1983.

Buniakovskii, V. *Taktika boia po opytu russko-iaponskoi voiny*. St. Petersburg, 1912.
———. *Taktika i tekhnika boevykh deistvii pekhoty (Kratkoe posobie nachal'nikam vsekh rodov voisk)*. St. Petersburg, 1912.
———, and I. Cherniavskii. *Pokhod, otdykh i boi po opytu tekushchei voiny*. Moscow, 1916.
Burne, A. H. *The Liao-yang Campaign*. London, 1936.
Bushnell, John. *Mutiny amid Repression: Russian Soldiers in the Revolution of 1905–1906*. Bloomington, Ind., 1985.
von Caemmerer, [Lieut.-General]. *The Development of Strategical Science during the 19th Century*. Trans. Karl von Donat. London, 1905.
Cheremisov, V., comp. *Russko-Iaponskaia voina 1904–1905 goda*. Kiev, 1907.
Curtiss, John Shelton. *The Russian Army under Nicholas I, 1825–1855*. Durham, N.C., 1965.
———. *Russia's Crimean War*. Durham, N.C., 1979.
Danilov, Iu. N. *Rossiia v mirovoi voine 1914–1915 g.g.* Berlin, 1924.
Danilov, N. A. *Lektsii po voennoi administratsii*. Leningrad, 1926.
Denikin, Anton I. *The Career of a Tsarist Officer: Memoirs, 1872–1916*. Trans. Margaret Patoski. Minneapolis, 1975.
Dobrorolski, Sergei [General]. *Die Mobilmachung der russischen Armee 1914*. Berlin, 1922.
Dragomirov, M. I. *Kurs taktiki, dlia gg. ofitserov uchebnogo pekhotnogo bataliona*. 2d ed. St. Petersburg, 1867.
———. *Sbornik original'nykh i perevodnykh statei M. Dragomirova 1858–1880*. 2 vols. St. Petersburg, 1881.
von Drygalski, A., C. von Zepelin, and C. F. Batsch. *Russland. Das Heer. Die Flotte*. Bd. 3, *Die Heere und Flotten der Gegenwart*, ed. C. von Zepelin. Berlin, 1898.
Durop, K. N., comp. *Taktika. Podrobnyi konspekt dlia voennykh i unkerskikh uchilishch*. 2 pts. St. Petersburg, 1886.
Elchaninov, A. G. *Vedenie sovremennykh voiny i boia*. St. Petersburg, 1909.
English, John A. *A Perspective on Infantry*. New York, 1981.
Epauchin [sic], [Colonel]. *Operations of General Gurko's Advance Guard in 1877*. Trans. H. Havelock. London, 1900.
Erickson, John, and E. J. Feuchtwanger, eds. *Soviet Military Power and Performance*. Hamden, Conn., 1979.
Fadeev, R. *Vooruzhennyia sily Rossii*. Moscow, 1868.
Fedorov, A. V. *Russkaia armiia v 50–70-kh godakh XIX veka*. Leningrad, 1959.
Fedorov, Vladimir G. *Vooruzhenie russkoi armii v krymskuiu kampaniiu*. St. Petersburg, 1904.
———. *Vooruzhenie russkoi armii za XIX stoletie*. St. Petersburg, 1911.
Fergusson, Thomas G. *British Military Intelligence, 1870–1914*. Frederick, Md., 1984.
Fisher, H. H., ed. *Out of My Past: The Memoirs of Count Kokovtsov*. Stanford, Calif., 1935.
Fuller, William C., Jr. *Civil-Military Conflict in Imperial Russia, 1881–1914*. Princeton, N.J., 1985.
Gaivoronskii, F. F., ed. *Evoliutsiia voennogo iskusstva: Etapy, tendentsii, printsipy*. Moscow, 1987.
Gareev, M. A. *Obshchevoiskovye ucheniia*. Moscow, 1983.
Garshin, Vsevolod. *From the Reminiscences of Private Ivanov and Other Stories*. Trans. Peter Henry, Liv Tudge, Donald Rayfield, and Philip Taylor. London, 1988.
Gazenkampf, M. *Moi dnevnik 1877–78 gg.* St. Petersburg, 1908.
Geisman, P. A., comp. *Kratkii kurs istorii voennogo iskusstva v srednie i novye veka*. 3 pts. St. Petersburg, 1893–96.
———. *Opyt izsledovaniia taktiki massovykh armii*. St. Petersburg, 1894.

314 / Bibliography

Gershel'man, Fedor. *Kavaleriia v voinakh XX veka*. St. Petersburg, 1908.
Gerua, A. V. *K poznaniiu armii*. St. Petersburg, 1907.
———. *Posle voiny: O nashei armii*. St. Petersburg, 1907.
Gerua, B. V. *Vospominaniia o moei zhizni*. 2 vols. Paris, 1969–70.
Glavnoe upravlenie General'nogo shtaba. *Vooruzhennye sily Avstro-Vengrii*. [Sekret'no]. 2 vols. St. Petersburg, 1912.
———. *Vooruzhennye sily Germanii*. [Sekret'no]. 2 vols. St. Petersburg, 1912–14.
Glinoetskii, N. P. *Istoricheskii ocherk Nikolaevskoi Akademii general'nogo shtaba*. St. Petersburg, 1882.
Golovin, N. N. *Sluzhba general'nogo shtaba*. St. Petersburg, 1912.
Golovine, N. N. *The Russian Army in the World War*. New Haven, Conn., 1931.
von der Goltz, Colmar Freiherr. *The Conduct of War*. Trans. G. F. Leverson. London, 1899.
Greene, F. V. *Report on the Russian Army and Its Campaigns in Turkey in 1877–1878*. New York, 1879.
———. *Sketches of Army Life in Russia*. New York, 1880.
Griffith, Paddy. *Forward into Battle: Fighting Tactics from Waterloo to Vietnam*. Chichester, Sussex, 1981.
Grishinskii, A. S., V. P. Nikol'skii, and N. L. Klado, eds. *Istoriia russkoi armii i flota*. 16 vols. Moscow, 1911–13.
Grulev, M. V. *Zapiski generala-evreia*. Paris, 1930.
Gudim-Levkovich, P. M. *Kurs elementarnoi taktiki*. 3 vols. St. Petersburg, 1887–90.
Gurko, V. I. *Features and Figures of the Past: Government and Opinion in the Reign of Nicholas II*. Ed. J. E. Wallace Sterling, Xenia Eudin, and H. H. Fisher. Trans. Laura Matveev. Stanford, Calif., 1939.
Hamilton, Ian. *A Staff Officer's Scrapbook*. London, 1905.
Higham, Robin, and Jacob W. Kipp, eds. *Soviet Aviation and Air Power: A Historical View*. Boulder, Colo., 1977.
Howard, Michael. *The Franco-Prussian War: The German Invasion of France, 1870–1871*. London, 1981.
———, ed. *The Theory and Practice of War*. Bloomington, Ind., 1975.
Ignat'ev, A. A. *Piat'desiat let v stroiu*. Moscow, 1988.
Ignatyev, A. A. *A Subaltern in Old Russia*. Trans. Ivor Montagu. London, 1944.
Jelavich, Barbara. *St. Petersburg and Moscow: Tsarist and Soviet Foreign Policy, 1814–1974*. Bloomington, Ind., 1974.
Kavtaradze, A. G. *Voennye spetsialisty na sluzhbe Respubliki Sovetov 1917–1920 gg*. Moscow, 1988.
Keegan, John. *The Face of Battle*. New York, 1980.
———. *Six Armies in Normandy*. New York, 1982.
Keep, John L. H. *Soldiers of the Tsar: Army and Society in Russia, 1462–1874*. Oxford, 1985.
Kennedy, Paul M. *Strategy and Diplomacy 1870–1945*. London, 1984.
———, ed. *The War Plans of the Great Powers, 1880–1914*. London, 1979.
Kersnovskii, A. A. *Filosofiia voiny*. Belgrade, 1939.
———. *Istoriia russkoi armii*. 4 pts. Belgrade, 1933–38.
Király, Béla K., and Gunther E. Rothenberg, eds. *War and Society in East Central Europe*, Brooklyn College Studies on Societies in Change No. 10. New York, 1979.
Kolenkovskii, A. *Manevrennyi period pervoi mirovoi imperialisticheskoi voiny 1914 g*. Moscow, 1940.
Krahmer, Gustav. *Geschichte der Entwicklung des russischen Heeres von der Thronbestegnung des Kaisers Nikolai I Pawlowitsch bis auf die neuste Zeit*. Leipzig, 1896–97.

Kuropatkin, A. N. *The Russian Army and the Japanese War*. Trans. A. B. Lindsay. 2 vols. New York, 1909.
Leer, G. A. *Metod voennykh nauk (strategii, taktiki i voennoi istorii)*. St. Petersburg, 1894.
———. *Opyt kritiko-istoricheskogo izsledovaniia zakonov iskusstva vedeniia voiny (Polozhitel'naia strategiia)*. St. Petersburg, 1871.
———. *Strategiia*. St. Petersburg, 1885.
———. *Zapiski strategii*. 3rd ed. St. Petersburg, 1877.
Levitskii, N. A. *Russko-Iaponskaia voina 1904–1905 gg*. 3rd ed. Moscow, 1938.
Lieven, D. C. B. *Russia and the Origins of the First World War*. New York, 1984.
Lobko, P. L. *Zapiski voennoi administratsii*. 6th ed. 3 pts. St. Petersburg, 1906.
Luvaas, Jay. *The Military Legacy of the Civil War: The European Inheritance*. Chicago, 1959.
MacKenzie, David. *The Lion of Tashkent: The Career of General M. G. Cherniaev*. Athens, Ga., 1974.
McElwee, William. *The Art of War: Waterloo to Mons*. Bloomington, Ind., 1974.
McNeal, Robert H. *Tsar and Cossack, 1855–1914*. London, 1987.
Maksheev, A. I., comp. *Istoricheskii obzor Turkestana i nastupatel'nogo dvizheniia v nego russkikh*. St. Petersburg, 1890.
Manikovskii, A. A., comp. *Boevoe snabzhenie russkoi armii v voinu 1914–1918 gg*. 3 pts. Moscow, 1920–23.
Martynov, E. I. *Iz pechal'nogo opyta Russko-Iaponskoi voiny*. 2d ed. St. Petersburg, 1907.
———. *Strategiia v epokhu Napoleona I i v nashe vremia*. St. Petersburg, 1894.
Mavrodin, V. A., and Val. V. Mavrodin. *Iz istorii otechestvennogo oruzhiia (russkaia vintovka)*. Leningrad, 1981.
May, Ernest R., ed. *Knowing One's Enemies: Intelligence Assessment before the Two World Wars*. Princeton, N.J., 1984.
Mayzel, Matitiahu. *Generals and Revolutionaries. The Russian General Staff during the Revolution: A Study in the Transformation of Military Elite*. Osnabruck, 1979.
Melikov, V. A. *Problema strategicheskogo razvertyvaniia po opytu mirovoi i grazhdanskoi voin*. Moscow, 1935.
Meshcheriakov, G. P. *Russkaia voennaia mysl' v XIX v*. Moscow, 1973.
Mikhnevich. N. P., comp. *Strategiia*. 3rd ed. St. Petersburg, 1911.
———. *Vliianie noveishikh tekhnicheskikh izobretenii na taktiku voisk*. St. Petersburg, 1893.
Miliutin, D. A. *Vospominaniia*. Reprint ed. Newtonville, Mass., 1979.
Miller, Forrestt A. *Dmitrii Miliutin and the Reform Era in Russia*. Nashville, Tenn., 1968.
Neznamov, A. A. *Iz opyta russko-iaponskoi voiny (Zametki ofitsera General'nogo shtaba)*. 2d ed. St. Petersburg, 1906.
———. *Osnovy sovremennoi strategii*. Moscow, 1919.
———. *Sovremennaia voina*. 2d ed. St. Petersburg, 1912.
———. *Trebovaniia, kotorye pred'iavliaet sovremennyi boi k podgotovke (obucheniiu) nachal'nikov i mass*. St. Petersburg, 1909.
Nojine, E. K. *The Truth about Port Arthur*. Trans. A. B. Lindsay. London, 1908.
Norregaard, B. W. *The Great Siege: The Investment and Fall of Port Arthur*. London, 1906.
Obzor deiatel'nosti voennogo ministerstva v tsarstvovanie Imperatora Aleksandra III-ogo 1881–1894. St. Petersburg, 1903.
Ocherk deiatel'nosti voennogo ministerstva za istekshee desiatiletie blagopoluchnogo tsarstvovaniia Gosudaria Imperatora Aleksandra Aleksandrovicha 1881–1890. St. Petersburg, 1892.
Opisanie russko-turetskoi voiny 1877–78 gg. na Balkanskom poluostrove. 9 vols. in 14 books. St. Petersburg, 1901–13.

Pankov, D. V., comp. *Razvitie taktiki russkoi armii.* Moscow, 1957.
Paret, Peter, ed. *Makers of Modern Strategy from Machiavelli to the Nuclear Age.* Princeton, N.J., 1986.
Peredel'skii, G. E., ed. *Otechestvennaia artilleriia. 600 let.* Moscow, 1986.
Pevnev, A. *Voiskovaia konnitsa i ee boevoe ispol'zovanie.* Moscow, 1926.
Pierce, Richard A. *Russian Central Asia.* Berkeley, Calif., 1960.
Problemy istorii feodal'noi Rossii. Sbornik statei k 60-letiiu Prof. V. V. Mavrodina. Leningrad, 1971.
Prokof'ev, E. A. *Voennye vzgliady dekabristov.* Moscow, 1953.
Puzyrevskii, A. K. *Desiat' let nazad. Voina 1877–78 gg.* St. Petersburg, 1887.
Reddel, Carl W., ed. *Transformation in Russian and Soviet Military History.* Washington, D.C., 1990.
Rediger, A., comp. *Komplektovanie i ustroistvo vooruzhennoi sily.* 4th ed. 2 vols. St.Petersburg, 1913.
———. *Uchebnye zapiski po voennoi administratsii.* St. Petersburg, 1888.
Rostunov, I. I., ed. *Istoriia pervoi mirovoi voiny 1914–1918.* 2 vols. Moscow, 1975.
———, ed. *Istoriia russko-iaponskoi voiny 1904–1905 gg.* Moscow, 1977.
———. *Russkii front pervoi mirovoi voiny.* Moscow, 1976.
———., ed. *Russko-turetskaia voina 1877–1878.* Moscow, 1977.
Russes et Turcs: La Guerre d'Orient. 2 vols. Paris, 1877–78.
Die Russische Armee. Berlin, 1912.
Russko-iaponskaia voina 1904–1905 gg. 9 vols. in 16 books. St. Petersburg, 1910–13.
Sakurai, Tadayoshi. *Human Bullets (Niku-Dan): A Soldier's Story of Port Arthur.* Trans. Masujiro Honda and Alice Bacon. Tokyo, 1907.
Sbornik voennykh razskazov sostavlennykh ofitserami-uchastnikami voiny 1877–1878. 3 vols. St. Petersburg, 1878.
Screen, J. E. O. *The Helsinki Yunker School 1846–1879.* Helsinki, 1986.
Shaposhnikov, B. M. *Vospominaniia. Voenno-nauchnye trudy.* Moscow, 1974.
Shavel'skii, G. *Vospominaniia poslednogo protopresvitera russkoi armii i flota.* 2 vols. New York, 1954.
Shatsillo, K. F. *Rossiia pered pervoi mirovoi voinoi (Vooruzhennye sily tsarisma v 1905–1914 gg.).* Moscow, 1974.
———. *Russkii imperializm i razvitie flota.* Moscow, 1968.
Showalter, Dennis E. *Railroads and Rifles: Soldiers, Technology, and the Unification of Germany.* Hamden, Conn., 1975.
Shvarts, N. *Ustroistvo voennogo upravleniia.* 2d ed. Moscow-Leningrad, 1927.
Skazaniia inostrantsev o Russkoi Armii v voinu 1904–1905 g.g. St. Petersburg, 1912.
Skalon, D. A., ed. *Stoletie Voennogo Ministerstva.* 48 pts. in 13 vols. St. Petersburg, 1902–13.
Sorokin, A. I. *Oborona Port-Artura.* Moscow, 1954.
Snyder, Jack. *The Ideology of the Offensive.* Ithaca, N.Y., 1984.
Stepanov, M. P. *Rushchukskii otriad v 1877–1878 gg. Istoricheskii ocherk.* St. Petersburg, 1888.
Stone, Norman. *The Eastern Front, 1914–1917.* London, 1975.
Strokov, A. A. *Istoriia voennogo iskusstva.* Moscow, 1965.
———. *Vooruzhennye sily i voennoe iskusstvo v pervoi mirovoi voine.* Moscow, 1974.
Suchomlinow, W. A. *Erinnerungen.* Berlin, 1924.
Sukhomlinov, V. A. *Vospominaniia.* Berlin, 1924.
Svechin, A. A. *Evoliutsiia voennogo iskusstva s drevneishikh vremen do nashikh dnei.* 2 vols. Moscow-Leningrad, 1927–28.
———. *Russko-iaponskaia voina 1904–1905 gg.* Oranienbaum, 1910.

———. *Strategiia*. 2d ed. Moscow, 1927.
———. *Strategiia v trudakh voennykh klassikov*. 2 vols. Moscow, 1926.
von Tettau, Freiherr. *Achtzehn Monate mit Russlands Heeren in der Manchschurei*. 2 vols. Berlin, 1907.
Thaden, Edward C. *Russia since 1801*. New York, 1971.
Tretiakov, N. A. *My Experiences at Nan Shan and Port Arthur with the Fifth East Siberian Rifles*. Trans. A. C. Alford. London, 1911.
Ushakov, K. P. *Podgotovka voennykh soobshchenii Rossii k mirovoi voine*. Moscow-Leningrad, 1928.
Verkhovskii, A. I. *Ocherk po istorii voennogo iskusstva v Rossii XVIII i XIX v.* Moscow, 1921.
Vinogradov, V. I. *Russko-turetskaia voina 1877–1878 gg. i osvobozhdenie Bolgarii*. Moscow, 1978.
Vitte, S. Iu. *Vospominaniia*. 3 vols. Moscow, 1960.
Voenno-morskaia strategicheskaia igra 1902 g. na temu "Zaniatie russkami silami verkhnogo Bosfora" [Ves'ma sekretno]. St. Petersburg, 1902.
Winton, Harold R. *To Change an Army*. Lawrence, Kans., 1988.
Wirtschafter, Elise Kimerling. *From Serf to Russian Soldier*. Princeton, N.J., 1990.
Wood, Oliver Ellsworth. *The Yalu to Port Arthur*. Kansas City, Mo., 1905.
Zaionchkovskii, A. M. *Lektsii po strategii, chitannye na voenno-akademicheskikh kursakh vysshego komsostava v Voennoi akademii RKKA v 1922–1923 gg.* 2 pts. Moscow, 1923.
———. *Mirovaia voina 1914–1918*. 2d ed. Moscow, 1931.
———. *Nastupatel'nyi boi po opytu deistvii Generala Skobeleva v srazheniiakh pod Lovchei, Plevnoi (27 i 30 avgusta) i Sheinovo*. St. Petersburg, 1893.
———. *Podgotovka Rossii k imperialisticheskoi voine*. Moscow, 1926.
Zaionchkovskii, P. A. *Rossiiskoe samoderzhavie v kontse XIX stoletiia*. Moscow, 1970.
———. *The Russian Autocracy in Crisis, 1878–1882*. Trans. and ed. G. M. Hamburg. Gulf Breeze, Fla., 1979.
———. *Samoderzhavie i russkaia armiia na rubezhe XIX-XX stoletii*. Moscow, 1973.
———, ed. *Spravochnik po istorii dorevoliutsionnoi Rossii*. 2d ed. Moscow, 1978.
———. *Voennye reformy 1860–1870 godov v Rossii*. Moscow, 1952.
Zander, G., comp. *Takticheskie primery dlia voennykh i iunkerskikh uchilishch*. 2d ed. Odessa, 1897.
Zheleznykh, V. I., ed. *Voenno-inzhenernoe iskusstvo in inzhenernye voiska russkoi armii*. Moscow, 1958.
Zhilin, P. A. *Istoriia voennogo iskusstva*. Moscow, 1986.
———, ed. *Russkaia voennaia mysl' konets XIX–nachalo XX v.* Moscow, 1982.
Zhuk, A. B. *Revol'very i pistolety*. Moscow, 1983.
———. *Vintovki i avtomaty*. Moscow, 1987.
Zolotarev, V. A. *Rossiia i Turtsiia. Voina 1877–1978 gg.* Moscow, 1983.
———. *Russko-turetskaia voina 1877–78 gg. v otechestvennoi istoriografii*. Moscow, 1978.

Articles

Adaridi, A. "Liubiteliam doktriny." *Russkii invalid*, no. 14 (17 January 1912), 3.
Adasovskii, [*Polkovnik*]. "Vliianie malokalibernogo oruzhiia i bezdymnogo porokha na deistviia artillerii v boiu." *Voennye besedy ispolnennyia v shtabe voisk Gvardii i Peterburgskogo voennogo okruga v 1892–1893 gg.* XIII (1893), 109 33.
Ageev, A. "Iz dnevnika N. P. Mikhnevicha." *Voenno-istoricheskii zhurnal*, no. 5 (May 1976), 69–75.

———. "Ofitsery russkogo general'nogo shtaba ob opyte russko-iaponskoi voiny 1904–1905 gg." *Voenno-istoricheskii zhurnal*, no. 8 (August 1975), 99–104.
———. "Voenno-teoreticheskoe nasledie A. A. Neznamova." *Voenno-istoricheskii zhurnal*, no. 11 (November 1983), 84–89.
———. "Voenno-teoreticheskie vzgliady N. P. Mikhnevicha." *Voenno-istoricheskii zhurnal*, no. 1 (January 1975), 90–95.
Askew, William C. "Russian Military Strength on the Eve of the Franco-Prussian War." *Slavonic and East European Review*, 30 (1951), 185–205.
Avdeev, V. A. "Voenno-istoricheskie issledovaniia v akademii general'nogo shtaba russkoi armii." *Voenno-istoricheskii zhurnal*, no. 12 (December 1987), 77–80.
———. "Voenno-istoricheskie issledovaniia v russkoi armii." *Voenno-istoricheskii zhurnal*, no. 3 (March 1986), 81–84.
B—v, A. "K 75-letiiu Nikolaevskoi akademii general'nogo shtaba." *Voennyi sbornik*, no. 12 (December 1907), 213–34.
Barbasov, A. "Russkii voennyi deiatel' N. N. Obruchev." *Voenno-istoricheskii zhurnal*, no. 8 (August 1973), 100–105.
Baumann, Robert F. "Universal Service and Russia's Imperial Dilemma." *War & Society*, 4, no. 2 (September 1986), 31–49.
———. "Subject Nationalities in the Military Service of Imperial Russia: The Case of the Bashkirs." *Slavic Review*, 46, no. 3/4 (Fall/Winter 1987), 489–502.
Baumgarten, A. "Artileriia v polevom boiu." *Voennyi sbornik*, 194, no. 8 (August 1890), 310–48.
———. "Artileriiskii zametki." *Voennyi sbornik*, no. 7 (July 1898), 122–46, no. 8 (August 1898), 442–60.
———. "K voprosu ob artileriiskikh massakh." *Voennyi sbornik*, 179, no. 2 (February 1888), 285–310.
Bellamy, Christopher D. "Antecedents of the Modern Soviet Operational Manoeuvre Group (OMG)." *RUSI*, 129, no. 3 (September 1984), 50–58.
———. "Heirs of Genghis Khan: The Influence of the Tartar-Mongols on the Imperial Russian and Soviet Armies." *RUSI*, 128, no. 1 (March 1983), 52–60.
———. "The Russian Artillery and the Origins of Indirect Fire." *Army Quarterly and Defence Journal*, 112, no. 2 (April 1982), 211–22, no. 3 (July 1982), 330–37.
———. "Seventy Years On: Similarities between the Modern Soviet Army ands Its Tsarist Predecessor." *RUSI*, 124, no. 3 (September 1979), 29–38.
Bezrukov, A. "Podderzhka sviazi mezhdu otdel'nymi chastiami na voine." *Voennye besedy ispolnenyia v shtabe voisk Gvardii i Peterburgskogo voennogo okruga za 1893–1894 gg.*, XV (1894), 109–44.
Bil'derling, A. "Dmitrii Alekseevich Miliutin." *Voennyi sbornik*, no. 2 (February 1912), 3–16.
Bogdanovich, M. I. (comp.). "Istoricheskii ocherk deiatel'nosti voennogo upravleniia v pervoe dvadtsati-piati-letie tsarstvovaniia Gosudaria Imperatora Aleksandra Nikolaevicha (1855–1880 gg.)." *Voennyi sbornik*, 132, no. 3 (March 1880), 5–58, no. 3 (April 1880), 257–82, 133, no. 5 (May 1880), 5–34.
Boldyreev, V. E., and S. A. Drutskoi, "Plevna." *Voennaia entsiklopediia*, XVIII, 455–69.
Bol'shakov, I. "Russkaia razvedka v pervoi mirovoi voine 1914–1918 godov." *Voenno-istoricheskii zhurnal*, no. 5 (May 1964), 44–48.
Bonch-Bruevich, M. "M. I. Dragomirov. Osnovy vositaniia i obrazovaniia voisk." *Voenno-istoricheskii zhurnal*, no. 3 (March 1973), 75–80.
Bond, Brian. "Doctrine and Training in the British Cavalry, 1870–1914." In Michael Howard (ed.), *The Theory and Practice of War*, 95–125.

Boretskii, A. "Ustavy i zhizn' (K polemike o edinoi voennoi doktrine)." *Russkii invalid*, no. 15 (19 January 1912), 4–5.
Borisov, D. S. "Nauchnyi podvig voennogo inzhenera." In V. I. Zheleznykh (ed.), *Voenno-inzhenernoe iskusstvo i inzhenernye voiska russkoi armii*, 173–84.
Borisov, V. "Znachenie Klauzevitsa v sovremennoi voennoi nauke." *Voennyi sbornik*, 244, no. 12 (December 1898), 241–47.
Borodkin, M. "Graf D. A. Miliutin v otzyvakh ego sovremennikov." *Voennyi sbornik*, no. 5 (May 1912), 1–16, no. 6 (June 1912), 1–20.
Brooks, E. Willis. "The Improbable Connection: D. A. Miljutin and N. G. Cernysevskij." *Jahrbücher für Geschichte Osteuropas*, no. 37, H. 1 (1989), 21–43.
———. "Reform in the Russian Army, 1856–1861." *Slavic Review*, 43, no. 1 (Spring 1984), 63–82.
Bushnell, John. "Peasants in Uniform: The Tsarist Army as a Peasant Society." *Journal of Social History*, 13, no. 4 (Summer 1980), 565–76.
———. "The Revolution of 1905–1906 in the Army: The Incidence and Impact of Mutiny." *Russian History*, 12, no. 1 (Spring 1985), 71–94.
———. "The Tsarist Officer Corps, 1881–1914: Customs, Duties, Inefficiency." *American Historical Review*, 86, no. 3 (June 1981), 753–80.
Chetyrkin, N. "Doktrina v ustave polevoi sluzhby." *Russkii invalid*, no. 41 (22 February 1912), 2–3.
———. "Ustav i doktrina." *Russkii invalid*, no. 260 (4 December 1911), 6.
Chichagov, M. "Bol'shoe kavaleriiskoe delo v amerikanskuiu voinu 1863 goda. Srazhenie pri Brendi." *Voennyi sbornik*, 159 (1884), 216–32.
John S. Curtiss. "The Army of Nicholas I: Its Role and Character." *American Historical Review*, 63, no. 4 (July 1958), 880–89.
D—v, N. V. "Nashi boevye instruktsii i nastavleniia." *Voennyi sbornik*, 241, no. 5 (May 1898), 81–102, no. 6 (June 1898), 336–57, no. 7 (July 1898), 81–107.
Dalinskii. "Edinenie i antagonizm v armii." *Russkii invalid*, no. 270 (17 December 1911), 5.
Degtiarev, [*Kapitan*]. "Boi roty in bataliona i primenenie artillerii v boiu po opytu russko-iaponskoi voiny." *Voennyi sbornik*, no. 9 (September 1906), 69–82.
Derevianko, I. V. "Mozg armii." *Voenno-istoricheskii zhurnal*, no. 10 (October 1989), 79–80.
———. "Russkaia agenturnaia razvedka v 1902–1905 gg." *Voenno-istoricheskii zhurnal*, no. 5 (May 1989), 76–78.
D'iakov, V. A. "O razvitii russkoi voenno-istoricheskoi mysli v poslednei chetverti XIX veka." *Voenno-istoricheskii zhurnal*, no. 5 (May 1959), 60–72.
Domanevskii, V. N. "Edinstvo voennoi doktriny." *Russkii invalid*, no. 274 (22 December 1911), 4.
———. "Edinstvo voennoi doktriny i polevoi ustav." *Russkii invalid*, no. 5 (6 January 1912), 5.
———. "K sporu o zlobdnevnom voprose." *Russkii invalid*, no. 23 (28 January 1912), 4–5.
———. "Samodeiatel'nost' nachal'nikov." *Russkii invalid*, no. 48 (1 March 1912), 5.
———. "Sluzhba general'nogo shtaba." *Russkii invalid*, no. 266 (13 December 1911), 3–4.
———. "Sluzhba general'nogo shtaba v shtabe divisii." *Russkii invalid*, no. 11 (14 January 1912), 3.
———. "Sluzhba general'nogo shtaba v shtabe korpusa." *Russkii invalid*, no. 31 (10 February 1912), 4–5.

Dragomirov, M. I. "Konspekt lektsii po taktike." *Voennyi sbornik*, no. 9 (September 1912), 1–28.

———. "Po povodu nekotorykh statei vyzvannykh poslednymi dvumia kampaniiami." *Voennyi sbornik*, 88, no. 12 (December 1872), 253–74, 89, no. 1 (January 1873), 89–106, no. 3 (March 1873), no. 4 (April 1873), 267–76.

Dragun. "Zametki o konnitse." *Voennyi sbornik*, 179, no. 2 (February 1888), 281–94.

Druzhinin, K. "Issledovanie strategicheskoi deiatel'nosti germanskoi kavalerii v kampaniiu 1870 goda, do srazheniia Mars-la-Ture vkliuchitel'no." *Voennye besedy ispolnennyia v shtabe voisk Gvardii i Peterburgskogo voennogo okruga v 1896–97 gg.*, XIX (1897), 5–22.

———. "Russko-turetskaia voina 1877–78 gg. na Balkanskom teatre." In A. S. Grishinskii et al. (eds.), *Istoriia russkoi armii i flota*, XI, 23–175.

———. "Voennoe delo v Rossii pri vstuplenii na prestol Imperatora Aleksandra II i pered voinoi 1877–78 gg." In A. S. Grishinskii et al. (eds.), *Istoriia russkoi armii i flota*, XI, 5–22.

Durop, K. "Takticheskoe obrazovanie armii v mirnoe vremia." *Voennyi sbornik*, 83, no. 2 (February 1872), 229–43.

E. U. "Uroki iaponskoi voiny." *Voennyi sbornik*, no. 8 (August 1908), 77–98.

Erickson, John. "The Soviet Military System: Doctrine, Technology and 'Style.'" In John Erickson and E. J. Feuchtwanger (eds.), *Soviet Military Power and Performance*, 18–43.

Fon-der-Khoven, A. I. "O magazinnykh ruzh'iakh." *Voennye besedy ispolnennyia v shtabe voisk Gvardii i Peterburgskogo voennogo okruga v 1885–1887 g.g.*, III (1887), 155–237.

Frolov, V. "Geroicheskaia oborona Port-Artura." *Voenno- istoricheskii zhurnal*, no. 9 (September 1984), 67–75.

———. "Russko-iaponskaia voina 1904–1905 gg. (Nekotorye voprosy voennogo iskusstva)." *Voenno-istoricheskii zhurnal*, no. 2 (February 1974), 83–90.

Fuller, William C., Jr. "The Russian Empire." In Ernest R. May (ed.), *Knowing One's Enemies: Intelligence Assessment before the Two World Wars*, 98–126.

G. "Popytka k vyiasneniiu veroiatneishego znacheniia malodymnogo porokha polevom boiu." *Voennyi sbornik*, 198, no. 4 (April 1891), 212–26.

Geisman, P. A. "Podgotovka k voine v shirokom smysle." *Voennye besedy ispolnennyia v shtabe voisk Gvardii i Peterburgskogo voennogo okruga v 1894–1895 godakh*, XVI (1895), 97–160.

Glinoetskii, N. "Zametki o polevykh poezdkakh ofitserov general'nogo shtaba." *Voennyi sbornik*, 87, no. 10 (October 1872), 249–67.

Gololobov, A. "Hasha akademiia general'nogo shtaba." *Voennyi sbornik*, 79, no. 5 (May 1871), 61–139.

Grigor'ev, V. "Doktrina—nauka, uchenie." *Russkii invalid*, no. 36 (16 February 1912), 4–5.

Gulevich, A. A. "Voina i narodnoe khoziaistvo." *Voennyi sbornik*, 239, no. 1 (January 1898), 60–75, 241, no.5 (May 1898), 17–57, no. 6 (June 1898), 257–310.

Ianushkevich, N. N. "Sostoianie vooruzhennykh sil Rossii so vremeni vvedeniia obshcheobiazatel'noi voinskoi povinnosti." In A. S. Grishinskii et al. (eds.), *Istoriia russkoi armii i flota*, XIII, 3–32.

Irvine, Dallas D. "The Origin of Capital Staffs." *Journal of Modern History*, 10, no. 2 (June 1938), 161–79.

Iunakov, N. "Nochnoi boi. Takticheskoe izsledovanie." *Voennye besedy ispolnennyia v shtabe voisk Gvardii i Peterburgskogo voennogo okruga v 1897–98 gg.*, XX, 67–121.

Jones, David R. "Administrative System and Policy-Making Process." In *The Military-Naval Encyclopedia of Russia and the Soviet Union*, II, 34–169.

———. "The Beginnings of Russian Air Power, 1907–1922." In Robin Higham and Jacob W. Kipp (eds.), *Soviet Aviation and Air Power: A Historical View*, 15–33.

———. "Advanced Guard (Avangard)." In *The Military-Naval Encyclopedia of Russia and the Soviet Union*, IV, 54–190.
Kaigorodov, [Kapitan]. "Artilleriiskiia massy na poliakh srazhenii." *Voennye besedy ispolnennyia v shtabe voisk Gvardii i Peterburgskogo voennogo okruga v 1887–1888 g.g.*, V (1888), 3–31.
Kartsov, P. "Traianov pereval." *Russkii vestnik*, 194, no. 1 (January 1888), 68–105.
Kavtaradze, A. G. "Iz istorii russkogo general'nogo shtaba." *Voenno-istoricheskii zhurnal*, no. 12 (December 1971), 75–80, no. 7 (July 1972), 87–92.
———. "Iz istorii russkogo general'nogo shtaba (1909–iiul' 1914 gg.)." *Voenno-istoricheskii zhurnal*, no. 12 (December 1974), 80–86.
———. "Voiskovoe upravlenie general'nogo shtaba russkoi armii." *Voenno-istoricheskii zhurnal*, no. 6 (June 1978), 77–82.
Kazakov, M. "Ispol'zovanie reservov v russko-iaponskoi voine 1904–1905 gg." *Voenno-istoricheskii zhurnal*, no. 4 (April 1971), 44–55.
Keep, John L. H. "Chernyshevskii and the 'Military Miscellany.'" In Inge Auerbach et al. (eds.), *Felder und Vorfelder russischer Geschichte i Studien zu Ehren von Peter Scheibert*, 111–33.
———. "The Military Style of the Romanov Rulers." *War & Society*, 1, no. 2 (September 1983), 61–84.
———. "The Origins of Russian Militarism." *Cahiers du Monde Russe et Sovietique*, 26, no. 1 (January-March 1985), 5–20.
———. "The Russian Army's Response to the French Revolution." *Jahrbücher für Geschichte Osteuropas*, Bd. 28 (1980), H. 4, 499–523.
Kenez, Peter. "Autocracy and the Russian Army." *Russian Review*, 13, no. 5 (July 1974), 201–5.
———. "A Profile of the Prerevolutionary Officer Corps." *California Slavic Studies*, 7 (1973), 121–58.
Kennedy, Paul. "Mahan *versus* MacKinder: Two Interpretations of British Seapower." In Paul Kennedy, *Strategy and Diplomacy 1870–1945*, 43–85.
Kipp, Jacob W. "Tsarist Politics and the Naval Ministry, 1876–1881: Balanced Fleet or Cruiser Navy?" *Canadian-American Slavic Studies*, 17, no. 2 (Summer 1983), 151–79.
Kipp, Jacob W., and Maia Kipp. "The Grand Duke Konstantin Nikolaevich: The Making of a Tsarist Reformer." *Jahrbücher für Geschichte Osteuropas*, Bd. 34 (1986), H. 1, 3–18.
Kipp, Jacob W., and W. Bruce Lincoln. "Autocracy and Reform: Bureaucratic Absolutism and Political Modernization in Nineteenth-Century Russia." *Russian History*, 6, pt. 1 (1979), 1-21.
Kolesnikov, A. A. "O vospitanii i obuchenii v russkoi armii." *Voenno-istoricheskii zhurnal*, no. 3 (March 1989), 82–84.
Kollega. "Iliuzii i predvziatost'." *Russkii invalid*, no. 49 (2 March 1912), 4.
Komissarov, A. M. "M. I. Dragomirov o voinskoi distsipline." *Voenno-istoricheskii zhurnal*, no. 6 (June 1989), 89–92.
Kondrat'ev, P. "Takticheskoe obrazovanie ofitserov." *Voennye besedy ispolnennyia v shtabe voisk Gvardii i Peterburgskogo voennogo okruga v 1896–97 g.*, XVIII (1897), 1–184.
Kopylov, V. N. "Miliutin kak voennyi deiatel'," *Voennaia mysl'*, no. 3 (March 1945), 59–72.
———. "Taktika russkoi armii v russko-turetskoi voine 1877–1878 gg." In D. V. Pankov (comp.), *Razvitie taktiki russkoi armii*, 188–227.
Korsun, N. G., and P. Kh. Kharkevich. "Taktikia russkoi armii v russko-iaponskoi voine 1904–1905 gg." In D. V. Pankov (comp.), *Razvitie taktiki russkoi armii*, 228–63.
———. "Taktika russkoi armii v pervuiu mirovuiu voinu 1914–1918 gg." In D. V. Pankov (comp.), *Razvitie taktiki russkoi armii*, 264–315.

Kozlov, S. "Voennaia nauka i voennye doktriny v pervoi mirovoi voine." *Voenno-istoricheskii zhurnal*, no. 11 (November 1964), 31–41.
Kr—kii, Ia. "O deistvii voisk v boiu na osnovanii Franko-germanskoi voiny 1870–1871 godov." *Voennyi sbornik*, 94, no. 11 (November 1873), 79–106.
Krasnov, P. "Lava." *Voennyi sbornik*, no. 3 (March 1911), 41–68.
Krivtsov, E. "Voennoe iskusstvo." *Russkii invalid*, no. 16 (20 January 1912), 5.
Leonidov, L. "Iz istorii russkoi voenno-istoricheskoi mysli 60-kh godov XIX veka." *Voenno-istoricheskii zhurnal*, no. 10 (October 1973), 95–100.
Leer. G. "Sintez taktiki." *Voennyi sbornik*, 179, no. 2 (February 1888), 243–80.
———. "Sovremennoe sostoianie strategii." *Voennyi sbornik*, 53, no. 1 (January 1867), 47–66, no. 2 (February 1867), 223–58, 54, no. 3 (March 1867), 3–28, no. 4 (April 1867), 99–140, 55, no. 5 (May 1867), 3–25, 57, no. 9 (September 1867), 25–65, no. 10 (October 1867), 219–70.
———. "Strategicheskii i takticheskii obzor Franko-prusskoi voiny (do Sedana)." *Voennyi sbornik*, 81, no. 10 (October 1871), 145–76.
———. "Teoreticheskie masshtaby." *Voennyi sbornik*, 61, no. 1 (January 1869), 77–89.
Liprandi, R. "Pereprava u Sistova." *Voennyi sbornik*, 250, no. 11 (November 1898), 48–55.
Lobov, V. N. "Russkii voennyi myslitel' i pedagog." *Voennaia mysl'*, no. 2 (February 1990), 43–51.
Luckett, Richard. "Prerevolutionary Army Life in Russian Literature." In Geoffrey Best and Andrew Wheatcroft (eds.), *War, Economy and the Military Mind*, 19–31.
Luvaas, Jay. "European Military Thought and Doctrine, 1870–1914." In Michael Howard (ed.), *The Theory and Practice of War*, 69–93.
McNeal, Robert H. "The Reform of Cossack Military Service in the Reign of Alexander II." In Béla K. Király and Gunther E. Rothenberg, eds., *War and Society in East Central Europe*, 409–21.
Mahon, John K. "Civil War Infantry Assault Tactics." *Military Affairs*, 25 (1961), 57–68.
Maksheev, F. A. "Neskol'ko slov o Dragomirovskom i Leerovskom periodakh nachal'stvovaniia akademiei." *Voennyi sbornik*, no. 12 (December 1907), 234–38.
Maslov, I. "Izsledovanie o sredstvakh k umen'sheniiu v voiskakh ot pekhotnogo ognia." *Voennyi sbornik*, 183, no. 9 (September 1888), 60–96, no. 10 (October 1888), 288–326.
———. "O svoistvakh pekhotnogo ognia i o sredstvakh k razvitiiu ego sily." *Voennyi sbornik*, 174, no. 3 (March 1887), 87–118.
———. "Osnovnaia ideia plana voiny." *Voennyi sbornik*, no. 7 (July 1905), 49–64, no. 8 (August 1905), 55–76, no. 9 (September 1905), 49–62, no. 10 (October 1905), 51–64, no. 11 (November 1905), 79–94.
Mayzel, Matitiahu. "The Formation of the Russian General Staff, 1880–1917: A Social Study." *Cahiers du Monde Russe et Sovietique*, 16, nos. 3–4 (July–December 1975), 297–321.
[Miliutin, D. A.] "Voennye reformy Imperatora Aleksandra II." *Vestnik evropy*, 17, no. 1 (January 1882), 5–35.
Menning, Bruce W. "A. I. Chernyshev: A Russian Lycurgus." *Canadian Slavonic Papers*, 30, no. 2 (June 1988), 190–219.
———. "The Army and Frontier in Russia." In Carl W. Reddel (ed.), *Transformation in Russian and Soviet Military History*, 25–38.
———. "The Deep Strike in Russian and Soviet Military History." *Journal of Soviet Military Studies*, 1, no. 1 (April 1988), 9–28.
———. "Train Hard, Fight Easy: The Legacy of A. V. Suvorov and His 'Art of Victory.' " *Air University Review*, 38, no. 1 (November–December 1986), 79–88.

----------. "Russian Military Innovation in the Second Half of the Eighteenth Century." *War & Society*, 2, no. 1 (May 1984), 23–41.
Meshcheriakov, G. P. "Vliianie boevogo opyta otechestvennoi voiny 1812 goda na razvitie taktiki v russkoi armii v pervoi polovine XIX veka." In L. G. Beskrovnyi (ed.), *1812 god*, 181–97.
M. G. "Ustroistvo i sluzhba general'nogo shtaba." *Voennyi sbornik*, 92, no. 7 (July 1873), 85–114.
Mikhnevich, N. P. "Partizanskie deistviia kavalerii v 1812 i 1813 godakh." *Voennye besedy ispolnennyia v shtabe voisk Gvardii i Peterburskogo voennogo okruga v 1887–1888 g.g.*, IV (1888), 263–311.
Morrill, Dan L. "Nicholas II and the First Hague Conference." *Journal of Modern History*, 46, no. 2 (June 1974), 296–313.
"Mysli o tekhnike voin budushchago." *Voennyi sbornik*, 211, no. 5 (May 1893), 35–55, no. 6 (June 1893), 255–71, 212, no. 8 (August 1893), 221–38, 213, no. 9 (September 1893), 32–55.
N. K. "Po povodu proekta ustava polevoi sluzhby i nastavleniia dlia deistviia v boiu otriadov iz vsekh rodov oruzhiia." *Voennyi sbornik*, no. 6 (June 1903), 72–85.
Neznamov, A. A. "Boevaia podgotvka armii." *Russkii invalid*, no. 38 (18 February 1912), 4–5.
----------. "Na zlobu dnia." *Russkii invalid*, no. 12 (15 January 1912), 4–5.
----------. "Operatsiia (voennaia)." *Voennaia entsiklopediia*, XVII, 130–33.
----------. "Plan voiny." *Voennaia entsiklopediia*, XVIII, 446–49.
Nikol'skii, V. P. "Voennoe delo v Rossii pri vstuplenii na prestol Imperatora Nikolaia II i pered voinoi 1904–05 g.g." In A. S. Grishinskii et al. (eds.), *Istoriia russkoi armii i flota*, 13, 33–41.
Noskov, A. "Komplektovanie armii vo vremia voiny," *Voennaia mysl'*, no. 9 (September 1938), 136–41.
"Obzor mnenii, vyskazannykh v nashei voennoi literature po raznym voennym voprosam za 1877–1879 gody." *Voennyi sbornik*, 132, no. 3 (March 1880), 173–210, no. 4 (April 1880), 410–23, 133, no. 5 (May 1880), 50–79, 134, no. 7 (July 1880), 159–75, no. 8 (August 1880), 259–79, 135, no. 9 (September 1880), 34–53, no. 10 (October 1880), 280–96, 136, no. 11 (November 1880), 280–96.
Orlov, N. A. "Nravstvennyi element v voennom dele." *Voennaia entsiklopediia*, XVII, 45–52.
Osipov, M. "Obzor voenno-nauchnykh trudov D. A. Miliutina." *Voenno-istoricheskii zhurnal*, no. 9 (September 1972), 102–7.
Parkin, P. "Vintovke S. I. Mosina 80 let." *Voenno-istoricheskii zhurnal*, no. 4 (April 1971), 122–26.
Pavlenko, N. "Iz istorii razvitiia teorii strategii." *Voenno-istoricheskii zhurnal*, no. 10 (October 1964), 104–16.
Petrov, A. N. "Otvet g. Maslovu." *Voennyi sbornik*, 256, no. 11 (November 1900), 62–67.
----------. "Zadachi sovremennoi strategii." *Voennyi sbornik*, 217, no. 5 (May 1894), 35–64.
Pintner, Walter M. "The Burden of Defense in Imperial Russia, 1725–1914." *Russian Review*, 43, no. 3 (July 1984), 231–59.
----------. "Russian Military Thought: The Western Model and the Shadow of Suvorov." In Peter Paret (ed.), *Makers of Modern Strategy*, 354–75.
"Po povodu takticheskoi podgotovki polevoi artilerii." *Voennyi sbornik*, 247, no. 5 (May 1899), 111–16.
Pototskii, [*General-maior*]. "Posledniia novovvedeniia i predlozheniia v polevoi artillerii." *Voennye besedy ispolnennyia v shtabe voisk Gvardii i Peterburgskogo voennogo okruga za 1892–1893 gg.*, XIV (1893), 67–85.

Potto, V. "Sovremennoe obrazovanie i vospitanie voisk." *Voennyi sbornik*, 105, no. 11 (November 1875), 49–89, no. 12 (December 1875), 413–71.
Ray, Oliver Allen. "The Imperial Russian Army Officer." *Political Science Quarterly*, 76, no. 4 (December 1961), 575–90.
Romanovskii, G. D. "Russko-Iaponskaia voina 1904–5 g.g.: Deistviia na Kvantumskom poluostrove. Oborona kreposti Port-Artur." In A. S. Grishinskii et al. (eds.), *Istoriia russkoi armii i flota*, XIII, 83–141.
Rostunov, I. I. "Boevye deistviia russkoi armii na Balkanakh v 1877–1878 gg." In G. L. Arsh, V. N. Vinogradov, et al. (eds.), *Russko-turetskaia voina 1877–1878 gg. i Balkany*, 10–25.
———. "Uroki russko-iaponskoi voiny 1904–1905 gg." *Voenno-istoricheskii zhurnal*, no. 2 (February 1984), 73–78.
Rothenberg, Gunther E. "Moltke, Schlieffen, and the Doctrine of Strategic Envelopment." In Peter Paret (ed.), *The Makers of Modern Strategy*, 296–325.
Savushkin, R. A. "K voprosu o vozniknovenii i razvitii operatsii." *Voenno-istoricheskii zhurnal*, no. 5 (May 1979), 78–82.
Schneider, James J. "The Theory of the Empty Battlefield." *RUSI*, 132, no. 3 (September 1987), 37–44.
Shatsillo, K. F. " 'Delo' polkovnika Miasoedova." *Voprosy istorii*, 42, no. 4 (April 1967), 103–16.
———. "Podgotovka tsarizmom vooruzhennykh sil k pervoi mirovoi voine." *Voenno-istoricheskii zhurnal*, no. 9 (September 1974), 91–96.
Shkurak, K., and N. Murzaev, "K istorii razvitiia russkikh polevykh i boevykh ustavov." *Voenno-istoricheskii zhurnal*, no. 11 (November 1962), 113–20.
Shor, D. I. "Inzhenernoe obespechenie osady Plevny v russko-turetskoi voine 1877–1878 gg." In V. I. Zheleznykh (ed.), *Voenno-inzhenernoe iskusstvo*, 126–46.
Sidorov, A. L. "Iz istorii podgotovka tsarizma k pervoi mirovoi voine." *Istoricheskii arkhiv*, no. 2 (1962), 120–55.
———. "Zheleznodorozhnyi transport Rossii v pervoi mirovoi voine i obostrenie ekonomicheskogo krizisa v strane." *Istoricheskie zapiski*, 26 (1948), 3–64.
Smychnikov, A. A., and Val. V. Mavrodin. "K voprosu o perevooruzhenii russkoi armii v seredine XIX veka." In *Problemy istorii feodal'noi Rossii. Sbornik statei k 60-letiiu prof. V. V. Mavrodina*, 235–56.
"Sovremennaia strategiia" [trans. by G. L. of General von Schlichting's *Osnovy sovremmenoi taktiki i strategii*]. *Voennyi sbornik*, no. 6 (June 1908), 241–62.
Stein, H. P. "Der Offizer des russischen Heeres im Zeitabschnitt zwischen Reform und Revolution (1861–1905)." *Forschungen zur osteuropäischen Geschichte*, 13 (1967), 346–507.
Stone, Norman. "Organising an Economy for War: The Russian Shell Shortage, 1914–1917." In Geoffrey Best and Andrew Wheatcroft (eds.), *War, Economy and the Military Mind*, 108–19.
Sukhotin, N. "Amerikanskaia konnitsa." *Voennyi sbornik*, 175, no. 5 (May 1887), 75–99.
Sul'kevich, M. A. "Takticheskaia podgotovka polevoi artilerii." *Voennyi sbornik*, 244, no. 11 (November 1898), 149–58.
Suvorov, A. N. "Voennaia igra starshikh voiskovykh nachal'nikov v aprele 1914 goda." *Voenno-istoricheskii sbornik*, vyp. 1 (1919), 9–29.
Svechin, A. A. "Bol'shaia voennaia programma." *Russkaia mysl'*, bk. VIII (August 1913), 19–29.
———. "Evoliutsiia operativnogo razvertyvaniia." *Voina i revoliutsiia*, no. 5 (May 1926), 3–26.
———. "Voennye deistviia v Man'chzhurii v 1904–1905 gg." In A. S. Grishinskii et al. (eds.), *Istoriia russkoi armii i flota*, XIV, 59–182.

Terekov, [*Podpolkovnik*]. "Voprosy vospitaniia i obucheniia voisk." *Voennye besedy ispolnennyia v shtabe voisk Gvardii i Peterburgskogo voennogo okruga v 1889–90 gg.*, VIII (1890), 131–92.
Towle, P. A. "British Estimates of Japanese Military Power, 1900–1914." In Philip Towle (ed.), *Estimating Military Power*, 111–38.
Turner, L. C. F. "The Russian Mobilisation in 1914." *Journal of Contemporary History*, 3, no. 1 (January 1968), 65–88.
———. "The Russian Mobilisation in 1914." In Paul M. Kennedy (ed.), *The War Plans of the Great Powers, 1880–1914*, 252–68.
Ukhach-Ogorovich, N. A. "Otsenka nabega na Inkou." *Voennyi sbornik*, no. 5 (May 1908), 61–70.
V. P. "Neskol'ko slov o general'nom shtabe." *Voennyi sbornik*, no. 5 (May 1871), 140–44.
Vasil'ev, A. V. "Strelkovoe oruzhie russkoi armii v pervoi mirovoi voine." *Voenno-istoricheskii zhurnal*, no. 7 (July 1987), 82–83.
Vinogradskii, A. N. "Russko-Iaponskaia voina. Obstanovka pered voinoi." In A. S. Grishinskii et al. (eds.), *Istoriia russkoi armii i flota*, XIV, 5–58.
———. "Taktika skorostrel'noi artilerii." *Voennyi sbornik*, 257, no. 1 (January 1901), 99–104.
Voide, K. M., ed. "Voina." *Voennyi sbornik*, 245, no. 1 (January 1899), 5–37, and 256, no. 12 (December 1900), 387–404.
Volotskii, N. "Teoriia veroiatnostei i boevoe snabzhenie artileri." *Voennyi sbornik*, no. 2 (February 1904), 139–52, and no. 3 (March 1904), 135–44.
———. "Teoriia veroiatnostei i boevoe snabzhenie patronami." *Voennyi sbornik*, no. 11 (November 1904), 81–100.
Von Wahlde, Peter. "Dmitrii Miliutin: Appraisals." *Canadian Slavic Studies*, 3, no. 2 (Summer 1969), 400–414.
———. "A Pioneer of Russian Strategic Thought: G. A. Leer, 1829–1904." *Military Affairs*, 35, no. 4 (December 1971), 1–8.
———. "Russian Military Reform: 1862–1874." *Military Review*, 39, no. 10 (January 1960), 60–69.
Voronov, P. "Nachalo Plevny." *Voennyi sbornik*, 252, no. 3 (March 1900), no. 4 (April 1900), 245–71.
Zaionchkovskii, A. M. "Edinstvo voennoi doktriny (Opasnosti i ubelichenii)." *Russkii invalid*, no. 280 (31 December 1911), 2–3.
———. "Eshche po povodu doktriny." *Russkii invalid*, no. 7 (10 January 1912), 3–4.
———. "Initsiativa chastnykh nachal'nikov v oblasti ikh strategicheskikh rabot." *Voennye besedy ispolnennyia v shtabe voisk Gvardii i Peterburgskogo voennogo okruga v 1895–1896 godakh*, XVII, 1–52.
Zaionchkovskii, P. A. "Perevooruzhenie russkoi armii v 60—70-kh godakh XIX veka." *Istoricheskie zapiski*, 36 (1951), 64–100.
———. "Vydaiushchiisia uchenyi i reformator russkoi armii." *Voenno-istoricheskii zhurnal*, no. 12 (December 1965), 32–36.
Zaitsev, L. A. "Voenno-pedagogicheskie vzgliady M. I. Dragomirova." *Voenno-istoricheskii zhurnal*, no. 9 (September 1985), 72–76.
Zakharov, V. A. "Sostoianie i razvitie russkogo voenno- inzhenernogo iskusstva i inzhenernykh voisk s nachala XX v. do Velikoi oktiabr'skoi sotsialisticheskoi revoliutsii." In V. I. Zhelezhnykh (ed.), *Voenno-inzhenernoe iskusstvo*, 147–72.
"Zapiski P. M. Gudim-Levkovicha o voine 1877–1878 goda." *Russkaia starina*, 124, no. 12 (December 1905), 513–604.
Zeddeler, [*fligel'-ad'iutant baron* L. L.]. "Pekhota, artilleriia i kavaleriia v boiu i vne boia

v Germano-frantsuzskoi voine 1870–1871 godov." *Voennyi sbornik*, 86, no. 7 (July 1872), 33–114.
Zhilin, A. P. "Bol'shaia programma po usileniiu russkoi armii." *Voenno-istoricheskii zhurnal*, no. 7 (July 1974), 90–97.
Zhilin, P. "Diskussii o edinoi voennoi doktriny." *Voenno-istoricheskii zhurnal*, no. 5 (May 1961), 61–74.
Zhilinskii, S. "Strategiia-nauka i strategiia-iskusstvo." *Voennyi sbornik*, 94, no. 11 (November 1873), 5–48.
Zolotarev, V. "Prichiny i kharakter russko-turetskoi voiny 1877–1878 gg." *Voenno-istoricheskii zhurnal*, no. 7 (July 1975), 78–82.

Encyclopedias

The Military-Naval Encyclopedia of Russia and the Soviet Union. 1978–84. 4 vols. incomplete.
Sovetskaia istoricheskaia entsiklopediia. 1965–70. 13 vols.
Sovetskaia voennaia entsiklopediia. 1932–33 ed. 2 vols. incomplete.
Sovetskaia voennaia entsiklopediia. 1976–80 ed. 8 vols.
Voennaia entsiklopediia. 1911–15 ed. 18 vols. incomplete.

Newspapers, Journals, and Periodicals

Air University Review
American Historical Review
Army Quarterly and Defence Journal
Cahiers du Monde Russe et Sovietique
California Slavic Studies
Canadian Slavic Studies
Canadian-American Slavic Studies
Canadian Slavonic Papers
Current History
Forschungen zur osteuropäeischen Geschichte
Istoricheskii arkhiv
Istoricheskie zapiski
Jahrbücher für Geschichte Osteuropas
Journal of Contemporary History
Journal of Modern History
Journal of Social History
Journal of Soviet Military Studies
Krasnyi arkhiv
Military Affairs
Military Review
Political Science Quarterly
RUSI [Journal of the Royal United Services Institute for Defence Studies]
Russian History
Russian Review
Russkaia mysl'
Russkaia starina
Russkii invalid

Russkii vestnik
Slavic Review
Slavonic and East European Review
Voennaia mysl'
Voenno-istoricheskii sbornik
Voenno-istoricheskii zhurnal
Voennye besedy ispolnennyia v shtabe voisk Gvardii i Peterburgskogo voennogo okruga
Voennyi sbornik
Voina i revoliutsiia
Voprosy istorii
War & Society

Index

Agapeev, A. P., military writer, 130
Aircraft: acquisition of, 233; in maneuvers, 269
Akhaltsykh Detachment, 78
Aleksandropol' Detachment, 78
Alekseev, Admiral E. I., Viceroy of Far East, 153; contradicts Kuropatkin, 154, 172
Alekseev, M. V., 220, 239, 241, 249
Alexander II, emperor, 58, 89, 272
Alexander III, emperor, 89, 287n.2
American Civil War, 40, 45–6
Ardahan, advance against, 78
Artamonov, N. D., and war plans, 52
Artillery: organization and strength, 26–27, 113–14, 144, 231; armament, 31–3, 106–07, 113, 231–32, 273; tactics, 46–7, 143–45, 258–59, 264–65; use of indirect fire with, 144–45; in maneuvers, 48, 49, 149, 267, 268; failure to coordinate with infantry, 85; in Large Program, 234; reserve augmentation of, 27, 113–14
Astaf'ev, A. I., 7–8, 38, 40
Automotive transport, 232–33
Auxiliary troops, organization and strength, 28–9, 114–15, 228, 232

Baiov, A. K., military historian, 206, 211
Balkan Crisis, 51–2, 233
Balloons: acquisition of, 108, 233; usefulness of, 203
Baltic Fleet: in Indian Ocean, 170; defeated at Tsushima, 195
Bariatinskii, A. I., 9, 15, 18, 21, 22, 96
Bariatinskii-Miliutin dispute, 15–16, 18, 21–2, 96–7
Battle: nature of, 134–35, 213–14, 276; stages of, 135, 137–38, 238–39
Batumi, advance against, 78
Bayazid, advance against, 78
Berdan, Hiram, 30–1
Berdan No. 2 rifle, or "Berdanka": characteristics of, 31; reequipping the army with, 104
Berlin, Congress of, 81
Beskrovnyi, L. G., military historian, 195
Bil'derling, A. A.: at Liaoyang, 176, 178; at Sha-ho, 179, 181; at Mukden, 186, 187
Birger, A. K., at Mukden, 190, 192, 193
Bismarck, Otto von, 91–2
Bloch, Jan S., 129–30, 209
Botkin, S. P., physician-writer, 64
Brooks, E. Willis, 116
Brusilov, A. A., 243, 249

Cadre and reserve system, 24, 25, 108–15 passim, 226, 227
Carr, E. H., 4

Cavalry: organization and strength, 27–8, 112–13, 230–31; tactics, 45, 145–47, 263–64; impact of technology on, 45; deep raiding function of, 45–6, 145–46, 230–31; in exercises and maneuvers, 46, 150, 267–68; and shock action, 146, 264; screening and security function of, 263
Central Asia, conquest of, 12
Chelishchev, Colonel, at Telish, 72–3
Chemulpo (Inchon), Japanese attack on, 157
Chenevix-Trench, F., 115
Cheremisov, V. A., 213
Cherniaev, M. G., 12, 52
Clausewitz, Carl von, 124–25, 204, 207
Combat experience: as component of military history, 2, 134, 204; foreign wars as source of, 40, 44; in *Military Collection*, 288–89n.28
Command and control: general problems of, 254–55, 260, 265–66, 269–70, 274, 276; in battle, 138, 265–66, 269–70
Commission on Terms of Service, 21, 22
Committee on Structure and Training of Troops, 39, 44, 88, 94, 136, 137
Conscription. *See* Universal military service
Cossack military service, 28, 114–15
Crimean War, 4, 40; impact of, 6–7, 8–9
Curtiss, John S., 4

Dal'nii (Darien), lease begun, 152
Danilov, Iu. N., 187, 218, 237, 240, 241, 245–46, 248, 250, 252
Danube, forced crossing of, 54–7
Decembrist Uprising, as sparked by military dissatisfaction, 7
Denikin, A. I., on senior commanders, 101
Deployments, Russian army, peacetime, 20–21, 115, 222, 224–26, 230
Derozhinskii, V. F., at Shipka, 65
Dobrorol'skii, Colonel Sergei, 239
Dobrovol'skii, M. V., at Lovech, 67–8
Dolni Dubnik, Russian attack on, 73–4
Dragomirov, M. I., 3, 6, 8, 38–9, 40, 41–2; forcing Danube, 54–7; wounded, 66, 85, 87, 103, 120, 135, 136, 139, 141, 142, 143, 149, 151, 204–05, 239, 270, 275; on "Tashkentsy," 281n.27
Dragomirov, V. I., 1914 war game, 254
Dual Alliance, Austro-German (1879), 91
Dubail, General Auguste, 242
Durop, K. N., tactician, 135, 146

Engel'gardt mortar system, 105–06
Engineer Troops (Sappers). *See* Auxiliary troops

Erevan Detachment, 78, 81
Erickson, John, 13, 38, 274
Ermolov, General N. S., 234
Erzerum, Russian advance on, 81
Exercises and maneuvers: during annual camps, 148–51; influence of Krasnoe selo on, 266–69; definition of, 148; regulations on, 48; duplicating effects of firepower in, 149

Fadeev, General R. A., 89, 96
Field administration, army, regulations on, 18–19, 98–100, 234–35
Field fortifications, 41, 85, 138, 259
Field gun, light quick-firing: adoption of, 106–07; impressions of in Russo-Japanese War, 203; pre-1914 emphasis on, 231
Field regulation (combat), army, 256–59
Firepower and shock action in battle, balance between, 39, 41, 42–3, 136, 137–38, 142, 260–62
Foch, Marshal Ferdinand, 251
Fock, General A. V.: at Dal'nii, 160; at Port Arthur, 164, 165
Fortress construction and armament, 20, 29, 117–19, 232, 248
Franco-Prussian War: impact on threat assessment, 19–20; impact on artillery development, 33; as source of combat experience, 40, 44
Franco-Russian Alliance, origins of, 92, 242, 245, 247; article III of convention, 245; pressures from, 247
Fuller, William C., Jr., 102, 117, 219
Future war, nature of, 2, 129, 208–09

Gadolin, A. V., Russian armorer, 32
Gareev, General M., 150
Garshin, Vsevolod, author of "Four Days," 82, 287n.80
Geisman, P. A., 206
General Staff: pre-reform understanding of, 15; failure to pursue Prussian model, 15–16; officers of the ("genshtabisty"), 38, 100, 102, 202–03, 207, 211, 236, 265. *See* Main Staff *and* Main Directorate of the General Staff
Gerngross, A. A.: at Telissu, 173–74; at Mukden, 193
Gershel'man, F. K., on lack of war planning (1901–02), 121
Gerua, A. V., 202, 207, 212, 255; on applied strategy and "operatika," 205
Gerua, B. V., 250, 251
Girs, N. K., Foreign Minister, 91
Glasko, F. F., at Telissu, 173–74
Golovin, N. N., 216
von der Goltz, 209, 212
Gorbatovskii, V. N., at Port Arthur, 164, 165
Gorchakov, A. M., Foreign Minister: and *recueillement*, 11; and origins of Russo-Turkish War, 51–2
Goremykin, F. I., military writer, 8
Gorlov, A. P., small arms developer, 30, 31
Gorni Dubnik: Russian attack on, 71–2; legacy of, 124, 136
Greene, Francis Vinton, American observer, Balkan Theater, 56–7, 65, 66, 73, 74, 76–77
Greger, Gorvits, and Kogan, supply contractors, 83

Grekh, B. N., on infantry tactics, 43, 44
Grekov, M. I., at Mukden, 188, 191
Grippenberg, General O. K., 185, 186
Guards regiments as hereditary elite, 102
Gudim-Levkovich, P. M., 69, 135
Guk, K. G., and indirect fire, 143
Gulevich, A. A., 130, 133, 209
Gunius, K. I., small arms developer, 30
Gurko, I. V., 149, 177; with Forward Detachment, 57–8, 59; on Sofia Road, 71–2, 73–4; in trans-Balkan campaign, 75, 78
Gurko, V. I.: and Military-Historical Commission, 216; as Young Turk, 216

Hamilton, Ian, at Battle of Yalu, 157–58
Heiman, General V. A., advance on Erzerum, 81
Higher Examination Board, 236
Hussein Khalil Pasha, at Kars, 80

Ianushkevich, N. N., 216, 250
Ignat'ev, N. P., and London Protocol, 52
Imeretinskii, A. K.: at Lovech, 67; at 3rd Plevna, 69, 70
Infantry: organization and strength, 24, 26, 109, 110–12, 228; tactics, 42–4, 136–39, 259–62; armament, 30–1, 104–05; in exercises, 49, 149, 267–68; in Large Program, 233; reserve augmentation of, 111–12, 226, 227
Inkou, Mishchenko's raid on, 184–85
Irvine, Dallas, describes capital staff, 15
Isakov, N. V., 35
Ivanov, General N. Iu., 243, 249, 252

Joffre, Marshal Joseph, 242
Jomini, 124–25, 126, 129, 204
Jones, David R., 246, 255–56
Junker schools, 34, 103

Karol, Romanian Prince, at Plevna, 69
Kars, Russian attack on, 78, 80–1; as example of night attack, 138
Kartsov, P. P., in trans-Balkan campaign, 75
Kashtalinskii, N. A., at Battle of Yalu, 158
von Kaufman, K. P., 12
Kaul'bars, A. V., 185; at Mukden, 186, 187, 189, 190–91, 192, 193, 194
Kawamura, General, at Mukden, 186, 187, 190
Keep, John L. H., 4
Keller, Count Fedor, at Motien Pass, 174
Kennedy, Paul, on geostrategy, 274
Kliuev, N. A., 241
Kobulety Detachment, 78
Kondratenko, P. I., at Port Arthur, 162–63, 164–65, 167–68
Kotsebu Special Commission on Military Administration, 96–7
Kozlov, S. N., 3; on nature of future war, 2
Kridener, N. P.: in advance to Nikopol, 60; at 2d Plevna, 62–3
Krnk, Sylvester, Bohemian armorer, 30
Krylov, E. K., at 3rd Plevna, 70
Kuprin, Alexander, 287n.80

Kuroki, General: in Korea, 157; at Fenghuangcheng, 172; at Liaoyang, 176, 177–78; at Mukden, 188–90
Kuropatkin, A. N., 220, 249, 276; early career, 92–4; as War Minister, 88; at Kursk maneuvers, 150–51; on strategic rail net, 117, 120; on unpreparedness for offensive war, 121; on Japanese as potential enemies, 153; Skobelev's alleged assessment of, 93–4, 154; arrival in Manchuria, 157; effects of Port Arthur siege on, 171–72; with dispersed forces in Manchuria, 172–73; opts for battle at Liaoyang, 175; during Liaoyang operation, 176, 178–79; at Sha-ho Operation, 179–81, 183–84; at Sandepu, 185–86; at Mukden, 186, 187, 189–90, 192, 193–94, 197; on Russo-Japanese War in retrospect, 202

Large Program for Strengthening the Army, 233–34, 248
Launits, M. V., at Mukden, 192, 194
Lazarev, I. D., in Caucasus, 80
Leer, G. A.: influence of Napoleon on, 36; on tactics, 40–1; origins of theory of strategy, 36; on implications of Franco-Prussian War, 36; as commandant of Staff Academy, 87; on strategy, 125–29, 131, 132, 133; declining influence in wake of Russo-Japanese War, 204, 205, 210; legacy of, 275
Liaoyang operation, 176–77
Linevich, N. P., 101, 185, 193
Lloyd, William, influence on Leer, 126
Lobko, P. L., and military administration, 98–9
Logistics: in war games, 252, 254; based on incorrect norms, 255–56. *See* Rear services
London Protocol (1877), 52
Loris-Melikov, M. T., in Caucasus, 78
Lovech, battle, 67–9
Lukomskii, A. S., as "Young Turk," 216

Machine guns: brief adoption of Gatling-type, 33; experimentation with, 107; Vickers-Maxim, 107–08, 232; in 1914 regulation, 260
Maevskii, N. V., armorer, 32
Maevskii steel gun system, adoption of, 105
Main Artillery Directorate, 106, 107, 255
Main Directorate of the General Staff (GUGSh): in 1863–65, 16–17; structure of, 218, 220, 221, 233, 234; and war planning, 241–42. *See* Main Staff
Main Staff: formation of, 16; organization and functions of, 17, 97–8; after 1905, 218, 220; unhurried work patterns in, 292n.127
Markarov, S. O., death in 1904, 155
Martynov, E. I., 128–29, 202
Maslov, I. P., military theorist, 135–36, 138, 142, 294n.49
Maslovskii, D. F., military historian, 131
Medical services: in Russo-Turkish War, 82; in Russo-Japanese War, 196
Mehemet Ali Pasha, 58, 60
Michael Artillery Academy, 35, 143
Michael, Grand Duke, in Caucasus, 78, 80–81
Mikhnevich, N. P., 127, 132, 136, 206, 208–11, 212, 213, 230, 234, 238, 242, 247, 278
Military Collection, beginning of, 10
Military district system, origins of, 13–14

Military-Historical Commissions: for Russo-Turkish War, 88, 94, 95; for Russo-Japanese War, 201
Military history: role of, 2–3, 130–32, 200–01, 275–76, 278; varying institutional support for, 88, 94, 95; "Russian" and "Academic" schools of, 208, 209, 211
Military schools (*voennye uchilishcha*), 34, 103
Military science, 2, 3, 277–78; nature of, 133
Military theory: and tactics, 134–36; and military history, 201; and practice, 205
Miliutin, D. A., 6, 7, 23, 220, 272, 277; on reform, 8–9, 10–11; early service career, 9–10; founds *Military Collection*, 10; relationship with A. I. Bariatinskii, 10, 21–2; on general staff organization, 15–16; War Ministry reforms of, 14–8; prescriptions for field exercises, 49, 50, 148; in the field with Imperial Suite, 74; on peace terms in 1878, 81; creation of Military-Historical Commission, 88; requests retirement, 89
Mishchenko, P. I.: with cavalry screen in Korea, 157; commands raid on Inkou, 184–85; receives mixed evaluations, 198, 277
Mobilization gap: origins of, 19–20; and rail net, 116–17; persistence of, 230, 273–74
Mobilization schedules, 115; No. 18 (1903), 100, 120, 121, 239–40; No. 19 (first variant), 240–41; No. 19 A and G, 242–43, 245–46, 247; No. 20, 243; No. 8 (Far East), 122
von Moltke, Helmuth: 124, 129, 204; comments on Russo-Turkish War, 58, 75
Mosin, S. M., armorer, 104–05
Mosin rifle, characteristics of, 104–05
Motien Pass, 173, 174
Mukden operation, 186–94
Muktiar Pasha: before Kars, 78; counterattacks and withdraws, 79–80; defends Kars, 80; flees to Erzerum, 81
Munitions and armament stores: in Russo-Turkish War, 81–2; in Russo-Japanese War, 196
Myshlaevskii, A. Z., 131, 220

Nagant, Leon, armorer, 104
Nagant revolver, characteristics of, 105
Nakamura, General, in night attack at Port Arthur, 170
Nanshan, battle, 159–60
Napoleonic paradigm: condemned, 8; retains persistence, 128, 129, 132
Naval capabilities, Russian: inadequate in 1877–78, 84; inadequate in 1904–05; required to support ground operations, 272–73
Naval reconstruction programs, 219–20
Nepokoichitskii, N. N.: chief of staff, Balkan Theater, 55; advises Kridener, 62; rumors of corruption, 83
Neznamov, A. A., 200, 202, 203, 206, 207, 208, 211–15, 230, 234, 238, 242, 247, 271, 276, 277, 278
Nicholas Academy of the General Staff (Nicholas Military Academy to 1855), 7, 16, 125, 126, 133, 275, 276; entrance requirements and curriculum of, 35–6, 38; curriculum reform at, 101–02; historical studies at, 95, 200, 206; post-1905 critique of, 203, 206–07, 211, 216
Nicholas I, emperor, 6; reaction to Decembrist Uprising, 7

Nicholas II, emperor, 216, 218, 219, 220, 238, 239, 248, 272; personality and convictions of, 89
Nicholas Nikolaevich (the Elder), Grand Duke, 21, 46; as Supreme Commander, Balkan Theater, 55, 57; launches trans-Balkan operations, 74; possible involvement in corruption, 83
Nicholas Nikolaevich (the Younger), Grand Duke, 146, 216, 217, 219, 220, 249–50, 251
Night attacks, 138–39, 262
Nikopol, reduction of, 60
Nodzu, General: prepares debarkation of 4th Army at Takushan, 158, 172; at Sha-ho, 181; at Mukden, 188, 191
Nogi, General: creates 3rd Army at Dal'nii, 160; operates on far approaches to Port Arthur, 161; deploys for initial assault on Port Arthur, 163; prepares 3rd storm, 169; prepares 4th storm, 170; at Mukden, 187, 188, 189, 192, 193, 194
Nozhin, Russian correspondent, describes battle for 203 Meter Hill, 171

Obruchev, N. N., 35, 145, 146, 217–18, 220; early career, 17–18; author of "Thoughts on the Defense of Russia," 19–20, 21; on professionalism of "genshtabisty," 38; in planning for the Russo-Turkish War, 52–3; plans counteroffensive for Caucasian Theater, 78, 79–80; as Chief of the Main Staff under Vannovskii, 90; supports Miliutin legacy in Kotsebu Commission, 97; proposes reorganization of Main Staff, 97, 98; insists on freedom of action in war planning, 119; labels mobilization an act of war, 119
Offense and defense, balance between, 124, 238–39, 257, 259
Officer corps: education of, 33–6, 38, 103; size of, 100; changing professional requirements for, 100–01; outlook and composition, 100–03
Oku, General: lands 2d Army near Dal'nii, 158; defeats Russians at Nanshan, 159–60; advances inland, 172; defeats Russians at Telissu, 173–74; at Sha-ho, 181; at Mukden, 188, 191, 192, 193
Operational art, 205, 207, 275, 301n.49
Operations: definition of, 127; nature according to Mikhnevich, 210; according to Neznamov, 213–14; theory for conduct of, 126–27, 205, 210, 212, 214–15
Opolchenie (militia), 26, 226
Oranovskii, V. A., 249
Organizational Commission, for universal military service, 21
Orlov, N. A., 134; characterizes aerial bombardment as absurd; at Liaoyang, 178, 179
Osman Pasha: leaves Vidin for Plevna, 60–1; defeats Russians at 1st Plevna, 61–2; defeats Russians at 2d Plevna, 63–4; sallies against VIII and IV Corps, 66; capitulates, 74
Oyama, Marshal: launches envelopment at Liaoyang, 166; assumes command of Japanese forces in Manchuria, 174–75; at Liaoyang, 177, 179; at Sha-ho, 179, 180, 181, 183; at Mukden, 186–87, 192

Palitsyn, F. F.: 218, 226, 227; chief of GUGSh, 239, 251
Palitsyn-Alekseev reform initiative, 220, 222, 225
Panjdeh incident, 91
Panslavism, 11, 51, 52
Pares, Bernard, 270
Paret, Peter, 1
Paris, Congress of, 51; terms of, 280n.2
Paul I, emperor, 7
Peking, Treaty of, 12
Peter the Great, emperor, 7
Petrov, A. N., military historian, 131
Pleve, V. K., on benefits for Russia of "a short glorious war," 155
Pleve, General P. A., 243, 249, 252
Plevna, battle, legacy of, 124, 136, 274–75, 277. *See* Russo-Turkish War
Port Arthur: lease begun, 152; surprise naval attack on, 155; description of defenses, 162–63; impact of siege on Russian field army in Manchuria, 171–72; influence operations, 197; uncertain legacy of, 277. *See* Russo-Japanese War
von Prittwitz, Max, 245
Prussian model: of General Staff, 15; of military administration, 96–7
Puzyrevskii, A. K., 130, 131, 134

Radetskii, F. F.: at Shipka-Sheinovo, 75
Railroad mobilization, necessity for, 292n.114
Railroad net, strategic: impact on war planning, 116–17, 120, 122, 195, 197; impact on peacetime deployments, 117
Rakaza, Colonel, at 2d Plevna, 63
Rear and front, relationship between, 130, 132–33
Rear services: in 1877–78, 82–3; in 1904–05, 195–96
Reconnaissance, failures in, 84
Rediger, A. F., War Minister, 219, 220, 221
Redl, Colonel Alfred, 246
Reinsurance Treaty, 92
Reitern, M. Kh., 21
Rennenkampf, P. K., 243, 244, 245, 249, 252, 254; at Sha-ho, 180; at Mukden, 187
Reserve troops, 24, 26–7, 109, 111–12, 226–27, 273
Romania: agreement for transit of Russian troops, 52; agreement to participate in war against Turkey, 69
Rostunov, I. I.: comments on Schedule No. 19, 240; on No. 19 A and G, 243, 273
Rothenberg, Gunther, 124
Rozhestvenskii, Z. P., defeated at Tsushima, 195
Rumiantsev, P. A., 131
Russian army: capacity for wartime expansion in comparison with other European powers, 11; strength and organization, 23–24, 26, 28–9, 108–09, 111–13, 114–15, 153–54, 226–28; transition to rifled weapons, 30–1; transition to smokeless powder weapons, 104–08; Miliutin reforms of, 11, 13–4, 23–4, 26–?; reforms of 1910, 222, 224–28, 230; corps organization, 22, 109, 227–28; impact of 1904–05 on, 217; budgetary constraints, 90–1, 219, 288n.16, 290n.74
Russo-Japanese War: Russian preparations for, 153–4;

war plans for, 153; Russian reliance on reserve troops for, 154; Japanese preparations for, 154–55; Japanese armaments for, 155; initial Japanese ground operations, 156–60; Japanese advance to Port Arthur, 160–61; 1st Japanese assault against Port Arthur, 164–66; 2d Japanese assault, 166–68; 3rd Japanese assault, 168–69; 4th Japanese assault, 170–71; Battle of Telissu, 173–74; Liaoyang Operation, 175–79; Sha-ho Operation; Battle at Sandepu, 185–86; Mukden Operation, 186–94; transport and supply in, 195–96; perceptions of lessons learned in, 202–03, 214, 276
Russo-Turkish War: events leading to, 51–2; Russian planning for, 52–3; Turkish preparation for, 53–4; stages of, 53; initial trans-Danube operations, 57–8, 60; 1st Plevna, 61–2; 2d Plevna, 62–4; Russian reversion to defensive, with additional mobilization, 64; Russian defense of Shipka Pass, 64–66; Russian attack on Lovech, 67–9; 3rd Plevna, 69–71; Russian attacks on Gorni Dubnik, Telish, and Dolni Dubnik, 71–4; fall of Plevna, Russian winter offensive, 74–8; initial Russian offensive in Caucasus, 78; Turkish offensive, 78–9; Russian counteroffensive, 78–81; Russian medical services in, 82; Russian rear services in, 82–3; assessments of Russian command and control in, 84–5; combined arms coordination in, 85; perceptions of lessons learned from, 275–76
Ruzskii, N. V., 243, 249, 252, 256

Sakharov, General V. V., proposed reorganization of Main Staff, 97–8
Samoilov, V. K., reports Japanese prepared for offensive operations, 153
Samsonov, A. V., 243, 244, 245, 249, 255; at Liaoyang, 178
San Stefano: Russian advance to, 78; Treaty of, 81
Sandepu, battle, 185–86
von Schlichting, Sigismund Wilhelm, 207, 212
Science, role in military theory, 125
Scouts, mounted, introduction of, 147
Sedan: as example, 124; influence on Japanese, 172, 186
Serbia, Russian officers in service of, 52
Sergei Aleksandrovich, Grand Duke, 150–151
Shakir Pasha, at Arab Konak Pass and Philippopolis, 75
Shakovskoi, A. I., in supporting attack at 2d Plevna, 62–3
Shaposhnikov, B. M., 207, 265
Shatilov, General, at Kars, 81
Sheinovo, battle, 75–6
Shil'der-Shul'dner, Iu. I.: advances on Plevna, 61; suffers defeat at 1st Plevna, 61–2
Shipka Pass: I. V. Gurko's attack on, 58; description of, 64–5;
Shtakel'berg, G. K.: at Telissu, 173–74; at Sha-ho, 179, 180, 181, 183
Shuvalov, P. A., 21
Skobelev, M. D.: at 2d Plevna, 62–3; at Lovech, 67–8; employs modified infantry assault tactics at 3rd Plevna, 70; at Sheinovo, 75–6; at Geok Tepe, 91

Sluchevskii, K. K., at Liaoyang, 176
Smith and Wesson Model 3 revolver, adoption of, 31
Snyder, Jack, 242
Sobolev, L. N., at Sha-ho, 183
Society of the Zealots of Military Knowledge, origins of, 95
Special (Secret) Conference (1873): on requirements for Russian army, 21–3; results of, 22
Staff rides, 35, 250
State Defense Council, 217, 218, 219, 220, 221
State Duma, 218
Stavka (Headquarters of the Supreme Command), 235, 245, 250
Stessel, A. M., 159; prepares defenses of Port Arthur, 161–62; Surrenders Port Arthur
Stoletov, N. G., at Shipka Pass, 60, 65–6
Stolypin, P. A., 219
Stone, Norman, 270
Strategic deployments and concentrations, 210, 213
Strategic offensive and tactical defensive, merits of, 86
Strategy: nature of, 127–28; two-pole (shock strike or exhaustion), 129–30, 210
Sukhomlinov, V. A., 205, 206, 221, 224, 225, 227, 230, 241, 242, 250, 272, 277; takes issue with Kuropatkin, 121; background, 221–22; and 1914 war game, 251, 252, 253–54
Sukhotin, N. N., 131, 139, 145, 146; stresses reconnaissance and deep raiding as cavalry functions, 46
Sukhozanet, N. O., War Minister, 9
Suleiman Pasha: counters Gurko at Eski Zagra, 58, 60; attacks at Shipka Pass, 65–6; suffers defeat at Philippopolis, 78
Suvorov, A. N., and 1914 war game, 254
Suvorov, A. V.: tradition of, 7, 8, 131; *Art of Victory* of, 216
Svechin, A. A., 125, 134, 216, 234, 278; on Russo-Japanese War, 201, 202; on operational art, 197, 205; on rear services in Far East, 195, 196
Sviatopolk-Mirskii, N. I., at Sheinovo, 75–6
·Sviatskii, V. N., on combined arms, 147

Tashihchiao, battle, 175
Technology: impact of change on military affairs, 2; on scale of battle, 274–75; on nature of tactics and operations, 123–24, 132–33, 135–36
Telegraph, tactical use in 1877–78, 84
Telish, battle, 72–3
Terminology (military), Russian precise definition, 300-01n.21
Threat assessment: 239–41, 246–47; by Obruchev, 19–20, 21; during the late 1880s and early 1890s, 92; leads Russians to discount Japanese, 152–53
Three Emperors' League, 13, 52, 91, 92
Togo, Admiral: attacks Port Arthur, 155; blockades Russian Pacific Squadron, 159
Topornin, D. A., at Mukden, 192, 193
Totleben, E. I., at investment of Plevna, 71
Training, according to Dragomirov system, 39
Trans-Siberian Railroad: as diversion of resources, 116;

strategic significance of, 274; carrying capacity of, 153, 292n.132; blamed for Far Eastern defeat, 195
von Traubenberg, Rausch, 252
Tretiakov, N. A.: at Nanshan, 159–60; at 203 Meter Hill, 170–71
Troop mobilization: poor projections of 1859, 9; achievements and shortcomings (1863–1872), 19; compared with Germany and Austria-Hungary, 19; before the Russo-Turkish War, 52, 53; to reinforce the Balkan Theater, 64; to reinforce the Caucasian theater, 78; post-1905 difficulties with, 222, 225, 230
Troyan Pass, 67, 75
Tsarist politics: as backdrop to military change, 2; as related to army, 89–90, 218–19
Tserpitskii, K. V., at Mukden, 193
203 Meter Hill: 163, 166, 167; defense of, 168; surrender of, 170–71

Unified military doctrine, 215–16, 236–37, 248
Universal military service, 222; Miliutin initially frustrated over, 13; Commission on Terms of Service, 22; deliberation over and planning for, 21, 22–3

Vannovskii, P. S., 220; as Chief of Staff, Eastern Detachment, 60; as War Minister, 88; background, 90; on requirements for financial stringency, 90–1
Vannovskii, V. P., assessments of Japanese armed forces (1900), 152–53
Vasil'ev, N. A., at Mukden, 192
Vessil Pasha, at Shipka-Sheinovo, 75–6
Vineken, A. G., 250–51
Vitgeft, V. K., sorties with Russian Pacific Squadron from Port Arthur, 163

War games, 250; of 1907, 251; of 1911, 251; of 1914, 251–55
War Ministry: reform requirements, 9; reorganization under Miliutin, 14; proposed reform under Alexander III, 96–7
War plan, 238; according to Mikhnevich, 210; according to Neznamov, 212–13
War planning, Russian: Obruchev's reflections on, 19–20; for war against Turkey, 52–3; essentially defensive under Obruchev, 116, 119; for war in the Far East, 153–54; under circumstances of late 1880s and early 1890s, 92; and mobilization schedules, 116, 120–22; during 1905-13, 239–43, 247
War scare of 1887, 91
Wells, H. G., 275

Yalu, battle, 157–58
"Young Turks" (reformist "genshtabisty"), 207, 216, 219, 220–21

Zabudskii, N. A., designer of quick-firing field gun, 106
Zaionchkovskii, A. M., military historian, 134, 204, 215, 216, 236–37, 241–42, 245, 259, 277, 278
Zaionchkovskii, P. A., military historian: on aging senior commanders, 101; on importance of concentration in theater of military actions, 120; on absence of plans for the western theater, 121; on maneuvers, 150
Zal'tsa, A. E., 243, 249, 252
Zarubaev, N. P., at Liaoyang, 176, 179
Zasulich, M. I., defeated by Japanese at Battle of Yalu, 157–58
Zeddeler, L. L., 135; criticizes Russian tactics in light of Franco-Prussian War, 44; wounded in infantry assault at Gorni Dubnik, 71; fate of, 286n.50
Zhilinskii, Ia. G., 242, 243, 245, 248, 252
Zimnicea, Russian forced crossing of Danube at, 55
Zotov, P. D.: as IV Corps commander at Pelishat-Sgalovna, 66; as Prince Karol's chief of staff, 69; predicts failure at 3rd Plevna, 69

Bruce W. Menning is an instructor of strategy with the Department of Joint and Multinational Operations at the U.S. Army Command and General Staff College, Fort Leavenworth, Kansas, where he has held the John F. Morrison Chair of Military History. He is a former Secretary of the Army Fellow.